Gravity and the Organism

Gravity and the Organism

Edited and with a Preface by

Solon A. Gordon and Melvin J. Cohen

The University of Chicago Press

Chicago and London

No part of this book may be reproduced by any mechanical, photographic, or electronic process, or in the form of a phonographic recording, nor may it be stored in a retrieval system, transmitted, or otherwise copied for public or private use without the written permission of the publisher, except for the purpose of official use by the United States government and on the condition that the copyright notice be included with such official reproduction.

International Standard Book Number: 0-226-30447-7
Library of Congress Catalog Card Number: 70-156302

The University of Chicago Press, Chicago 60637
The University of Chicago Press, Ltd., London

© 1971 by The University of Chicago
All rights reserved. Published 1971
Printed in the United States of America

CONTENTS

Preface ix

Introduction
    The Organism and Gravity 1
        A. H. BROWN

Part I. Physical Bases
1. The Physics of Gravity and Equilibrium in Growing Plants* 13
    G. D. FREIER and F. J. ANDERSON
    Discussion: *Audus, Freier, Gualtierotti, Hertel, Jenkins,*
        Pollard, Weis-Fogh, Westing, Wilkins
2. Physical Determinants of Receptor Mechanisms* 25
    E. C. POLLARD
    Discussion: *Audus, Banbury, Brown, Cohen, Galston, Gordon,*
        Hertz, Kaldewey, Pickard, Pollard, Shen-Miller
3. Oscillatory Movements in Plants under Gravitational Stimulation 35
    A. JOHNSSON
    Discussion: *Hertz, Johnsson, Shen-Miller, Wilkins*
4. Aspects of the Geotropic Stimulus in Plants 41
    R. HERTEL

Part II. Gravity Receptors in the Plant
5. Gravity Receptors in Lower Plants* 51
    A. SIEVERS
    Discussion: *Audus, Etherton, Galston, Gualtierotti, Larsen,*
        Shen-Miller, Sievers, Weis-Fogh, Westing
6. Gravity Receptors in Phycomyces 65
    D. S. DENNISON
    Discussion: *Brown, Dennison, Hertel, Johnson,*
        Leopold, Weis-Fogh, Westing
7. The Susception of Gravity by Higher Plants* 73
    P. LARSEN
8. The Susception of Gravity by Higher Plants: Analysis of Geotonic Data for Theories of Georeception 89
    B. G. PICKARD

---
*Lead paper.

9. A Case against Statoliths   97
    A. H. WESTING
    Discussion: *Audus, Larsen, Laverack,*
                   Pickard, Wilkins

Part III. Correlation in Plant Geotropism

10. Hormone Movement in Geotropism*   107
    M. B. WILKINS
    Discussion: *Audus, Brown, Galston, Gordon, Hertel, Larsen,*
                   Pickard, Pollard, Westing, Wilkins

11. On Hormone Movement in Geotropism   127
    H. RUFELT
    Discussion: *Brown, Cohen, Gordon, Rufelt*

12. Hormone Movement in Geotropism: Additional Aspects   131
    A. C. LEOPOLD and R. K. de la FUENTE
    Discussion: *Audus, Galston, Leopold, Weis-Fogh, Wilkins*

13. Linkage between Detection and the Mechanisms Establishing Differential Growth Factor Concentrations*   137
    L. J. AUDUS
    Discussion: *Audus, Davis, Hertel, Leopold, Pollard*

14. Bioelectric Phenomena in Graviperception*   151
    C. H. HERTZ
    Discussion: *Etherton, Galston, Gordon, Hertz, Kaldewey,*
                   Schrank, Weis-Fogh, Westing, Wilkins

15. A Vibrating-Reed Electrometer for the Measurement of Transverse and Longitudinal Potential Differences in Plants   159
    L. GRAHM

16. Geotropic Curvature of Avena Coleoptiles as Affected by Exogenous Auxins   163
    A. R. SCHRANK

Part IV. Invertebrate Responses to Gravity

17. Stabilizing Mechanisms in Insect Flight*   169
    D. M. WILSON
    Discussion: *Brown, Cohen, Galston, Weis-Fogh, Wilson*

18. Flying Insects and Gravity   177
    T. WEIS-FOGH
    Discussion: *Brown, Cohen, Gualtierotti, Howland, Weis-Fogh*

19. Proprioceptive Gravity Perception in Hymenoptera*   185
    H. MARKL

20. Gravity Orientation in Insects: The Role of Different Mechanoreceptors   195
    G. WENDLER
    Discussion: *Gualtierotti, Hertel, Howland, Jenkins,*
                   Markl, Wendler, Westing

21.	**Primitive Examples of Gravity Receptors and Their Evolution*** G. A. HORRIDGE Discussion: *Audus, Cohen, Galston, Gordon, Horridge,* *Laverack, Lowenstein, Lyon, Weis-Fogh*	203
22.	**Gravity Receptors and Gravity Orientation in Crustacea*** H. SCHÖNE	223
23.	**The Integrative Action of the Nervous System in Crustacean Equilibrium Reactions** W. J. DAVIS Discussion: *Brown, Cohen, Davis, Galston, Gualtierotti, Horridge,* *Hoshizaki, Johnson, Laverack, Schöne, Wilson*	237

Part V. Vertebrate Responses to Gravity

24.	**Functional Anatomy of the Vertebrate Gravity Receptor System*** O. LOWENSTEIN Discussion: *Lowenstein, Mittelstaedt*	253
25.	**The Gravity Sensing Mechanism of the Inner Ear*** T. GUALTIEROTTI Discussion: *Cohen, Gualtierotti*	263
26.	**Semicircular Canal and Otolithic Organ Function in Free-swimming Fish** H. C. HOWLAND Discussion: *Howland, Lowenstein*	283
27.	**Central Nervous Responses to Gravitational Stimuli*** W. R. ADEY	293
28.	**Vestibular Influences on the Brain Stem** H. SHIMAZU	311

Part VI. Plant and Animal Gravimorphism

29.	**Gravimorphism in Higher Plants*** L. S. JANKIEWICZ Discussion: *Galston, Leopold*	317
30.	**Geoepinasty, an Example of Gravimorphism** H. KALDEWEY Discussion: *Galston, Kaldewey*	333
31.	**Plant Responses to Chronic Acceleration*** S. W. GRAY and B. F. EDWARDS Discussion: *Audus, Banbury, Edwards, Etherton, Galston,* *Gordon, Gray, Howland, Kaldewey, Larsen,* *Leopold, Pickard, Smith, Weis-Fogh, Wilkins*	341
32.	**Chronic Acceleration of Animals*** A. H. SMITH and R. R. BURTON Discussion: *Edwards, Gualtierotti, Lowenstein,* *Smith, Weis-Fogh, Wunder*	371

33. **The Effects of Chronic Acceleration of Animals: A Commentary** 389
    C. C. WUNDER
    Discussion: *Brown, Edwards, Johnson, Schöne, Sparrow, Weis-Fogh, Westing, Wilkins, Wunder*

Part VII. "Space" Oriented Studies

34. **Simulated Weightlessness Studies by Compensation*** 415
    S. A. GORDON and J. SHEN-MILLER
    Discussion: *Audus, Brown, Gordon, Howland, Johnsson, Kaldewey, Westing, Wilkins*
35. **Growth Responses of Plants to Gravity** 427
    C. J. LYON
    Discussion: *Galston, Larsen, Lyon*
36. **Effect of Net Zero Gravity on the Circadian Leaf Movements of Pinto Beans** 439
    T. HOSHIZAKI
    Discussion: *Brown, Galston, Hoshizaki, Wilkins*
37. **The Experiments of Biosatellite II*** 443
    J. F. SAUNDERS, O. E. REYNOLDS and F. J. deSERRES

Part VIII. Summations

38. **Responses to Gravity in Plants** 453
    A. W. GALSTON
39. **Gravity and the Animal** 469
    O. LOWENSTEIN

List of Participants 473

Preface

In 1967, from September 18 to 20, a Symposium on Gravity and the Organism was held at Sterling Forest, Tuxedo, New York. It was sponsored by the Space Science Board of the U.S. National Academy of Sciences and supported by the National Aeronautics and Space Administration. The proceedings of that symposium, updated in 1970 by the participants, form the contents of this volume.

The symposium focused on the influence of gravitation (or its acceleration equivalent) upon the properties and behavior of living matter. Research on how biological systems respond to gravity has increased with the growing interest in extraterrestrial environments. Concepts about gravity sensors, particularly those of the plant, have been changing, and research has expanded on the physiological and morphological sequelae of sensor activation. A cogent reason for the symposium was the opportunity it would afford for discussion and appraisal of these recent developments. Another purpose was to promote interaction between the zoologist and the botanist concerned with the mechanisms by which organisms respond to gravitational force, and the physicist who has thought about momentum flows in gravitational fields. Briefly dwelling on the physics of gravity and equilibrium, the symposium moves from the nature and activation of the sensors of gravitational force and direction to the physiological, morphological and behavioral consequences of such activation. Throughout there recurs the consideration of an anatomical matrix which has evolved from, and in turn modulates, the pervasive force of gravity.

The program was organized on a format of lead and discussant papers. Lead speakers were asked to establish a foundation for the topic assigned, recognizing the diversity in backgrounds of the participants, and to bring their discussions to the speculative level. Formal discussants were asked to comment on and extend the presentation of the lead speaker, introducing their own work in supplement. The open discussions that followed the formal presentations have been included in the present volume, as have the closing summations of content and implication by a botanist and a zoologist.

We hope that this exchange between scientists of diverse disciplines, considering a single stimulus and response at various levels of molecular organization to that stimulus, has led and will lead to a broadening of awareness and an enhancement of understanding.

September 1, 1970

SOLON A. GORDON
MELVIN J. COHEN

The Organism and Gravity: An Introduction

Allan H. Brown, *University of Pennsylvania*

In this symposium we shall explore some of the more interesting reactions of plants and animals to the gravitational component of their environment. In concept the symposium is interdisciplinary, and in this introduction I hope I can avoid a scientifically parochial outlook on the research topics to which we are giving our attention. Keeping in mind what we all have in common, both in background and in interests, I hope I can identify certain generalizations that most of us can accept with regard to gravity and the organism. In addition, I think it may be useful to point out where differences in the experimental materials we select, in the techniques we use for research, or in our underlying biological concepts may divide us, perhaps on more than trivial grounds.

The title of my general introduction is a deliberate inversion of the name chosen for this symposium. I wish, thereby, to emphasize that the organism comes first, that our viewpoint is that of experimental biology, and that we are concerned principally with learning more about organisms by understanding the ways in which gravity is important to them.

It is almost strictly true that every environmental factor which varies in intensity or quality or direction—either with time or with geographic location—is exploited biologically for purposes or orientation, navigation, as an energy source, or in some other fashion. Gravitational attraction is no exception. All plants and animals experience it. They evolved under its influence. They have learned to cope with it and to make use of it—learned in the evolutionary as well as the ontogenetic sense.

In keeping with long-standing physiological tradition, as we explore the biological role of an environmental factor, we feel a need to know how our favorite experimental organisms respond to changes in stimulus direction, stimulus intensity, and stimulus quality (although it is doubtful that the last can be made a variable in the case of gravity). Such basically descriptive information is often requisite to our goal of understanding underlying biological mechanisms. In general we can claim that past researches on the biological role of gravity have been modestly fruitful. We know that most organisms use the *direction* of the gravitational force. This is quite obvious in the case of many mobile species, but perhaps it can be demonstrated even better by some immobile forms—particularly by the developmental patterns of both higher and lower plants, which often achieve remarkably precise orientation with respect to the gravitational field direction. We should keep in mind, however, that in some organisms sensing of direction may be a special result of a balanced sensing of intensities.

In favorable examples it has been shown that the test organisms sense an accelerating force, loosely called g, as a vector, whether acceleration arises from gravitational, centrifugal, or other inertial forces, separately or in combination. The experimental evidence strongly supports the conclusion that the *intensity* of g can be the organism's guide to function and to development. Nevertheless we often do not understand just how detection of g is accomplished. In some instances we may feel the answers are clear, but we cannot generalize with confidence; with many groups of organisms we are still very much at the stage of case-by-case exploration.

When accelerations are made intermittent rather than continuous, summation effects of stimuli sometimes become apparent, and it then is possible to speak of the *frequency* of the stimulating factor. If the repetitive stimuli are made to recur often enough, this vibration—still measured conveniently in g units—can induce important biological effects. For repetitive g stimuli in the audiofrequency range many organisms have marvelously adapted sensors. We are not accustomed to thinking of any of the well-studied biological detectors of sound energy as having much to do with sensors for earth's gravity. This may be our misfortune, because both are special kinds of bioaccelerometers; in identifying biological g sensors and characterizing their detection mechanisms, it might be helpful to take into account the way organisms are affected by vibration.

How should we be studying g-related biological phenomena? In recent years it has become increasingly popular to emphasize that biological topics are profitably treated on several levels of organization—those of the population, the organism, the cell, the organelle, and the molecule. Phenomena in which we here share a common interest are chiefly those we observe at the level of the organism—tropism, taxis, growth, and development. We also seek to study these things at the cellular and, preferably, at the subcellular or molecular levels, but this is not always possible. One can argue that the depth of our understanding of gravity-related phenomena in physiology and development is closely predicted by the level of biological organization for which we currently design our experiments: the higher that level the less we understand. Here we must admit that many of the most intriguing gravity-related biological phenomena, especially in the plant kingdom, are currently being studied mainly at the level of the organism or organ. The point is worth noting because it suggests where the greatest potential for research in this field may lie.

More specifically, I suggest that we have found it useful to think in terms of a *stimulus-response system*, and to direct our efforts first toward exploring diverse properties of this system, later toward analysis of functional mechanisms as they are identified or postulated. Most investigations have depended heavily on detailed morphological information about the test material, especially since g-perception and response functions are often highly localized. Logically the first task is to identify and characterize structually the relevant sensorium. In this task, anatomical and cytological studies have by no means exhausted their potential.

If we first ask where in the living structure the bioaccelerometers lie, our next question is sure to be how do they function. What basic principle of detection is employed by the system of sensors? Moreover, is acceleration detected by one mechanism or by several? Intracellular stratification of organelles or other inclusions, stress on plasma membranes, distortion of an endoplasmic reticulum, displacement of cell parts or of portions of multicellular structures—all have been invoked as the mechanical basis for g detection. Only in a few instances does the answer seem obvious. Here it may be noted that the zoologist or neurophysiologist often has a better right than the botanist to be confident that he understands the principle of g detection in his particular experimental material. I need not labor the point that frequently our ignorance of detector function is more impressive than our knowledge.

Where morphological examinations do not provide attractive bases for speculation, experimental physiology is called upon to delineate the properties (especially the kinetic properties) of the g

sensor, as a clue to the mechanism of detection. The research questions are quantitative and precise. What is the sensitivity of the bioaccelerometer? Are intensity of stimulus and duration of application reciprocally related for the production of a standardized response? If so, for what range of these variables? Is there a g threshold below which the sensor is unaffected no matter how long the stimulus is applied? Is there a minimal (presentation) time required for the sensor to detect a brief stimulus? If so, is the presentation time a function of stimulus intensity? What function? Is g stimulation detected on an all-or-none basis, or is the detector output graded in accord with the intensity of the stimulus? If the sensor responds differently to stimuli of differing intensities, does g preception proceed linearly or logarithmically, or does it obey some other law? Can the detector be saturated? What is its dynamic range? Does the g-sensing system exhibit a memory function?—that is, can summation of stimuli be demonstrated? Is the calibration of the bioaccelerometer constant or does it adapt to different chronically imposed g levels? Is the sensor's capacity to detect g an indigenous property, or is only the ability to acquire that capacity implanted in the genome? Accordingly, what (if any) aspects of g detection become modified if the test individual should develop in a field greater or less than the earth's 1 g?

Analytically, the next category of research questions relates to *transduction* and *transmission* of g information in specific stimulus-response systems. It is here that we find a striking difference between animal and plant reactions to gravitational stimuli—namely, the much greater rapidity with which animals respond. This in itself does not mean that the difference is fundamentally related to the detection systems. But as we examine and compare the more fully studied reactions of higher animals and plants, it seems that the mechanisms are quite disparate. Among animals, insects' mechanical sensors of the relative position of body parts, inertial changes of fluid (as in the vertebrate inner ear), and mass displacement in statocysts of many forms all involve relatively rapid transduction of the mechanical stimulation into altered nerve-cell activity. Beyond that stage the information-transfer and response process ceases to be unique. In the higher plants transduction commonly involves a hormonal component, and the mechanisms for eliciting biologically appropriate responses include stimulus-induced alterations in the production, transport, and growth responses to local concentrations of particular chemical messengers. Just as the rapid responses of animals to changes in orientation and acceleration are chiefly the province of the neurophysiologist, so the slower growth responses of plants are primarily the domain of the plant-hormone specialist. The plant system serves for communication of many kinds of information not specifically related to g. Working out in detail this communication system and the responses which it orders has formed the basis for one of the most active and important fields of research in plant physiology over the past four decades; it has conferred scientific immortality on the oat plant comparable with that enjoyed by the cat, frog, mouse, squid, Drosophila, *Escherichia coli*, and Chlorella.

An obvious difficulty in conferences which bring together botanists and zoologists is that of understanding each others' experimental material. Perhaps we can overcome the difficulty, or even put it to our mutual advantage. For comparative purposes it may be useful for us to focus specifically on the gravity-sensitive organs encountered in a general survey of living systems.

Animals. A very widely distributed but by no means ubiquitous g-sensing device of animals is the *statocyst*. This organ consists of a small mass (*statolith*) surrounded by a fluid of lower density and anchored by a tension-sensitive transducer which communicates with an associated neuron as the statolith undergoes displacement. In spite of its many variants in different taxa, the statocyst is a surprisingly conservative structure. Comparison of this neuromechanical transducer with a man-made strain gauge is quite apt.

Experimental manipulation of especially accessible statocysts (involving neurosurgery, recording of electrical activity, removal of the statolith, etc.) and detailed anatomical examination at cellular and subcellular levels continue to reveal fresh possibilities for imaginative investigations. Animals make use of their statocyst-derived information on accelerations in many different ways. Combining results of behavioral studies with comparative neurophysiological and anatomical information on statocyst structure and function has been, and will surely continue to be, a fruitful interdisciplinary study of the influence of gravity on animals.

Plants. Only a limited number of test species have so far been investigated in depth. It follows that a comparative physiological approach has not yet been fully exploited. The sensing devices which plants use for gravity perception are not well understood, even though the regions of a plant which are most sensitive to gravity are often highly localized, and readily accessible, and though in some instances they have been well described at several morphological levels. There is no clear concensus among plant physiologists about the mechanism by which the "typical" gravity sensor operates, even though the consequences for growth of the plant are the subject of a voluminous literature justifying quantitative and often very precise predictions about the way a plant will respond to gravity.

At present there is a paucity of concepts and even of speculations pertinent to the plant gravity sensor. Conditioned, apparently, by knowledge of animal examples, plant physiologists have shown a strong desire to locate and identify plant statoliths—to invoke a sensing mechanism dependent on the stratification of particulates. But what is displaced by gravity? In the long history of this search for statoliths in plant cells, sedimenting starch grains (amyloplasts) within the cytoplasm of cells in the most sensitive plant regions have emerged perennially as the most likely candidates. The evidence which can be marshaled to support the case for the starch grain is impressive but unfortunately not without deficiencies. Here I wish only to point out that, even if we should all agree that displacement of amyloplasts or other particles is, at least in some instances, involved in the plant's detection of the gravitational vector, we do not thereby understand any better the mechanism connecting statolith movement and the organism's response.

Before going on to other systems, I want to speak of a certain quantitative feature of a higher plant's response to gravity and about an experimental procedure that has been extensively exploited by plant physiologists. It does not seem to be equally appropriate for experimentation with animals, and therefore some zoologists may not be familiar with its advantages. If a growing plant—almost any higher plant will do—is displaced with respect to the vertical, some tens of minutes later it will alter its growth in such a way as to restore its original orientation, with the longitudinal axis in coincidence with the gravitational vector. If it is displaced only briefly and then restored to its original orientation well before the growth response can set in, it still responds anyway. By varying the duration of the displacement we can obviously determine the minimal time, the *presentation* time, necessary for a detectable growth response. Depending on what plant we use and whether we are considering the root or the shoot, presentation times range from about 0.3 to 10 min or even longer—a surprisingly long time when compared to most animal systems. Because the presentation time is so long, it is possible to incline the plant and rotate it about its long axis, slowly enough to avoid complications due to centrifugal acceleration, yet rapidly enough to sweep the gravitational vector completely around the plant axis well within the minimal time for a response to a fixed displacement from the vertical to show itself. As one might guess, the plant as it grows then continues to maintain its axis in the new orientation (even 90° to the vertical) as long as the slow rotation (say, about 1 rpm) is continued. It is as if the plant were "unaware" that it is not vertically oriented, since its growth shows no righting tendency.

The apparatus for imposing the displacement and the rotation is called a *clinostat*, and it is as common in laboratories where plant physiologists study gravitational responses as is the oscilloscope in laboratories of neurophysiology.

As we review experimental results obtained with the clinostat, it is well to keep in mind this question: Is the plant's experience of being on a rotating clinostat a simulation of exposure to a reduced gravitational field or is it something distinct? With so much attention being paid to space flight as a means of attaining what is colloquially referred to as *zero gravity* or *weightlessness*, can the plant physiologist validly boast that he has been working with a $0g$ simulator for about a century? Depending on our assumptions, we can argue that a clinostat does or does not simulate weightlessness. Solon Gordon prefers to describe clinostat treatment as a *compensation* or *nullification* of the effect of the directional component of the gravity force vector, and one cannot quarrel with that terminology. For myself, I cannot help thinking of the constant lateral stimulation tracing its path around and around the plant axis, as something quite different from the situation of a vertically oriented plant (which is not exposed to a fluctuating pattern of stimulation) or of a plant maintained in an orbiting space capsule, where in its state of free fall it receives no g stimulation at all. These three sets of conditions are conceptually distinct, and I should expect, when appropriate comparisons have been made, that 90°-clinostat-grown plants (which we know differ in some respects from vertical control plants) will be found also to differ from plants grown in earth orbit.

Protista seem to be unable to detect gravity directly. Where it has been claimed on morphological grounds that a particular protozoan or microalga is sensing gravity, the evidence is more suggestive than compelling. Naturally one must be careful not to impute gravity perception to an organism which orients itself by a gradient of light intensity or hydrostatic pressure. Even though some protozoan structures have been called analogues of metazoan organs with undisputed capabilities for detecting acceleration, I am ignorant of any incontestable example of gravity sensing in microorganisms.

In this connection it can readily be demonstrated that sensing mechanisms based on passive stratification of particulates or phases of different density must encouter severe design limitations as miniaturization is attempted. With reasonable assumptions about density differences, viscosity of the continuous phase, and particle size, a lower limit is predicted for the size of a particle that will settle out, against the dispersing tendency of Brownian motion. It is interesting that a size threshold for graviperception thus estimated falls very roughly at a linear dimension of about 10 μm. In other words, organisms at the size range of bacteria are predicted to be incapable of sensing in microorganisms.

We should be aware that this prediction is not quite categorical; the loophole is that a sensing mechanism which depends, for example, on differential stretching of upper and lower plasma membranes rather than on stratification of cell contents, would be subject to a different size limitation. That limit might indeed be lower than the limit set by the requirement for escaping the consequence of Brownian motion. Nevertheless, it is at least not unreasonable to maintain that very small organisms (including all bacteria) have no means of sensing gravity or, if motile, of responding to it. Perhaps we tend to believe this as much because we cannot see any advantage in gravity perception for microorganisms as because we think a sedimentation sensor on this scale is impossible in principle and have not yet found any other kind.

We should also recognize that the failure of an organism to sense gravity, or to use gravitational information to influence behavior, is not the same as the organism's lacking any functional interaction with gravity. If we pose the right experimental questions we might discover interactions—unfamiliar and therefore scarcely predictable—which are not part of the normal "response to environment" patterns worked out by the physiologist, psychologist, or embryologist.

I must not put too much emphasis on problems of g detection, for I believe other aspects of stimulus-response systems warrant equal attention. How does the test organism use its g information? Here we are not concerned with gravity perception *per se*. We naturally think first of the most obvious systems: geotropism in plants, animal orientation—namely, systems that reveal capacities for relatively rapid responses. There are in addition those (often slower) biological effects of g which are better described as responses to chronic stress. Especially in larger organisms, chronic g stress can be viewed as a challenge to the organism. Among the results of prolonged stress most easily revealed are structural changes; the *form* of the test organism is thus conditioned by g. Where these changes have been studied experimentally by restricting orientation in special ways or by increasing g stress in centrifuges, the responses often have been patently adaptive. Relatively rapid behavior responses or geotropic responses may leave the organism in an essentially normal condition, but with chronic g stress, organisms respond more slowly, toward altered morphological end points. These phenomena are sometimes labeled *geomorphism*, but since geologists have preempted that term for a different phenomenon, we usually refer instead to *gravimorphism*.

At the present stage of our thinking, the research questions that can be attacked by developmental biologists investigating gravimorphism often require them to manipulate the strength or the direction of the gravitational field. For chronic exposures to altered g intensity, the centrifuge and the satellite are the principal (in fact the essential) research tools.

The condition of increased g has been called *hypergravia*. Because its effects are often adaptive, if not destructive, they are more easily predictable than those of subnormal g or *hypogravia*. Especially at the lower limit of g, the condition of weightlessness, we feel least secure in predicting biological consequences. Organisms may use g in subtle ways just because it is there to be exploited; unfortunately, here we are deprived of intuition based on teleology, which so often has served to suggest a fruitful experimental approach to a poorly understood phenomenon. Throughout all of evolution there can have been no advantage for an organism to adapt to hypogravia. Is terrestrial life g-related in some manner unlikely to be discovered except through g deprivation tests of long duration?

The limiting case of $0\,g$ may be an especially interesting test condition. One cannot study (on the earth) physiological or developmental processes in the absence of any gravitational stimulus. The student of graviperception or gravimorphism can alter the direction of the earth's acceleration vector with respect to the sensing organism; he can decrease it momentarily (in free fall); he can increase it on a centrifuge; but he cannot turn it off. Crudely put, this is analogous to a student of vision being allowed to experiment with the position of a bright light source, to dim or shut out the light for only the briefest of moments, to make it brighter, but never to turn it off entirely for long periods to make dark baseline observations. Although still left with many interesting experimental capabilities, our hypothetically restricted student would be sure to feel a nagging frustration at not being able to extend his investigations by using both darkness and dim light for prolonged experiments. After all, is not darkness the proper control condition for an experiment involving the effects of light on a photosensitive subject? For our present discussions the condition analogous to darkness is $0\,g$.

Fortunately there is a way to escape this frustration by prolonging the free-fall condition in an earth-orbiting satellite. It is now late 1967, the year of man's first exploitation of the earth satellite environment exclusively for gaining biological information. We have just heard about the first successful recovery of a biosatellite capsule. I think we biologists all feel a sense of historic importance associated with the use of this new facility, an orbiting "laboratory" which provides prolonged free-fall conditions. Personally I have no doubt that eventual full exploitation of orbiting facilities will provide valuable or even dominant contributions to future conferences such as this. The NASA biosatellite program is only a start. I am pleased that NASA has accepted its obligation to make this utterly *new* facility, the earth-orbiting satellite environment, available for biological experimentation. A prime motivation for organizing the conference on gravity and the organism was to provide an authoritative and comprehensive survey of background information against which research in future biological satellites should be designed.

In broad terms what kind of biological researches can best be carried out in space? Can we learn about the role of gravity in the life of an organism by exploiting a condition in which g stimulation is absent? The traditional method of physiology for exploring autoregulation has also been one of the simplest in concept: stress the organism by challenging it with an environmental influence outside the normal range, and observe those changes whereby the test subject accommodates to the stress. This method is eminently suited to the study of an organism's responses to hypo- and hypergravia.

With the sophisticated biosatellite facilities, manned as well as unmanned, which we may expect to have available in the future, there are a number of research questions for which we may hope to find answers. Some questions are specific and the answers may be decisive; others are necessarily general, since new phenomena are to be examined and early research must be exploratory. The following questions may serve as examples.

1. What is the threshold for g perception without a 1-g background limitation? We are not yet sure how sensitive biological g detectors may be. In space the question might be answered unambiguously. For certain higher plants, attempts have been made in several earth-based laboratories to measure a g-detection threshold. Different methods were used, but all were necessarily indirect. The results have spread over 4 or 5 orders of magnitude. A definitive set of measurements subject to unambiguous interpretation probably can be obtained only in space.

2. For plant geotropic responses, reciprocity between stimulus intensities and presentation times has been claimed for a limited range of these variables. The measurements cannot be extended below 1 g except in a space laboratory, where it will be possible to define the limits within which reciprocity obtains.

3. How does an organism adapt to g levels either above or below $1\,g$? To answer this kind of question for hypogravic conditions obviously requires a space laboratory. The topic, while fundamentally interesting, has practical overtones in the case of manned space flight of long duration. For periods of the order of two weeks man has adapted to a $0\text{-}g$ environment (although not without some difficulties). He has readapted to $1\,g$ apparently without undesirable lasting sequelae. We have no confidence that the same will be true for much longer missions, and some physiologists think readaptation to the earth's $1\,g$ may become increasingly difficult the longer the $0\text{-}g$ experience.

4. Is the establishment of biological polarity critically g-dependent? The influence of g on certain aspects of function, and especially on growth pattern, may be chiefly to provide an environmental clue for orienting some process, probably at the subcellular level. This is a possible but still unproved link between g perception and the origin of polarity in the developing organism. The thought is not new that perhaps morphological polarity cannot be initiated under weightlessness, even though once established it may persist. Whether or not this is true, it seems evident that we can confirm or disprove it only by observing biological development in the absence of g as an orienting clue.

5. How is the determination of form influenced by g? To date, studies of gravimorphism have been able to test only the condition of $1\,g$ and upwards, but in a satellite a "$0\text{-}g$ morphology" will become accessible for examination. Extrapolation back down from results obtained under hypergravia are suspect. Space experiments may show us what properties of developmental systems depend on g.

6. To what degree does so-called nullification of gravity by the clinostat produce a condition biologically equivalent to free fall? In a very limited way this type of question already has been posed in some experiments on board Biosatellite II.

As with any new research tool or method it is easier to suggest applications than to predict the results. Whenever we have been able to exploit a really new experimental or observational approach, or to examine a hitherto unexplored range of an important environmental variable, the results have been rewarding. I think we are justified in this same expectation for biosatellite research on sensing and responding to "unfamiliar" values of g. In particular it seems to me likely that prolonged deprivation of gravitational stimulus will bring about major adjustments within organisms and reveal functional relationships which are not even suspected. Thus we may be due for some surprises.

I should like to venture another prediction—better call it an educated hunch. As more and more experiments on gravity perception and response are designed and carried out with diverse experimental organisms in satellite environments, the so-called $0\text{-}g$ condition will not by any means provide all the scientifically worthwhile results. I expect that studies of effects of g between 0 and 1 will demonstrate that there is much to be learned by exploring that particular range of the g variable. Accordingly we may look forward to future biological satellites with on-board provision for applying centrifugal force. This I submit will not be just for the purpose of having appropriate $1\text{-}g$ controls in orbit (which, however, will be necessary for some kinds of experiments) but even more to provide greater than 0- but less than $1\text{-}g$ exposure for the test subjects.

In conclusion, I want to reemphasize that one of the conceptual bases for the conference on gravity and the organism was that by an interdisciplinary consideration of our many specific researches, common themes and (we hoped) new ideas for furthering particular investigations would emerge. We should seek out the generalizations which seem valid across disciplinary lines, and we should at least try to understand the research approaches of other participating biologists. We are united only by broad problems which employ a diverse array of experimental organisms, ranging widely in both plant and animal kingdoms. Although we are all physiologists, we are not much more alike in our experimental methods than in the taxonomic affinities of our experimental organisms. We have much to learn from each other. That this conference will be fruitful interdisciplinary experience is more than a hopeful assumption; it is a challenge. We must learn by sharing our diverse research experiences and each of us must profit by comparing other prejudices with his own.

I Physical Bases

1
The Physics of Gravity and Equilibrium in Growing Plants

G. D. Freier and F. J. Anderson, *University of Minnesota*

Nature of Gravitational Force

Newton found that the gravitational force, F, between the earth of mass M, and a particle of mass m, was $F = GMm/r^2$ when their centers were separated a distance r. In this equation G is the gravitational constant and is necessary because force, mass, and distance are defined in other operations. Particles have weight because of this force, but in carrying out the operation of weighing we may or may not arrive at a value for the gravitational force. In the operation of weighing a particle we find what force is necessary to bring the particle to equilibrium, i.e., to reduce the acceleration to zero in our frame of reference. If our frame of reference is accelerated with respect to the frame of reference of another observer, the second observer will say that the particle is not in equilibrium, and he will say that the gravitational force has not been balanced in the operation of weighing.

The observation of weighing may also be carried out by dropping the object of mass m, and measuring the acceleration, $\vec{g}$. The weight is then said to be $m\vec{g}$, but $\vec{g}$ will be different for observers who are accelerated with respect to each other. The fact that the acceleration due to gravity, $\vec{g}$, varied between the pole and the equator because of rotation of the earth about its axis did not change Newton's belief that the force of gravity was always GMm/r^2. He knew that different points on the earth were accelerated differently and that different forces at different points must be combined with GMm/r^2 to give the observed accelerations. This then leads to "weightlessness" in a satellite, as no force additional to the gravitational force is required to produce the required accelerations. The true cause of weight is the gravitational force, and it will persist in satellite motion about the earth. To say that things are weightless in a satellite leads to some misunderstanding in that we may feel that we have removed the gravitational force. On the contrary, the gravitational force still exists, but the other forces which produce equilibrium on the surface of the earth have been removed. It is commonly stated that objects in orbit about the earth are weightless, and hence forces due to the gravitational field can no longer be detected. Experiments done on a single-point mass indicate that the particle behaves as if it were in an inertial system, and the principle of equivalence tells us that the experiments can in no way differentiate between gravitational forces and inertial forces. All that is required for an inertial reference frame is that a particle at rest in this frame remains at rest, and a particle in motion would tend to remain in uniform motion if no force acts on the particle. The point-mass particle in a satellite fulfills this condition.

Things become more complicated when the satellite has a finite size and the motions of more particles are considered. An astronaut doing experiments in an orbiting sealed satellite would make some strange observations if he extended the experiments over a finite region of the satellite.

Suppose that experiments are done with two-point masses. If they are weightless in the satellite, the observer should be able to place them at any point in the satellite and they would not fall as they would in an earth-bound laboratory. Figure 1 shows a satellite revolving about the earth without rotating. (For simplicity we will assume the orbits to be circular.) If the two masses A and B are released side by side as shown, they will simply be two particles in the same orbit and thus follow each other around the earth. When the satellite is at position II of figure 1, an observer on the floor, X, would see the masses with their positions interchanged. The observer in the satellite would see the two masses rotating about each with no force between them to produce the necessary central acceleration. (The gravitational forces between the particles is much too small to account for this motion.) The presence of the gravitational field seems to produce some strange motion of the particles. If the satellite rotated with the same angular velocity as found for the relative motion of the two particles described above, the observer would note no motion of the particles, but the presence of the gravitational field would then place a rather definite boundary condition on the rotational motion of the satellite.

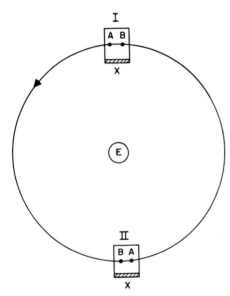

Fig. 1
A satellite in a circular orbit about the earth. The observer in the satellite sees the point masses A and B exchange positions in moving from position I to position II.

A second experiment may be one in which the observer in the satellite places the masses A and B above each other as shown in figure 2. The two particles are now in different orbits, and Kepler's third law tells us that the square of their periods of revolution about the earth is proportional to the cube of the radii of their respective orbits. Particle A in the larger orbit will have a longer period of revolution than particle B, and since it has farther to go in completing its orbit in a longer time, it will certainly get behind B. If the center of mass of the satellite is initially between the release points of A and B, and if the satellite does not rotate, the observer on the floor, X, will see the particles moving toward opposite walls where they collide with the wall. The observer will see the particles bouncing back and forth between the walls because he is in the gravitational field of the earth. Rotating the satellite will not alleviate this strange observation.

If the particles are bolted into position within the satellite, there will be stresses and strains in the bolts to maintain this fixed position. A finite-sized satellite and its contents with all objects being weightless will still suffer strange stresses and strains because it is in the earth's gravitational field.

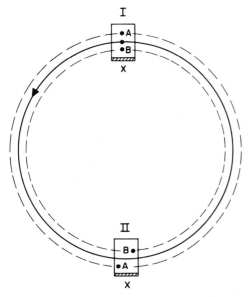

Fig. 2
A satellite in a circular orbit about the earth. The observer in the satellite sees the point masses A and B move from side to side in the capsule.

These simple experiments indicate that a single-mass point cannot reveal the presence of a gravitational field but that a complex of particles will do so. How can these effects be reconciled?

Einstein pointed out that the inertial properties of a small system are obtained from the space-time structure in its immediate vicinity and moves on a straight line in its local inertial frame. There is no evidence of gravitation when one studies only the motion of a single particle. However, the space-time "straight" line is non-Euclidean and has curvature, and this curvature is determined by the gravitational fields of all other particles in the universe. If two particles are studied, they each move along their "straight-line" geodesics—which may curve differently and thus give rise to the strange motions described above. The concept of particles moving on "straight" lines in space-time which is actually curved is analogous to two observers starting at different points on the earth's equator and walking "straight" north with equal speeds. Their motion on a curved surface brings them strangely closer together as if some force of attraction existed between them.

In general relativity, even light does not travel on straight lines in the Euclidean sense but follows a curved non-Euclidean line whose curvature reflects the presence of mass particles distributed throughout the universe. Similarly, the motion of every particle is determined by the curvature given to its "straight-line" motion by all other particles in the universe. In order to study a system of particles, we must study the curved space-time motion of each particle and see how these curved lines approach and recede from each other owing to their different curvatures.

In contrast to this, Newton did not use the local inertial frame for each particle, but he posited an absolute space to which all motions could be referred. This absolute space is usually chosen as Euclidean space with coordinate reference frames attached to the fixed stars. In this absolute coordinate system the motions of all particles may be described. From these motions the accelerations of particles may be determined, and, if the mass of the particle is known, the net force on the particle can be determined. From these forces Newton could delineate those which were due to gravity, and he thus formulated his law of universal gravitation. These gravitational forces remain mysterious, but they produce the observed accelerated motions that particles throughout the universe exhibit when subjected to them.

Forces on Particles in Growing Plants

If we examine the word *geotropism*, we find that it means "earth bending," and since we are applying the word to growing plants, some motion is involved. When a physicist observes any kind of directed motion, he is inclined to look for forces which might produce that motion. For example, if he observes positive charges moving in one direction and negative charges moving in the opposite direction, he will say that there is an electric field permeating the medium in the direction in which the positive charges move. If there is a force $\vec{F}$, acting on a particle of mass m, to give it an acceleration $\vec{a}$, he will apply the law $\vec{F} = m\vec{a}$ if he can, but if it is observed that $\vec{a} = 0$, he cannot conclude that there are no forces acting. There may be two or more forces in equilibrium. In the case of electric charges in a resistive medium permeated by an electric field, he will observe that the average acceleration of the charges is indeed equal to zero. He explains the situation by saying that the field force gives energy and momentum to the particle, and this energy is then lost by collisions with other particles in the medium. The charges appear to move with uniform motion while the field forces are in equilibrium with kinetic collision forces. Energy and momentum are taken from the electric field and given to the medium where they become evident by a rise in temperature.

In plant growth the situation is a little more complicated. In a gravitational field we observe that shoots grow up and roots grow down, but all forces between the elements of the plant and the earth are attractive and in a single direction. There are no positive masses which are accelerated in one direction and negative masses accelerated in the other direction similar to the electrical case. We explain the situation by saying that the shoots and roots receive similar stimuli, but they respond oppositely.

The plant, however, is a physical system, and if we invoke a sufficient number of physical laws, we should be able to trace the physics all the way from the stimulus to the response. In this paper, however, we will concern ourselves primarily with an inspection of the forces which produce the stimulus. The plant's adjustment to a gravitational field necessarily brings other forces into play, since the elements of mass in the plant are approximately in equilibrium during the plant's growth. In studies of geotropism we must be as concerned with these equilibrium forces as we are with the force of weight produced by gravity.

If one considers a particle in a plant on the surface of the earth, this particle will be acted upon by a body force of gravitation. But the particle is normally in equilibrium, so there must be other forces acting on the particle. If the orientation of the plant system changes, the orientation of the body force with respect to the plant changes, and when the particle in the plant comes to equilibrium in the new position, the equilibrium forces will also have shifted.

There are two main classes of forces which can be called into play; namely, body forces and surface forces. Gravitational forces are perhaps the best example of body forces, in that the force of gravity acts on every element of mass in the particle to give it the characteristic we call weight. An inertial force is also a body force in that an element of force must be transmitted to every element of mass to give the particle its acceleration. When a body is in free fall, every element of mass is acted on by the body force of weight to give that element the acceleration commonly called the acceleration due to gravity. If electric charge is distributed uniformly throughout the particle, and if the particle is in an electric field, we again have an example of body force.

Surface forces are usually classified further as two kinds, those which produce normal surface stresses and those which produce tangential surface stresses. Since a stress is a force per unit area, it is necessary to integrate the surface stress over the surface of the particle in order to find the resultant force on the particle. In a fluid a normal pressure force integrated over the surface of the particle will give the particle buoyancy. This requires that the pressure vary around the particle, and this variation in pressure is usually caused by a body force in the fluid. A body floating in a fluid is in equilibrium under the body force of weight and the surface force of buoyancy.

In a fluid the tangential stresses are usually produced by viscosity, and this viscous stress must also be integrated over the surface of the particle in order to give the resultant force. A body in free fall at terminal velocity is in equilibrium under the body force of weight and the surface forces of buoyancy and viscosity.

If a particle interacts directly with other particles, there may be contact forces having both normal and tangential components. These again are surface forces which act only over a fraction of the area of the particle.

The contact type of force may come from many impulses during interactions with other particles. This type of force then requires a concentration gradient so that there are more impulses on one side than on the other in order to produce a net force which may hold the gravitational body force in equilibrium. One then says that the kinetic diffusion is in equilibrium with the gravitational field.

Experiment

An experiment was done to determine if any of these equilibrium-producing forces could be eliminated as stimulus-producing forces. Oat seedlings were forced to grow between blotter paper and a clear plastic sheet in a vertical plane which rotated about a horizontal axis normal to the plane (figure 3). A particle in the plant will then be at the equilibrium distance, r_0, from the axis and rotate with some constant angular velocity, ω. In the frame of reference of the rotating plant, an observer would have the gravitational force rotating in the sense opposite to that of the angular velocity, ω, to give a periodic driving force. The mass, m, would also experience a constant centrifugal body force of $m\omega^2 r_0$ directed radially outward.

It is assumed that the equilibrium position, r_0, of any mass element is determined by both these forces being zero. When these forces act, the mass, m, may suffer a displacement, z, and while making the displacement, it will have a velocity dz/dt. The mass will also experience an acceleration d^2z/dt^2 in acquiring its velocity.

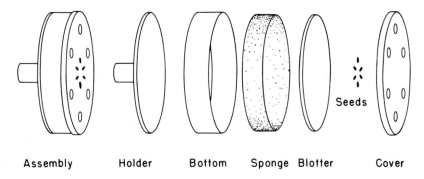

Fig. 3. Exploded view of the container in which oat seeds were grown while the container rotated about a horizontal axis.

The forces from normal surface stresses will, in general, create a restoring force proportional to the displacement given by kz, where k is an average elastic constant. The forces from tangential stresses will probably have their origin in viscous phenomena which would make this force proportional to the velocity of the particle or equal to $\lambda dz/dt$, where λ is an undetermined constant. Since all the motion is confined to a vertical plant, the plane may be considered as a complex number plane, so the driving force due to gravity may be written as $mge^{j\omega t}$, where t is the time the system has rotated with the angular velocity, ω, and $j = \sqrt{-1}$. The equation of motion of the mass is then

$$m\frac{d^2 z}{dt^2} + \lambda \frac{dz}{dt} + kz = m\omega^2 r_0 + mge^{j\omega t}.$$

The steady-state solution for the equation is

$$z = a + be^{j\omega t},$$

where

$$a = \frac{m\omega^2 r_0}{k}$$

and

$$b = \frac{g}{\left(-\omega^2 + \frac{k}{m}\right) + j\frac{\lambda}{m}\omega}$$

There are two important points to note in this solution. (1) The centrifugal force, $m\omega^2 r_0$, can bring about the constant displacement a to hold the particle in equilibrium as mg would do if the system were not rotating. Furthermore, a must be proportional to r_0 so that it can lead to a test of variable sensitivity. (2) If there is viscosity ($\lambda \neq 0$), then the displacement will not be in phase with the driving force, and this phase difference will depend on the direction of rotation. If the geotropic response of the plant depends on the displacement b and its related phase angle, it seems possible that the growth should depend upon the direction of rotation.

When seeds are planted in this rotating system so that the first growth of the shoot is outward, the shoot will be growing in the direction of the body force, $m\omega^2 r_0$, and will respond by attempting to reverse its direction of growth. It can turn to its right or to its left in reversing the direction, and if it is sensitive to the phase angle described above, there should be a preferential direction of turning for a given direction of rotation. No such preference was indicated in the experiments. The time average of the displacement, $be^{j\omega t}$, is, of course, zero, but the phase angle will remain constant throughout the growth. Many possible explanations can be given. It may be that the displacement, b, is completely dominated by the displacement a; it may be a test to show that viscous forces are not important, or it may be that too many simplifying assumptions have been made in the analysis.

The analysis can easily be modified to the case of a plant growing in a satellite moving in an elliptical orbit about the earth. In this case, the forces $m\omega^2 r_0$ and mg are both present but exactly cancel each other. All terms on the left side of the previous equation of motion are then zero. When the satellite is in orbit, the weight mg is still present, but there is no further need for surface forces to hold it in equilibrium. To make this statement true for the entire flight of the satellite, every point of the satellite must describe a true elliptical orbit, or there should be no rotation about the center of mass of the satellite in a coordinate system determined by the fixed stars. When a plant is grown in orbit, it still has weight, and the forces which are changed are those which hold the particles in equilibrium when the plant is at rest on the ground.

The remaining part of our experiments were designed to test the magnitude of the geotropic sensitivity. The oat seeds were planted at a distance r_0 from the axis of rotation such that $\omega^2 r_0$ was less than g, and the seed was oriented so that the shoot initially had an outward growth. As the shoot grew outward, the tip would reach a region where $\omega^2 r$ equaled a certain fractional value of g, and then it would turn through some angle less than 180°, so that subsequent growth would not be directly toward the axis. If a leaf did not appear, the shoot would turn a second time after crossing the central region. If a leaf had started to appear, the growth would continue outward in whatever direction it had at the time of formation.

Once a day the rotating system was stopped sufficiently long to add water and take a picture. The analysis of the growth was made from the photographic negative which could be studied on a moving stage microscope. The coordinates of about 20 random points along each shoot or root were measured with respect to the axis of rotation. Inspection of pictures from day to day showed that there was negligible buckling of the plant as it increased in length. Most of the measurements were made from a picture after 4 or 5 days of growth. The calculations and graphing of results were done with the aid of a computer.

Analysis

The analysis followed closely the calculation outlined by Gibbs and Wilson (1901, p. 121) or by H. B. Wills (1940, p. 56). From the coordinates of the 20 measured points we could calculate the relative displacements $\overrightarrow{\Delta S_i}$ of each point with respect to its preceding neighbor and then calculate the total displacement from the center as $\vec{r} = \vec{r}_0 + \sum_i \overrightarrow{\Delta S_i}$ where $\vec{r}_0$ is the initial displacement of the point where the growth started. The magnitude of $\vec{r}$ would be the radial distance at any given point from the origin. The arc length along the shoot or root was calculated as $S = \sum_i |\overrightarrow{\Delta S_i}|$, and this value of S rather than the time was used as an independent variable in all further calculation.

In order to get the curvature of the growth trajectory, it was necessary to calculate the rate of change of a unit tangent vector to the plant at each measured point along the path. The unit tangent vector was calculated as $\vec{t} = \overline{\Delta S_i}/\Delta S_i$. The next step was to calculate the change in the unit tangent vector in going from point to point as $\overline{\Delta t}/\Delta S_i$. When this is divided by its magnitude, the result is a unit vector which is normal to the unit tangent vector and directed toward the center of curvature of the path at the point in question. The magnitude of $\overline{\Delta t}/\Delta S_i$ gives the rate of turning along the growth path, and the direction of this cross product, $\vec{t} \times \overline{\Delta t}/\Delta S_i$ was used as an index of whether the turning of the plant was parallel or antiparallel to the direction of rotation. The curvature of the path is defined as $C_1 = |d\vec{t}/dS|$ and will be expressed in units of reciprocal centimeters.

Typical measurements of r for a shoot are shown in figure 4 and for a root in figure 5. The measurements of the curvature, C_1, for a shoot are shown in figure 6.

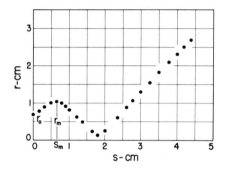

Fig. 4. Radial distance of a shoot from the center of rotation as a function of the shoot length with $\omega = 22.5$ rad/sec.

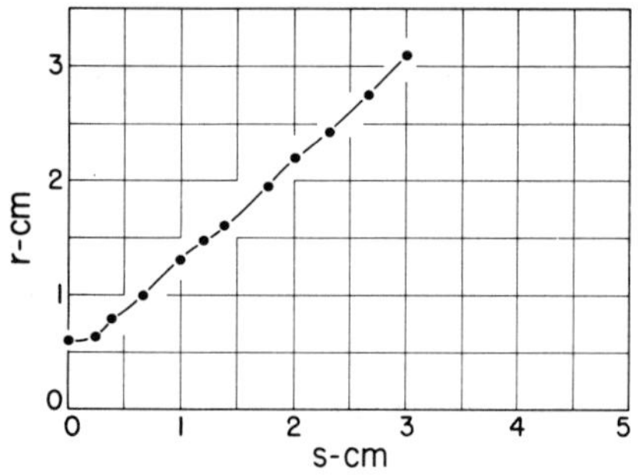

Fig. 5. Radial distance of a root from the center of rotation as a function of the root length with $\omega = 22.5$ rad/sec.

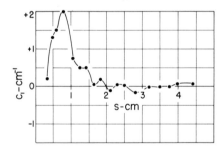

Fig. 6
Curvature or rate of change of a unit tangent vector per unit of shoot length as a function of the shoot length in centimeters.

The salient features of the graphs for the shoots, as shown in figure 4, appear to be the maximum and minimum for the radial distance as a function of path length. When the shoots were planted at the initial position, r_0, and directed so the first growth was toward increasing r, they would start turning and pass through a maximum of $r_m \cong 1$ cm when $S = S_m$. The rate of rotation was such that the acceleration was $0.52\,g$ at 1 cm.

The direction of turning of the shoot was parallel or antiparallel to the direction of turning with equal frequency; that is, graphs similar to figure 6 for other plants would have negative values with equal probability. The large values of C_1 would occur at the maximum values of r.

The first turn of the plant growth would not bring the shoot to a direction where it was growing directly toward the axis. After passing through the maximum at $r \cong r_m$, it would grow across the central region to give a finite value for the minimum value for r as shown on figure 4. The shoot apparently was unable to sense the smaller accelerations at very small values of r.

The roots would always show almost exclusive radial outward growth with negligible curvature. They were always growing into a region of increasing central acceleration and increasing central stresses. Since roots naturally grow opposite to these stress forces, the experiment was not a very good test for the response of the roots to these stresses.

In order to test further the sensitivity of the shoot to central accelerations, the rate of rotation was reduced so that the central acceleration, $\omega^2 r$, was equal to the acceleration due to gravity at $r = 3.20$ cm. A typical growth record for a shoot under these conditions is shown in figure 7. When this is compared with figure 4, it is seen that the turning point is further from the axis of rotation.

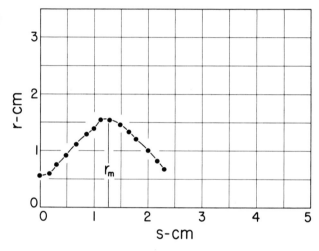

Fig. 7
Radial distance of a shoot from the center of rotation as a function of the shoot length with $\omega = 17.5$ rad/sec.

When the results for the faster rotation rate are compared with those for the slower rotation rate, we find that the average value of r_m changes from 1.19 to 1.49 cm while the value of r (for the central acceleration to be equal to the acceleration due to gravity) changes from 1.93 to 3.20 cm. If the shoots had shown the same sensitivity in both cases, the shoots should have gone through a maximum value of $r_m = 0.9$ cm of the faster rotation. This suggests that the stimulus depends on the time integral of the stimulating force rather than the magnitude of the stimulus, as found by Rutten-Pekelharing 1910).

For this experiment we may calculate this time integral as

$$\int \frac{\omega^2 r}{g} dt = \frac{\omega^2}{g} \int_0^{S_m} \frac{r}{V} dS$$

where V is the velocity of growth and S_m is the value of S at $r = r_m$. We did not carefully measure V in these experiments, but if the average value of V were assumed during the growth, it would have to be smaller by a factor of 1.3 for the faster rotation if the total impulse to produce the turning of the growing section of the plant were to be the same in both cases. Since different plant cells are involved in the above integral, it seems that perhaps the lower limit should not be zero but should be adjusted to some finite value depending on a lower limit of sensitivity. The lower limit to sensitivity also seems to be borne out by the fact that the shoots do not grow directly toward the axis of rotation after they have turned around. As the growth approaches the central region, the shoot does not appear to be able to detect where the axis is located.

It was assumed throughout the experiment that the stresses which balanced the weight were averaged to zero. The gravitational forces for varying orientations of the plants have been studied extensively by Gordon (1963). In our experiments we tacitly assumed that the period of the variation in the gravitational stress was much shorter than any characteristic time associated with the growth process, in which case the average value of the gravitational stress should then be zero for our rotating system.

An interesting check on the averaging process could be made by growing a seed in an "upside down" pendulum as described by Den Hartog (1947, p. 412). In this system the inertial forces produce a stable system with the bob of the pendulum oscillating above the support point rather than below it. If a seed were planted in the bob, which way would the shoots and roots grow?

Summary

In studying geotropic responses in a satellite experiment attention should be directed toward the forces which normally hold the system in equilibrium, as the gravitational force will change very little from its value at the surface of the earth. The forces which normally produce equilibrium tend to be reduced to zero in satellite motion. Even in the weightless environment of a satellite a finite-sized system will have to sense very small stresses to hold its geometric configuration so that threshold values of stimuli should be considered.

An experiment was done to study what forces other than gravitational force may be involved in growing plants to maintain equilibrium. It was concluded that plants have a lower limit to their ability to sense gravity and that equilibrium forces are probably not due to viscous forces.

Acknowledgement

We wish to thank the Numerical Analysis Center of the University of Minnesota for computer time used in the analysis.

References

Den Hartog, J. P. 1947. *Mechanical Vibrations*. New York: McGraw-Hill.

Gibbs, J. W., and Wilson, E. B. 1901. *Vector Analysis*. New Haven: Yale University Press.

Gordon, S. A. 1963. In *Space Biology*. Proceedings of the Twenty-Fourth Biological Colloquium. Corvallis: Oregon State University Press.

Rutten-Pekelharing, C. J. 1910. *Rec. Trav. Neer.* 7: 241.

Wills, A. P. 1940. *Vector and Tensor Analysis.* New York: Prentice Hall.

Discussion

WESTING: In your illustrations, Dr. Freier, it seemed that the seeds were all oriented similarly. Have you tried seeds in other orientations?

FREIER: We have tried various orientations of the seeds. If the root starts to grow outward, it will continue to grow straight out. For a random orientation of the seed, the root will, in general, make some minor turns to get around the seed between the blotting paper and the cover. Once it is around the seed, it will grow straight out along the radius. If the shoot, on the other hand, starts growing inward, it will grow towards the center but not necessarily pass directly through the center of the system. It can grow across the central region and then to the far side to reach again approximately the same maximum radial distance.

GUALTIEROTTI: If you maintain the centripetal acceleration constant, changing both the radius and the rotation speed, then you should be able to determine selectively the effect of the speed of rotation. Moreover, there should be a way of measuring the time constants of the phenomenon.

FREIER: We did not investigate that. Were we to make the rotation slower and slower, then eventually we should see the effect of time constants in the action of *g*. We did not do this type of experiment.

WEIS-FOGH: You have quite small chambers and the diameter of the shoot is large relative to the radius where you placed the plant. Have you ever tried to do this on a really large scale, where the diameter of the shoot is relatively small? As soon as curvature occurs, you have a difference on the two sides.

FREIER: We think that the diameters of the shoot are less than the errors in measurement of the radius, and that any changes in acceleration across the diameter of a shoot could not be detected in this experiment.

WILKINS: I think there may be a complicating factor in your data. You are dealing with an oat coleoptile which is growing from 0 to 3 or 4 centimeters. Up to the time it's about 0.75 to 1 cm long there's a great deal of cell division going on. After that only cell extension occurs. Now it may well be that the capability for bending during the growth up to 1 cm in length is extremely limited. It would be interesting to do your experiment starting with a coleoptile that is already 1 cm long. Between 1 cm and 3 cm in length we know that growth is fairly linear with time and is due primarily to expansion of cells.

FREIER: I agree. It was felt that this was not a problem because we took pictures on every day, from day 1 to day 5. The data were analyzed from the 4th and 5th day pictures, but the coordinates, as read on the previous days, were superimposed on pictures of the last day. There was a negligible buckling of the plant, and it did not change its shape from day to day. If there was expansion of the cells, it would have to shove the whole plant along its own growth axis.

AUDUS: I would like to make two points. One is that there is no strong evidence, for roots at least, that the whole sensitivity of the root to gravity changes very dramatically in the early stages of growth. There is evidence, for instance, that during the first few millimeters of growth you may in fact get a negative geotropic response, which is followed later by the normal positive response. This very much complicates one's interpretation of your data. The other remark I'd like to make concerns other movements that occur in plants, not only under conditions of simulated zero gravity on a clinostat but also under normal gravitational conditions. These are intrinsic oscillations that we call circumnutation, which are presumably due to some feedback mechanism in growth control. Now it seems to me that the existence of circumnutation obviates the possibility of showing that the plants can detect these space changes: you will get intrinsic oscillations, presumably not due to a field of any kind, which might overshoot—quite at random, one way and then the other—the position where it experiences these space changes.

FREIER: This is possible.

HERTEL: I would like to support Dr. Audus's suggestion. For example, auxin transport in a very small coleoptile is atypical, and not comparable with the transport in coleoptiles that are 1 cm long. I also think that Dr. Wilkins' objection is valid. The plant couldn't possibly make a complete 180° turn at one *point*. Thus I believe that the type of experiment presented by Dr. Freier does not give an accurate measure of the geotropic sensitivity.

FREIER: I think it does reflect the sensitivity by the fact that if we change the rate of rotation, we find a different turning distance at which the shoot again seeks the center.

AUDUS: There is also an objection based on the fact that the systems are changing. That is, the position of the sensor is changing, and therefore the forces operating on the sensor likewise change.

FREIER: Yes, and I think that is what we are trying to test. The sensors are predominantly at the tip. The position of the tip is changing, and hence $\omega^2 r$ is also changing. When $\omega^2 r$ gets big enough, then the tip senses the acceleration, and if $\omega^2 r$ isn't big enough, then the tip doesn't sense the accelerating forces.

JENKINS: In a 200-mile orbit of a biosatellite there is 95% of the gravity at the surface of the earth. The forward acceleration vector counterbalances exactly the earth's gravity and results in the state of free fall. Do you want to call this free fall and not weightlessness? And do we ever have zero gravity?

POLLARD: Free fall is a fine term. Zero gravity would be approached a long way from the earth if you suppose a finite size for the universe. It becomes a semantic problem.

2
Physical Determinants of Receptor Mechanisms

Ernest C. Pollard, *Pennsylvania State University*

There are two reasons for this paper, one largely negative and the other, I hope, positive. Suppose I take the negative first. A few years ago, as a member of the Bioscience Committee of NASA, I encountered the desire to examine problems of weightlessness by observing the behavior of bacteria in spacecraft. Such experiments are attractive, since a culture can be tucked away in a remote corner, but they seemed to me to be very dubious and, in reality, very expensive. So I performed (1965) a theoretical analysis of the probable usefulness of small cells in studying the effects of gravity. I concluded that the processes of diffusion are so potent in covering small distances that it would be pointless to expect any effect of gravity to introduce any perturbation in the action of diffusion unless the cell exceeded about $10\,\mu$ in diameter. Since this is an important conclusion, I propose to outline the reasoning involved and to show the results of some experiments on bacteria at $50{,}000\,g$ that provide experimental support for the theory.

The process of diffusion, or, what is the same thing, Brownian movement, is characterized by violent, aimless motion with an extremely high collision frequency. The actual molecular motion takes place at the speed of thermal agitation, which is governed by the relation

$$\frac{1}{2} mv^2 = \frac{3}{2} kT, \tag{1}$$

where m is the mass of the diffusing particle, v its velocity, k Boltzmann's constant, and T absolute temperature. Substitution of numbers, even for a relatively massive particle like a ribosome of "molecular weight" 5,000,000 gives the following calculation:

$$\frac{1}{2} \times \frac{5 \times 10^6}{6.03 \times 10^{23}} v^2 = \frac{3}{2} \times 1.4 \times 10^{-16} \times 300$$

$$v = 120 \text{ cm/sec}$$

Thus, even such a relatively clumsy object moves with a speed of over 1 m/sec, and a small molecule like indoleacetic acid travels at over 250 m/sec. So, even though the motion is aimless, it is very fast, and if the distances to be traversed are not great, the rapid random motion wins handily over the relatively slow, purposeful motion induced by gravity.

In order to see this more quantitatively we can consider the time it takes to diffuse across a cell $1\,\mu$ in diameter. We can use a diffusion constant, as measured in cellular material by Lehman and Pollard (1965), of 10^{-6} cm^2/sec and obtain the following:

$$\delta^2 = 2 Dt, \tag{2}$$

where δ is the distance diffused, D the diffusion constant, and t the time taken. For $\delta = 10^{-4}$ cm we find $t = 5 \times 10^{-3}$ sec. Thus, the time involved is only 1/200 sec. If we contrast the time taken to move the same distance under gravity for an object the size of a ribosome, we have

$$\frac{4}{3}\pi r^3 (\rho\text{-}1)g = 6\pi\eta r v, \tag{3}$$

where r is the radius of the ribosome, ρ its density, 1 the density of water, g the acceleration of gravity, η the viscosity, and v the velocity. Taking η to be 10^{-2} poise, $\rho\text{-}1 = 0.5$ and $r = 10^{-6}$ cm, we find v is about 10^{-8} cm/sec. So the time required to travel 1 μ, 10^{-4} cm, is 10^4 sec, two million times longer. This is a cogent reason for not expecting to see gravity effects in bacteria.

If the cell is 100 times larger, and the object which has to diffuse is 10 times larger, the situation is altogether different. Then we find a 10,000 times longer diffusion time, and 100 times faster sedimentation under gravity, so that the factor of a million is not now present. We can predict that the effect of gravity will be comparable to diffusion, and we accordingly expect to see effects due to gravity.

Experiments on Bacteria at 50,000 g

To reinforce this conclusion some experiments performed in our laboratory by Brian Walker and Mrs. Bonnie Nichols are briefly described. The concept of the experiment is shown schematically in figure 1. The ability of the bacterial cell to synthesize DNA, RNA, and protein is observed by measuring the uptake of ^{14}C-labeled thymine, uracil, and leucine into the trichloroacetic acid (TCA) insoluble fraction. In order to subject the cells to high-speed centrifugation without forming a pellet and changing their condition drastically, the cells must be suspended in an isodense suspension. The polysaccharide FICOLL (Pharmacia) permits doing this without involving the cell in any osmotic problems, as the average molecular weight of the particles of FICOLL exceeds 500,000. Roughly 35% solutions were found to be quite satisfactory. The expectation is that a depression in the amount of synthesis of the three macromolecular constituents will be observed, as is shown in the second graph in figure 1.

Calculations made on the basis of eq. (3) assuming a radius of 2,000 Å, a density difference of 0.1, a viscosity of 10 poise, the macroscopic viscosity estimated by Lehman and Pollard (1965), indicate that at 50,000 g the DNA, assumed to form a nuclear body in the cell, should sediment at 4×10^{-5} cm/sec. Thus, this nuclear body should reach the end of the cell in a few seconds. Polysomes should sediment in much the same way. Here the viscosity assumed, again on the basis

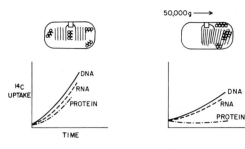

Fig. 1
Scheme of the centrifugation experiment. The normal cell, with its DNA and polysomes normally distributed can synthesize DNA, RNA, and protein as indicated. The measure of such synthesis is incorporated into the TCA insoluble fraction of ^{14}C counts. The centrifuged cell, with distortion of its synthetic machinery should show a reduced synthesis as indicated on the right.

of the experiments quoted, is less, namely, 1 poise, the radius assumed is 400 Å, and the density difference is 0.2. While the polysomes will be expected to sediment in much the same way as the DNA nuclear body, the smaller size suggests that there will be back diffusion in the case of the polysomes but not in the case of the nuclear body. Thus the polysomes will form a gradient toward the end of the cell. This gradient is shown schematically in figure 1.

The results given in tables 1, 2, and 3 are self-explanatory, and it can be seen that with the possible exception of the leucine uptake, corresponding to protein synthesis, there is no effect.

TABLE 1. Thymine Uptake ^{14}C Counts, 50,000 g TCA Insoluble Fraction

	0 min	30 min	70 min	108 min
Control	104	290	472	657
Centrifuged	110	243	481	726
Ratio control/centrifuged	0.95	1.20	0.98	0.91

Conclusion: DNA synthesis is unaffected.

TABLE 2. Uracil Uptake ^{14}C Counts, 50,000 g

	0 min	35 min	70 min	110 min
Control	450	1350	2800	5000
Centrifuged	440	1750	3800	5250
Ratio control/centrifuged	1.0	0.8	0.75	0.95

Conclusion: RNA synthesis is unaffected. Temperature variation could account for the increase in the centrifuged case.

TABLE 3. Leucine Uptake ^{14}C Counts, 50,000 g TCA Insoluble Fraction

	0 min	42 min	80 min
Control	205	2040	3150
Centrifuged	202	1350	2940
Ratio control/centrifuged	1.0	1.5	1.07

The low value of uptake at 50,000 g repeated, at 40 minutes or so in four experiments. The repetition is probably significant. Conclusion: some aspect of protein synthesis may be slightly influenced by 50,000 g, but in the main there is no effect.

This result includes some surprising elements, for the cell should be definitely stratified, and one would not expect a stratified cell to function as a normal cell. It is possible that the cells were "tumbling" during the centrifugation, so that they presented different ends to the gradient. Calculations based on the measured viscosity of the FICOLL solution of 10 poise show that tumbling is very unlikely, and to give an experimental support to this conclusion some experiments were carried out with filamentous forms of *Escherichia coli*, the bacterium used, whose length is 20 or more times the diameter, and these, at 10,000 g, showed no difference from normal cells.

A series of experiments with *B. cereus*, a larger cell, at 10,000 g showed that the uptake of ^{32}P into the TCA insoluble fraction was unaffected also. We are led to the idea that the internal organization of the bacterial cell is fastened to the membrane in some way. The existence of a growing point on the membrane, through which the DNA has to pass, has been postulated (Hanawalt and Haynes 1967). If this were fairly substantial in size, as expected—for it certainly has more than one enzymatic function—it could hold the nuclear body in place. Then the only predicted effect of distortion would be in the polysomes, the region of protein synthesis. This may have been observed.

In any event, these experiments, together with the theory, suggest that gravity effects at 1 g are unlikely in bacterial cells.

Physical Considerations Regarding a "Gravity Receptor"

This second or positive section of the paper owes its origin to a remark made by Thimann at a meeting. He said that we do not know the nature of the "gravity receptor." Since this was the first time that I as a physicist had realized the necessity of a biological concept of a "receptor," the train of thought given here was started. It should be said that many physical factors that influence cells hardly can be said to have receptors. Thus, temperature affects enzymes and active transport mechanisms, transcription, and DNA replication, to name only a few, and presumably only one particular element serves as the temperature "receptor" for any one cell in any one range. A mutant cell can have a different behavior because of one mutation. On the other hand, ultraviolet light is absorbed most readily in the nucleic acids, and so these can be thought of as the receptors. Since Newton's law of universal gravitation contains the famous words "*Every* body attracts *every* other body," one would expect that the whole cell, or indeed the whole system, may prove to be the receptor. It may well be that this is the simple truth, and if biological experiments were designed by physicists, I suspect that this would be the initial working hypothesis.

When one takes a second look, the bland statement above looks less persuasive. The cellular organelles have different sizes and densities; they perform different functions, they form part of a biological ensemble that can adapt to develop a slow but large response. It becomes reasonable then to examine the effect of the gravitational field on various components of the cell, and when this is done candidates are seen for the role of receptor.

First, there is one very important aspect to consider. If, indeed, the location of an organelle is significant, then it must be certain that there the distribution of the organelles is such that they can be influenced by gravity. The first test to apply is that of the Boltzmann distribution under gravity: the distribution of an isothermal atmosphere, the distribution of colloidal grains as classically studied by Perrin. The relation is

$$n/n_0 = e^{-mgh/kT} = e^{-V(\rho-1)gh/kT}$$

where n/n_0 is the ratio of the number in the distribution at a height h to that at height 0, V is the volume of the organelle, ρ its density, g the acceleration of gravity, and kT is the familiar product of Boltzmann's constant and the absolute temperature.

This experimental type of relation is sharply varying. If the term in kT is large so that the exponent is nearly 0, the ratio is very close to unity, meaning that there is uniform mixing. This holds for air molecules in the atmosphere, where h has to be hundreds of meters for an appreciable deviation. On the other hand, if the numerator in the exponent is large compared to kT, there is a sharp distribution, completed in millimeters. Now we can be fairly sure that for distances of the order of $1\,\mu$ the biochemistry of the cell takes place by diffusion as we have stated, and as discussed more fully in Pollard (1961, 1963). The synthetic processes occur by what has been termed "random collision and specific selection," and the properties of Brownian movement make this very effective. It is beyond $1\,\mu$ that the distribution needs to alter with the gravitational field. So for a limiting case we can require that for h to be $1\,\mu$ we need $V(\rho\text{-}1)gh$ to be of the order of kT. Assuming that $\rho\text{-}1$ is 0.5, we find that the volume of this distributed element is about 10^{-12} ml^3, about right for a mitochondrion but too large for a ribosome. Any dense object larger than this, which if spherical has a radius of 6,000 Å or $0.6\,\mu$, will certainly be distributed under gravity and so is a candidate for a gravity receptor.

Clearly the whole nucleus can serve as such. One has to ask whether the position of the nucleus matters. This opens up the second line of thought about a receptor: it must have a function that depends on how it is distributed. The aspect of the nucleus that is concerned with the generation and diffusion of particles into the cytoplasm certainly has such a function. Because of this it is very worthwhile to examine the character of the nucleolus as playing the role of receptor. The examination is interesting. The size and density of the nucleolus, which can easily be $5\,\mu$ in diameter and which can be very dense, both suggest that it is considerably distributed in a gravitational field. If, as Perry (1966) suggests, the nucleolus produces ribosomal RNA and causes this to diffuse into the cytoplasm, then clearly a strong asymmetry in the position of the nucleolus will cause a concentration gradient of ribosomes and a consequent concentration gradient of protein synthesis. Thus, the nucleolus cannot be ignored as a possible receptor.

Stress on Membranes: Physically Weak Membranes

Hydrostatic stress on weak membranes can cause a change in the separation of the component molecules that form the membranes. This has been discussed (Pollard 1965), and the basis for assessing the relative importance of hydrostatic stress on such membranes was some experiments of Weibull which suggested that the protoplast membrane has an elasticity of 10^8 dynes/cm^2. This is also the sort of figure deduced for the membrane of *Phycomyces blakesliana*. However, discussions with individuals concerned with artificial lipid membranes strongly suggest that this figure is much too great and that there may be membranes in cells that are genuinely of a lipoprotein form and that might have considerably less strength. Such physically weak membranes are excellent candidates for gravity effects. The kind of thing expected is shown in figure 2. For the purpose of making a point an elongated cell made of separate units of lipoprotein is hypothesized. In the one instance this cell is on its side, and it can be seen that the type of distortion undergone by the membrane is completely different from what occurs when the cell is on end. Placing the cell on end results in a longitudinal modification of the lipid membrane, and this might easily lead to considerable excess diffusion, or transport, of a small molecule. Thus, where there is any suspicion that the small molecule, such as indoleacetic acid, is involved, one of the first regions to suspect is a region that is rich in these physically weak membranes.

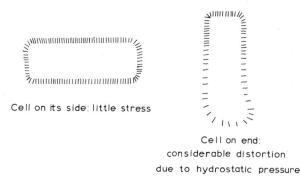

Fig. 2
Elongated cell made of separate units of lipoproteins, showing the effect of hydrostatic pressure on the lower part of the membrane when the cell is on end. Increased permeation would be expected.

Enzymatic Transport

If there is any reason to suspect that a small molecule is transported by means of an enzyme, then it is of some interest to see whether this transport process could be influenced by gravity. There is no doubt that our understanding of a permease mechanism is all too little. However, if it does correspond to a physical change of structural form, such as the shortening of the muscle fibril or the transition from random coil to helix, then such physical motion is bound to cause alterations in the water entrained with it, and this in turn is bound to cause change in the entropy of activation of the system. The change in the entropy of activation should be observed as a temperature coefficient. We would expect gravity effects to have strong temperature coefficients, and if these were observed, it would be presumptive evidence that an active transport mechanism was involved in the process.

Since such a membrane has to have a certain amount of rigidity and since the degree of stress in cells is not great, one would tend to predict that not much effect should be observed on active transport systems.

By observation, the available concentrations when the cell is considered as a fully open system are often too low. Such cells must contain borderline compartmentation. When this borderline compartmentation occurs (for example, by the use of the increased concentration within a segment of the endoplasmic reticulum), we would expect the effect of hydrostatic stress on the membrane to be considerable, and gravity effects should be observed. Thus, when there are many weak membranes, and when these membranes exert some effect in confining particular metabolites, an excellent situation for a gravity-affected system is present.

As a word of final philosophy, speaking more as a biologist than as a physicist, I would like to say that the cell seems to use for control positive devices that work over a considerable concentration range. *If* there is a clear-cut decisive development, that development (we may suspect) would be hard to influence by weightlessness, because it would almost certainly take place regardless of a concentration change. Hence it would not seem wise to look at sharp, clear developmental changes. If, on the other hand, much variety is possible, and if it is certain that there is a concentration dependence, then one almost would be tempted to say there is certain to be a gravity effect.

Summary

Theoretical reasons are given why the effects of weightlessness should not be studied in bacterial cells. The rapidity of Brownian movement renders diffusion more efficient than convection. Experiments at 50,000 g that support this are described.

Suggestions regarding a gravity receptor are made. One is that the whole nucleus might act as receptor; a second is that the nucleolus might be important; and a third is the implication of physically weak membranes such as those in the endoplasmic reticulum. If in any way active transport involving enzymes is part of the gravity-mediated process, then temperature effects of weightlessness should be studied. It is suggested that cell components that function over a large concentration range will not be involved with gravity, but that effects that occur in considerable variety and that are influenced by concentration should be gravity-dependent.

Acknowledgments

The help of Mrs. Bonnie Nichols and Brian Walker in the centrifugation experiments is gratefully acknowledged.

References

Hanawalt, P. C., and Haynes, R. H. 1967. *Scientific American* 216: 36-43.

Lehman, R. C., and Pollard, E. C. 1965. *Biophys. J.* 5: 109-19.

Perry, R. P. 1966. *In* National Cancer Institute Monograph 23, ed. W. S. Vincent and O. L. Miller, Jr., pp. 527-47. Bethesda, Md.: National Institutes of Health.

Pollard, E. C. 1961. *J. Theoret. Biol.* 1: 328-41.

———. 1963. *J. Theoret. Biol.* 4: 98-112.

———. 1965. *J. Theoret. Biol.* 8: 113-23.

Discussion

PICKARD: There are perhaps two objections to the nucleolus hypothesis. The first is that the viscosity of the surrounding nucleic acids is very high, so that the movement of the nucleolus would be expected to be quite slow. The second is the experimental observation that in the geotropism of the wheat coleoptile the nucleoli are distributed at random.

COHEN: How do you check the position of the nucleolus?

PICKARD: In glutaraldehyde-fixed material.

COHEN: I am suspicious of this criterion and would have preferred observations on living material. Don't you think there are great changes in viscosity with any fixative and that this might affect the position of the nucleolus?

PICKARD: Of course, viscosity changes during fixation, but glutaraldehyde is a relatively mild fixing agent and properly used does not move the organelles from their normal position as far as electron microscopists can tell.

COHEN: How do you know? Observations on living material would be less equivocal evidence.

BANBURY: We have on film the motion of actively streaming cytoplasm which shows marked displacements on the introduction of glutaraldehyde.

AUDUS: I think the nucleus and nucleolus are automatically ruled out, simply because of the phenomenon of cytoplasmic circulation. This streaming is so active that any kind of differential diffusion from the nucleus would be completely swamped by the circulation. My feeling is that gravity sensing could only take place in those regions of the cell that are relatively static, that is, those bordering on the cell wall.

POLLARD: I hate to stick my neck out, but you are discussing something you can see versus something you cannot see. You cannot see diffusion. You can see streaming. Any calculation you make on streaming versus diffusion puts streaming a very poor second.

AUDUS: Yes, but if the nucleus and other organelles are being carried around, you are not going to get any substantial concentration difference.

GORDON: I see no reason to assume unidirectional displacements by the fixative. It seems to me, therefore, that a pragmatic approach would be the statistical. Can significant differences in position or orientation of the nucleus (nucleolus) be correlated with reorientation of the cells in a 1-g field? Such correlations seem to be essential from the physical considerations discussed by Pollard, though not in themselves proof of a sensor function.

COHEN: I would like to support Pollard's ideas about the possible localization of the nucleolus and ribosomes. We have been studying similar processes in nerve cells of insects. If the axon of an insect nerve cell is injured, there is a rapid change in the distribution of cytoplasmic RNA in that cell. We have frequently seen a dimple in the nuclear membrane adjacent to the concentrations of RNA. The ultrastructure of this area indicates that new endoplasmic reticulum (ER) has been deposited and that ribosomes have studded the cisternal walls to form rough ER. This structure is largely absent in the normal insect neuron. Although this has nothing to do with gravity, it bears on what you were saying about localized effects. There seems to be a localization of ribosomes and ER at a particular region of the nuclear membrane.

POLLARD: There may be no gravity receptor either.

COHEN: There probably is no direct effect of gravity on neurons, but nerve cells satisfy many of your criteria for a gravity receptor. It also raises the possibility that high g may directly affect nerve cells.

KALDEWEY: Concerning the comment of Audus, who said that the nucleus is ruled out as a georeceptor because of being moved through the cell with the cytoplasmic streaming, we may ask whether the speed with which the nucleus is moved is the same at every locus within the cell. If we assume that the nucleus moves faster in the upper half than in the lower half of the cell, the duration of its presence in each half would be different. Consequently, more nuclear substances could diffuse into the lower half than into the upper half of the cell. Furthermore, if the particles diffusing out of the nucleus are small enough to be fixed in the part of the cell where they are liberated, this would result in a polarity within the cell.

GALSTON: I believe that the evidence on the tropistic responses of multicellular organs rules out the nucleus, or any part of the nucleus, as a possible gravity receptor. The fact of the matter is that the stem or root responding to gravity is a multicellular organ. If one observes the distribution of nuclei across that horizontally placed organ one sees no constant position of the nucleus with respect to the direction of the gravitational field. If you want to suppose that the nucleoli which have fallen down to the bottom of the nucleus are going to produce a gradient responsible for the curvature, you have got to find some way to summate this stimulus for the entire multicellular mass with randomly placed nuclei.

SHEN-MILLER: Concerning possible gravity sensors, we might consider the dictyosomes, or Golgi apparatus. Several years ago Audus and his coworkers published on the geosensitivity of the root cap. Removal of the cap completely eliminated the geotropic sensitivity of the root. Now, one of the differences between the cap cells and the cells behind the cap is in their content of dictyosomes: there are about 400 dictyosomes per cap cell as compared to 10 in the cells behind the cap. Dictyosomes produce vesicles which move toward the cell wall region and fuse with the cell membrane, and their contents have been found in the cell wall matrix. Tropism is the result of a differential wall synthesis or wall stretching or both, and dictyosomes are organelles that are actively involved in cell-wall formation.

AUDUS: I'd like to make one comment concerning membranes and the effect of starch statolith sedimentation. In root tips of *Vicia faba*, we've done a statistical survey of the distribution of organelles in the cell in relation to the sedimentation of these plastids, and we find that there is no significant change in the distribution of mitochondria or of the Golgi, but there is a significant shift in the ER. There tends to be an accumulation of ER on the upper side of the cell brought about presumably by a mechanical effect of the statoliths sedimenting to the bottom of the cell.

BROWN: Dr. Pollard, I should like to return to the model of a tension-sensitive membrane system as a possible g sensor. Could you calculate the smallest g force which could be detected by such a system? Alternatively could you estimate the lower size limit for the organism where differential stresses on membranes would no longer make it possible for the cell to sense 1 g?

POLLARD: That's very difficult to answer because of the problem of determining how weak the membrane is. If in fact the membrane is as weak as those ascribed to the red cell, then I would say that the cell with a number of these would be very sensitive indeed.

BROWN: Suppose you would like to find the sensor in a very small cell, a bacterial cell, would your model offer any hope for this?

POLLARD: I think not. Mostly because there are not many membranes in the bacterial cell—only one.

BROWN: Perhaps it takes only one. You're making the assumption because you find some limit in sensitivity in terms of cell size. It seems to me that the limit is going to be very, very small.

HERTZ: Another clue to the size of the statoliths may be obtained from the presentation time, the exposure duration that elicits a barely detectable georeaction of the plant. The length of the presentation time depends on organ, species, and temperature; for example, in the oat coleoptile it is about 2 min at 20°C. This relatively accurate definition of the perception time invites speculation about its physical origin. Along the lines just described by Pollard, and elsewhere by Audus and by Gordon, presentation time can be calculated by using Stokes' law for the sedimentation of particles in a viscous medium and Einstein's relation for the average motion of a particle subject to Brownian movement. Our calculations showed that sedimentation of mitochondria would be compatible with the presentation time for the oat plant and could be responsible responsible for the geoperception.

Assuming that the perception mechanism involves particles sedimenting in a uniform fluid, the relation between the presentation time t_p and the gravitational force f should be given by $t_p f^2$ = constant. Johnsson (*Physiol. Plantarum* 18 [1965], 945) has made a careful study of this relationship, using both the geoelectric effect and geotropism as indicators. Using a centrifuge to vary the force applied, he found that the two variables followed very closely the form $t_p f$ = constant. I believe that this result excludes the possibility that particles the size of mitochondria or smaller play an important part in the *perception* process.

However, because of our insufficient knowledge of the physical constants of the cell contents, inferences about the precise minimum dimensions of the statolith from simplified physical models probably have only limited value. Protoplasm undoubtedly is not a Newtonian fluid—its viscosity is probably not constant and it is likely to be thixotropic. Furthermore, protoplasmic streaming in the cell may also influence the movement of statoliths.

3
Oscillatory Movements in Plants under Gravitational Stimulation

Anders Johnsson, *Lund Institute of Technology*

In connection with the two preceding papers it should be pointed out that in the calculations on sedimentation of particles in cell protoplasm, all so-called non-Newtonian behavior of the cell content is neglected. We should be very careful before we assume the cytoplasm to be a Newtonian fluid, that is, one in which the viscosity is independent of the applied force. Only if the liquid is Newtonian will the falling velocity of a particle be proportional to the applied body force.

In experiments on Elodea leaves no displacement of the chloroplasts was reported after several minutes of centrifugation at about 900 g. Increasing the centrifugation time, however, caused the chloroplasts to move abruptly (Virgin 1951). I think such types of behavior must be kept in mind when calculating particle displacement within cells.

When detecting a phase angle between a force vector and a displacement vector, we must imagine a sensing system in which both quantities are independently but simultaneously recorded and then compared. I think this is a very interesting aspect of gravity perception in the plants that Professor Freier worked with. However, so far as I know, two such distinct sensing systems have not been reported to exist in plant cells. Therefore I find it symptomatic that the experimental results Professor Freier reported point to a simpler gravity-sensing system than a phase-detecting one.

In investigating gravity-perception systems in plants one must keep certain facts in mind.

1. Gravity perception and the consequent action must be related to the direction in which the gravitational force (or centrifugal force) is applied. Nothing justifies the assumption that a gravitational stimulation is perceived in the same way when applied longitudinally as when applied laterally. The sensitivity to gravitational force and the plant's response to it therefore might very well be different in the two directions, a possibility that should not be overlooked in experimental evaluations.

2. Investigations of a plant's reaction to gravitational force should include data on how long a certain force has acted. The theory that the plants react to the integrated force, that is, that

$$\text{reaction} = \text{constant} \times \int_0^t f \cdot dt,$$

is not valid under all circumstances—intermittent stimulation, for example (Günter-Massias 1928). I feel strongly that it is advisable to use constant forces in such experiments to justify use of the established formula. When constant force is used, the reaction should, of course, be proportional to the force $\times$ time. An appreciable reaction occurs if the force is applied longer than the so-called presentation time, t_p. It has been shown that in the neighborhood of 1 g, $f \times t_p$ = constant for laterally applied forces (Johnsson 1965).

If we measure the reaction due to a lateral gravitational stimulation as indexed by the maximum deflection angle, we obtain the relationship

$$\text{max. curvature} = \text{constant} \times \log\left(\frac{t}{t_p}\right) \qquad (1)$$

In the present instance, this means that the reaction does not follow a simple integrated force function. Again, the conclusion is that experimental test forces should be kept constant.

In equations proposed to characterize movements of plants (as after gravitational stimulation) the time lag between stimulus and response is frequently omitted. I shall try to show how profoundly this time lag can take part in the growth movements of the hypocotyls of the sunflower plant, Helianthus.

These hypocotyls exhibit oscillatory movements from the vertical (or circumnutations) having a period time T of about 2 hours at room temperature. The oscillations are approximately sinusoidal in one dimension (even when the amplitude slowly increases during longer experiments). We may therefore describe the oscillations in one dimension as

$$a = a_0 \sin \omega t, \qquad (2)$$

where ω is the angular frequency and a the angle with respect to the vertical.

Let us examine the movements in one dimension (the two-dimensional movement being composed of two perpendicular sinusoidal movements) and denote the rate of curvature at a time t as $da(t)/dt$. It is reasonable to assume (Israelsson and Johnsson 1967) that this quantity is approximately proportional to the difference in concentration of active auxin between the both sides of the plant stem. Moreover, we can assume this in turn to be proportional to the sine of a at an earlier time t_0:

$$\frac{da(t)}{dt} = -k \sin a(t - t_0). \qquad (3)$$

Here t_0 denotes the time lag between stimulus and response and is seen in figure 1.

The equation is a simple one, probably not justified under all circumstances, but it will prove useful in the following treatment. We can consider small angles and first study

$$\frac{da(t)}{dt} = -ka(t - t_0). \qquad (4)$$

Remembering that the oscillations are sinusoidal, we may put this condition in the equation to obtain

$$2\pi \frac{t_0}{T} = \frac{\pi}{2}. \tag{5}$$

Thus we can check the equation by looking for the linear relation between t_0 and T. There is a discrepancy between the experimental and the theoretical values of t_0/T, but a linear relationship is found between the two quantities. This is seen in figure 2, where t_0 and T vary with the temperature.

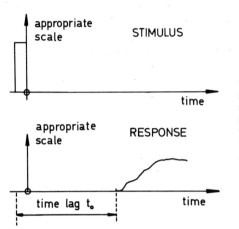

Fig. 1
Time lag between stimulus and response in a stimulus-response system. The gravitational stimulation of the Helianthus hypocotyl consisted of tilting the plant with respect to the vertical. The response is the geotropic curvature. The time lag t_0 can be interpreted as the geotropic reaction time and is of the order of 25 min at room temperature.

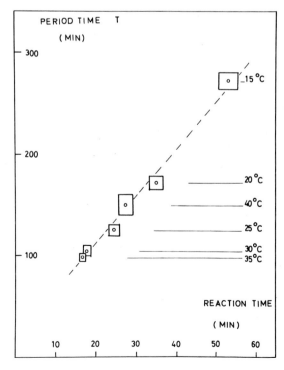

Fig. 2
Period time, T, for circumnutations of Helianthus hypocotyls plotted as a function of the reaction time at various temperatures. The standard deviations of the means are represented by the squares.

When we extend the equation to account for the plant's ability to "remember" a stimulation, the observed ratio t_0/T can also be accounted for. The equation is then somewhat more complicated:

$$\frac{da(t)}{dt} = -k \int_1^\infty f(x) \sin a(t - xt_0) \, dx. \tag{6}$$

It is obvious that when we know $f(x)$ we can predict the plant's response to a gravitational stimulation. From investigations on circumnutations (Israelsson and Johnsson 1967) reasonable values of k, $f(x)$ and t_0 have been approximated; $f(x)$ has been assumed to have an exponential decay. In figure 3 the equation has been used to calculate the response to a 10-min geotropic stimulation. It is seen that the experimental curve fits the theoretical relatively well. (The amplitudes differ, since the theory for the oscillations has been derived under the assumption that only the upper part of the stem is stimulated, whereas in the experiment the whole stem is stimulated.) More recent work (Johnsson and Israelsson 1968, 1969) has supported the validity of the model.

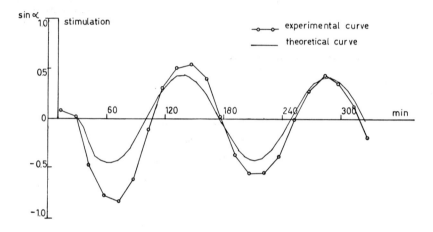

Fig. 3. Theoretical curvature (based on eq. [6] in text) and experimental curvature of Helianthus hypocotyls after a 10 min period in horizontal position ($\sin a = 1.0$).

If we have no regulating function, as under free-fall conditions, it could be argued that the output should be determined by the noise in the system. Thus

$$\frac{da(t)}{dt} = \text{constant} \times e(t), \tag{7}$$

where $e(t)$ is a noise-generating function. When $e(t)$ represents a white noise generator, the mean value of the output should be zero, that is, $\bar{a} = 0$. The quantity $\bar{a}^2$ should increase linearly with time. Under clinostat conditions this has, in fact, been found to be the case (Johnsson 1966, 1968).

In conclusion then, the equations presented describe satisfactorily circumnutations, response to short gravitational stimulations, and behavior on the clinostat.

Summary

In living organisms detectable response to a gravitational stimulation often lags behind the stimulus, that is, a time lag t_0 must elapse before reaction is detectable.

This time lag must be introduced into the equations describing gravitational compensating movements in plants. In Helianthus large oscillatory movements (nutations) occur and these can be described satisfactorily by a differential equation.

The equation can describe oscillations in the system, and these then correspond to the observed nutations. By introducing a "memory" function with respect to gravitational stimulation, experimental findings appear to be in good accordance with the theory.

References

Günter-Massias, M. 1928. *Z. Botan.* 21: 129-72.

Israelsson, D., and Johnsson, A. 1967. *Physiol. Plantarum* 20: 957-76.

Johnsson, A. 1965. *Physiol. Plantarum* 18: 945-67.

–––. 1966. *Physiol. Plantarum* 19: 1125-37.

–––. 1968. *Studia Biophys.* 11: 149-54.

Johnsson, A., and Israelsson, D. 1968. *Physiol. Plantarum* 21: 282-91.

–––. 1969. *Physiol. Plantarum* 22: 1226-37.

Virgin, H. 1951. *Physiol. Plantarum* 4: 255-357.

Discussion

WILKINS: When studying geotropic bending of horizontally placed shoots, I have observed that the bending is not always a smooth, continuous process, but that a kind of oscillating or fluctuating movement is sometimes superimposed. These oscillations seem to have a relatively high frequency, at least much higher than the frequencies calculated by you. Is there any explanation for these fast oscillations?

SHEN-MILLER: Relatedly, investigating geotropism of clinostat-grown seedlings, I find about a 30-min periodicity in the curvature development. And, as Dr. Hertel pointed out, comparable fluctuations were observed by Newman.

JOHNSSON: We have shown that solution of eq. (6) yields frequencies of five times the fundamental and even higher. The oscillations you mention would probably come out of the equation. The oscillation described here is only the simplest solution, and one that characterizes adequately the periodicity of nutations.

4
Aspects of the Geotropic Stimulus in Plants

Rainer Hertel, *Michigan State University*

Different Phases of Stimulation According to a Statolith Hypothesis

If we disregard the interesting reactions of plant organs to gravity-induced, external deformations (see chap. 6), the primary forces and displacements operative in geotropism are effective within the externally undisturbed tissue and probably within the individual cell (including the cell wall). Thus, considerations of the physical forces and their first consequences may tentatively be limited to the single cell.

Among the physical forces responsible for geotropism in plants, only the very first and biologically trivial action of gravity does not require knowledge or assumptions concerning the nature of the geosensor: the gravitational (or centrifugal) acceleration necessarily acts on masses. Questions about the nature of other forces involved ("equilibrium forces," see chap. 1), such as whether the masses acted upon must undergo significant displacement to be effective, are dependent on the actual mechanism of stimulation. This, however, is unknown for higher plants.

The "neutral" description of the geotropic reaction chain (exposure–susception–reception) (see, for example, Rawitscher 1932) has not provided a meaningful subdivision of physical forces and transduction. If, however, a statolith particle mechanism (Haberlandt 1900; Nemec 1900) is assumed from the beginning, one can distinguish three distinct phases of possible "primary" stimulation which follow an alteration of the direction of gravitational acceleration, as by moving a plant organ from the vertical to the horizontal position: (1) displacement of statoliths in the cell; (2) action of the particles at the new equilibrium position (such as the lower membrane); (3) return to symmetrical distribution (after the plant is returned to the vertical position or placed on a clinostat).

Phase 1, under the assumptions stated, may or may not be relevant for the existence of a presentation time; the reciprocity law (gt = constant) has been verified, at least for threshold stimuli (Johnsson 1965; Shen-Miller 1970).

Phase 2 can be studied apart from phase 1 by finding conditions where displacement is completed rapidly or where subsequent biochemical processes are blocked. Since effects such as lateral auxin transport (see chap. 10) continue long beyond the time needed for displacement of the supposed statoliths (that is, long past completion of phase 1), stimulation most likely occurs during phase 2. If this is so, it can then be asked whether cellular asymmetry is achieved by contact forces (pressure), by electrical forces, or by chemicals produced from the sensor, and also whether the actual displacement (phase 1) has any effect by itself. Of phase 3 the statolith hypothesis predicts that stimulation will, under "physiological" conditions on a clinostat, continue and slowly decrease, regardless of whether asymmetric distribution or pressure of the sensor is effective. (But for qualification of this argument, see Richter 1914.)

Testing Correlations between Displacement of Particles and Geotropic Bending

Comparisons of geotropic presentation time with displacement rates of movable starch, in various plants and organs with fast geotropic reactivity, have demonstrated striking correlations (Hawker 1932; chap. 7, this volume). For further study of the possible geosensor a zero response method was developed. This avoids the measurement of presentation time, which is difficult to achieve, as well as the need to draw conclusions from the degree of bending, which may be controlled by several factors.

A special centrifuge (fig. 1; detailed description in Ouitrakul and Hertel 1969) permits exposure of excised corn coleoptiles (about 2 cm long) to forces up to 40 g. The coleoptiles are mounted, 8 per Plexiglas stand, on agar over needles provided with soft tubing at the tips. The apparatus is covered with a hood made of hollow aluminum bars. If cooling is desired, the hood can be filled with water which is frozen and allowed to thaw slightly; this provides a temperature of 2-4°C. Rapid change to room temperature can be achieved by simply switching covers.

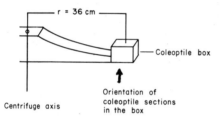

Fig. 1
One Arm of the Coleoptile Centrifuge. From Ouitrakul and Hertel 1969, by permission.

In a typical experiment coleoptiles were centrifuged at 2-4°C and 20 g (force normal to organ axis) and subsequently transferred to 25°C and to a clinostat (1 rpm; organ axis and previous force vector in plane of rotation). Curvature was measured 90 min after transfer to 25°C. Sample coleoptiles were fixed quickly at the end of centrifugation and examined by standard microscopic techniques for starch distribution in the cells.

All sets were centrifuged for 40 min in the cold in one direction (force to the "left"). A control was centrifuged for another 40 min, force downward, to return the starch grains to the basal part of the cell; this control developed no significant bending when transferred to the clinostat, indicating that no sizable perception had taken place during the relatively short time in the cold (see Brauner and Hager 1958) or that the stimulation had been "wiped out" by the vertical centrifugation. Test coleoptiles were centrifuged, with force to the "right" (opposite to their initial, 40-min treatment) for 0, 10, 15, 20, 25, and 40 min, and then transferred to the clinostat.

The 0- and 10-min sets developed curvature to the "right" (10° and 4°, respectively) the 20-, 25-, and 40-min sets showed clear bending to the "left" (-5°, -8°, -9°), whereas the 15-min set showed no significant curvature (zero response = 0.4 ± 1°). The distribution of amyloplasts across the cells was symmetrical in this 15-min set, but asymmetrical in all others, more to the "left" in the 0- and 10-min sets, more to the "right" in the 20-, 25-, and 40-min sets.

These preliminary data seem to confirm that starch grains have the appropriate displacement rate to account for the behavior of the statolith.

Recent experiments on starch-depleted roots (Iversen 1969) and studies with mutant corn coleoptiles (Hertel, de la Fuente, and Leopold 1969; Filner et al. 1970), summarized below, have provided further support for the starch-statolith hypothesis.

In coleoptiles of the amylomaize corn mutant (AM), the amyloplasts are much reduced in size in comparison with the wild type corn (WT), permitting a comparison of geotropic responsiveness as related to lateral displacement of amyloplasts and lateral transport of auxin. The amyloplasts of AM showed 30-40% less lateral redistribution in response to horizontal exposure as compared with WT. Under geotropic stimulation the lateral transport of auxin in the direction of growth was 40-80% less, and the geotropic curvature by the coleoptiles was also significantly less in the mutant as compared with WT (Hertel, de la Fuente, and Leopold 1969).

In contrast to the geotropic effect, phototropically induced lateral auxin asymmetry was not significantly different in WT and AM. Eleven other single-gene endosperm starch mutants of corn were compared with their corresponding normals. In *all* pairs, when a difference is geosensitivity of lateral auxin transport was present, it could be correlated with a parallel difference in amyloplast sedimentation. For example, sugary 1 ("67") had an amyloplast asymmetry index of 0.32 and a 13% gravity effect on auxin transport; the paired wild type had both a greater amyloplast asymmetry (0.61) and a greater gravity effect on transport (23%). Correlations between gravity effects on auxin transport and amyloplasts have also been demonstrated in comparisons of apical and basal sections of corn, oat, and sorghum coleoptiles (Filner et al. 1970).

The idea that starch plastids are involved in geostimulation is not incompatible with the findings of Pickard and Thimann (1966). Their results with starch-depleted coleoptiles prove that in higher plants massive displacement of large starch grains is not necessary to obtain a georeaction. The *start* of the bending, however, was greatly delayed in these coleoptiles, although there was no lag in elongation. This may indicate a different type of geostimulation. Pressure of other particles (or "empty" plastids) on the lower membrane, without major displacement, may also be operative (see the argument between Haberlandt [1902] and Jost [1902]). Furthermore, it is possible that a structure (perhaps on the inner surface of the plasma membrane) is "mechanically" sensitive to both total cell weight and to statoliths, and the latter, if present, would dominate and be responsible for any strong and fast reactions.

Action of Possible Statoliths in New Equilibrium Position

In studying phase 2, the action of statoliths, it may be asked whether, after the statoliths have been displaced, the amount of g force has a significant influence on the degree of stimulation. If the amount of force does have a significant effect, statoliths probably act by pressure or some similar mechanism; if it does not, they probably act by their mere presence. An experiment by Richter (1914), using roots which were briefly exposed horizontally and subsequently rotated with different orientation on the clinostat (the previous direction of gravity was either in or perpendicular to the plane of rotation), indicated that the postulated statoliths had to exert pressure, on the side they were closer to, to be effective during clinostat treatment.

This suggestion is supported by the following experiment (fig 2; see also Ouitrakul and Hertel 1969; Filner et al. 1970), which assumes an amyloplast statolith mechanism.

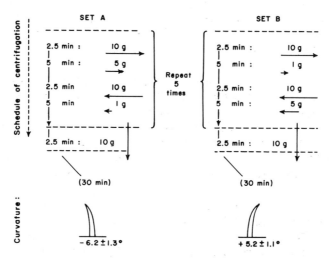

Fig. 2. Effect of Different *g* Forces on the Bending of Corn Coleoptiles. Two sets of 16 excised coleoptiles each were subjected to the schedule of centrifugation described in the figure. Samples for microscopy were taken after each of the 4 steps of the first cycle and at the end of the centrifugation program. The arrows at the symbols for *g* forces indicated the vector with respect to the coordinates of the coleoptiles schematically drawn in the figure. Curvature is given as degrees of angle (mean of 11 coleoptiles ± standard error of the mean). From Filner et al. 1970, by permission.

Excised corn coleoptiles were centrifuged for 2.5 min as described previously, but at 25°C and 10 *g* (exposure 1). Microscopy showed a complete displacement of movable starch to one side of the cells. The test treatment consisted in exposing the coleoptiles to a force (5 *g* or 1 *g*) in the same lateral direction for 5 min (exposure 2). Centrifugation for 2.5 min was then started in the opposite direction (exposure 3), compensating for the first exposure, followed by 5 min at 1 *g* or 5 *g*, (exposure 4). The sequence was repeated 5 times. At the end, the starch was centrifuged back to the basal cell wall and the curvature was measured on photographs after an additional 30 min of standing in the vertical, normal position.

The results indicate that even after displacement of the suspected statolith, 5 *g* is more effective than 1 *g*. In controls the first and third exposures were extended to 5 min to assure even more complete displacement; the results were essentially like those in figure 2. Different *g* forces (10 as against 1, 5, or 3; 3 as against 1; and 10 as against 5) were applied in the second and fourth exposures; the direction of curvature was always away from the direction of the greater force. To exclude the possibility that the result was due to thigmo- or traumatotropism caused by the inside needle, coleoptiles in other controls were exposed to the same treatment in glass tubing or under agar; the direction of curvature was the same as that in figure 2.

These findings suggest that the pressure, not the mere asymmetric distribution of the statoliths is effective—that a statolith might, for example, have to be pressed into a sensitive layer above the membrane.

Occasionally it has been implied that any statolith must produce some biologically active compound (see chap. 2). The data reported here support rather the original hypothesis of Haberlandt (1900), who proposed a "mechanical" action of the starch statolith on the lower side.

Features of Auxin Transport in Corn Coleoptiles

An analysis of the physical forces responsible for the biochemical processes leading to geotropic bending could also be attempted in the reverse order, by close investigation of biochemical processes in the response chain. Such knowledge might then point to the type of physical forces affecting the mechanism under study.

A change in the direction of auxin movement is an early consequence of the gravitational stimulus (see chap. 10). This lateral auxin transport is mediated by pumping mechanisms that are very similar if not identical to those of the active, longitudinal transport (chap. 12; but see also Burg and Burg 1966), and the latter is sensitive to gravity just as lateral transport is (Ouitrakul and Hertel 1969). Better understanding of lateral transport may suggest possibilities for a transduction mechanism.

The active and controlling steps of auxin transport are thought to occur at the plasma membrane. Evidence suggests that the individual cell is the unit of transport, that the pumping processes are repeated from cell to cell (Leopold and de la Fuente 1967), and that the exit step out of each cell (secretion) is the partial process controlling auxin movement (Hertel and Leopold 1963*a*).

The hypothesis has been proposed (Hertel and Leopold 1963*a*) that the geosensors act on auxin movement in a rather direct manner by favoring secretion at the physically lower membrane of each cell, resulting in an overall lateral transport through the organ (see chap. 13). Under this hypothesis, the transduction problem may perhaps be stated as follows: How can the pressure of the geosensor favor a membrane configuration which allows an increased auxin secretion?

Studies by Hertel and Flory (1968) have demonstrated two additional characteristics of auxin transport.

1. The active and specific pumps probably "handle" the auxin molecules by *noncovalent* interactions; the auxins indole-3-acetic acid (IAA) and napthalene-1-acetic acid (NAA), which is transported in a similar manner, move as free molecules.

A search for a transient auxin-X complex which was chasable, as required for a transport-carrier intermediary, gave negative results.

NAA was labeled in its carboxyl group with the heavy isotope ^{18}O and passed through the coleoptile transport system. If any covalent interaction occurred during passage, it would most probably take place at the carboxyl group and result in a loss of ^{18}O. The NAA arriving in the receptor blocks was isolated, purified, cocrystallized with excess ^{16}O NAA and analyzed for ^{18}O.

It was found that there was no loss of ^{18}O from the auxin transported through the tissue. This result makes the existence of covalent interactions in the transport process unlikely.

2. Auxin stimulates its own transport ("cooperativity"). The transport of the auxin 2,4-dichlorophenoxyacetic acid is usually sluggish; it is increased several times, however, if some IAA is added. Earlier observations (Hertel and Leopold 1963b) on the maintenance of IAA transport capacity by pretreatment with low levels of IAA can be understood on the same lines, indicating the possibility for amplification in the auxin transport system.

Summary

√ To discuss the physical forces involved in graviperception, several assumptions about the nature of the sensor are made. Accepting the hypothesis of a cellular statolith particle mechanism, a subdivision of the geostimulation is proposed.

A method is described to test correlations between microscopically visible cell particles and the direction of geotropic bending.

√ Within the framework of the starch statolith hypothesis, evidence is presented that the sensor does not stimulate by mere presence at the lower membrane but by some "mechanical" action dependent on gravitational force.

√ By analyzing auxin transport it is hoped to obtain clues concerning transduction and the type of physical forces involved in geoperception. The transport mechanism is likely to operate through specific noncovalent chemical interactions at the membrane. The observed "cooperativity" in this transport may provide possibilities for amplification.

Acknowledgments

This work was supported by the U.S. Atomic Energy Commission under contract no. AT (11-1)-1338. It is a pleasure to acknowledge the intelligent assistance of Ingeborg Schmidt and the cooperation of R. Geyer in designing and building the coleoptile centrifuge. Suggestions of F. Bertossi concerning the zero response method are greatly appreciated.

References

Brauner, L., and Hager, A. 1958. *Planta* 51: 115.

Burg, S. P., and Burg, E. A. 1966. *Proc. Nat. Acad. Sci.* 55: 262.

Filner, B.; Hertel, R.; Steele, C.; and Fan, V. 1970. *Planta* 94: 333.

Haberlandt, G. 1900. *Ber. Deut. Botan. Ges.* 18: 261.

———. 1902. *Ber. Deut. Botan. Ges.* 20: 189.

Hawker, L. E. 1932. *Ann. Botany (London)* 46: 121.

Hertel, R., and Leopold, A. C. 1963a. *Planta* 59: 535.

———. 1963b. *Naturwissenschaften* 50: 695.

Hertel, R., and Flory, R. 1968. *Planta* 82: 123.

Hertel, R., de la Fuente, R. K.; and Leopold, A. C. 1969. *Planta* 88: 204.

Iversen, T.-H. 1969. *Physiol. Plantarum* 22: 1251.

Johnsson, A. 1965. *Physiol. Plantarum* 18: 945.

Jost, L. 1902. *Biol. Zb.* 51/1: 161.

Leopold, A. C., and de la Fuente, R. K. 1967. *Ann. N. Y. Acad. Sci.* 144/1: 94.

Nemec, B. 1900. *Ber. Deut. Botan. Ges.* 20: 241.

Ouitrakul, R., and Hertel, R. 1969. *Planta* 88: 233.

Pickard, B. G., and Thimann, K. V. 1966. *J. Gen. Physiol.* 49: 1065.

Rawitscher, F. 1932. *Der Geotropismus der Pflanzen.* Jena: Gustav Fischer.

Richter, E. 1914. *Ber. Deut. Botan. Ges.* 32: 302.

Shen-Miller, J. 1970. *Planta* 92: 152.

II Gravity Receptors in the Plant

5
Gravity Receptors in Lower Plants

Andreas Sievers, *Bonn University*

Gravity receptors are to be found only in organisms which show a specific gravity-induced reaction. The gravity responses of the lower plants are rather diverse. Some flagellates, for example, Euglena and Chlamydomonas, are negatively geotactical, and the spermatozoids of the brown alga Fucus are positively geotactical (see Haupt 1962). In centrifuged eggs of Fucus the rhizoid develops at the side where plastids and other plasmic granula are removed. Several sessile algae, for instance, Vaucheria, Caulerpa, and Chara, have positively orthogeotropic rhizoids, just as do some of the mosses (see Banbury 1962). The dorsiventrality of the gemmae of the liverwort Marchantia can be induced by gravity (see Halbsguth 1965). The sporophytes of the hepaticae and the gameto- and sporophytes of the musci grow negatively geotropically. Rhizoids and roots of ferns probably are only weakly sensitive to gravity. The graviperception of their fronds depends on their developmental stage. Most hyphae of fungi are nongeotropic. Many sporangiophores, for example, those of Mucor, Phycomyces, and Pilobolus, show a negatively geotropic growth (see Banbury 1962).

It is not my task to give a complete survey of the gravity responses of lower plants. The examples mentioned above are enough to demonstrate that many lower plants are able to perceive gravity. What contrivances in these plants enable them to perceive gravitational or centrifugal force as a stimulus?

As with gravity receptors of animal forms, many investigators have searched for statoliths in the plant. There is a fundamental difference, however, in the manner of operation between the animal and the plant statolith. An animal statolith is always extracellular in the statocyst, and stimulates specific sensory cells by changing its position. Plant "statoliths", however, are located within the cell as inclusion bodies. Therefore, a change of their position must immediately interfere with preceding or "normal" functions of the cell in which they are enclosed. The susception and the perception of the stimulus provoked by acceleration must be located within the very same cell.

A brief reference to the so-called statolith starch, often found in cells of mosses, for instance (see Banbury 1962), should be made. On reorientation these starch grains always settle to the lower side of the cell, as can be seen in the cells of the root cap and starch sheath of the higher plant. Many observations lead to the idea that these "falling" starch grains operate as statoliths, but this is not proved with certainty. We do not know how normal cell functions are altered by positional shifts of these grains. Therefore the term "statolith starch" should be used with reservation.

In only two instances have the gravity receptors of lower plants been analyzed in detail, in the sporangiophores of Phycomyces and in the rhizoids of Chara.

The Sporangiophores of Phycomyces

The sporangiophores of the fungus *Phycomyces blakesleeanus* react negatively orthogeotropically. Dennison (1961) demonstrated that they have two different kinds of gravity receptors. He stimulated the sporangiophores by centrifugal force, and at 2 to 5 g the sporangiophores showed the expected curving, in air as well as in a fluid of a greater density. Dennison concluded that in this case the geoperception is mediated by displacement of cellular particles. He suggested that the vacuole moves to the "upper" part of the cell, whereas the cytoplasmic layer becomes thicker at the "lower" part, which also increases in length. To a certain degree this implies that the vacuole acts as a statolith, a function that remains to be substantiated, particularly in view of the probably participation of cytoplasmic organelles in the growth of the fungus cell wall.

A second receptor can be observed as the plant reacts to a sudden change of centrifugal force. In this instance the greater displacement of the relatively heavy sporangium at the tip of the cell first causes a passive bending of the whole sporangiophore. This induces a transient, active curvature of the small growing zone in the opposite direction. Even a mechanical deformation, without any change in centrifugal force, causes the same kind of short-term curving. Consequently the passive bending of the sporangiophore is the beginning of a true tropistic reaction.

The Rhizoids of Chara

The outer nodal cells of the alga *Chara foetida* often produce rhizoids, which are tubelike cells with a diameter of about 30 μ (fig. 1). In a length of some 300 μ the cell apex contains a dense granular cytoplasm lacking normal vacuoles. The so-called *Glanzkörper*, which operate as statoliths (Buder 1961), lie about 20 μ above the cell tip. Their chemical nature is still unknown, but they are not starch grains. In normally growing rhizoids there are 30 to 60 statoliths equally distributed in a position that is relatively stable in the axial direction of the cell, but unstable in the radial.

The rhizoids grow only at the paraboloidal part of the tip, at a rate of about 100 μ/hr or faster. They are positively orthogeotropic and especially suited for geotropic experiments because of their high sensitivity to gravity. If they are turned to the horizontal from their normal vertical position, the cell tip curves downwards. Figure 2 demonstrates that curvature can be seen after about 22 min. Two to 3 hr later the cell tip returns to the vertical direction and again grows normally. Only the basal cylindrical part of the rhizoid, which has ceased growing, remains in horizontal orientation. The cell regulates the direction of growth by means of such curving movement only in a small part of its growing tip.

After some minutes of horizontal exposure, the statoliths settle to the lower flank of the cell tip (fig. 2), where they at last lie spread over a length of 20 to 30 μ. This lateral displacement shows that they are heavier than the other plasmatic components. In the again normally growing cell tip the statoliths are redistributed as they were before reorientation.

Geoperception depends on the presence of the statoliths in the cell tip. Buder (1961) displaced them from the tip to the basal part of the cell by careful centrifugation. These centrifuged cells grew at normal rate, but without geotropic response. Some hours after centrifugation the rhizoid apex began to regenerate new statoliths. Simultaneously with this regeneration, the ability of the rhizoid to respond geotropically was restored. Thus the greater the number of statoliths, the

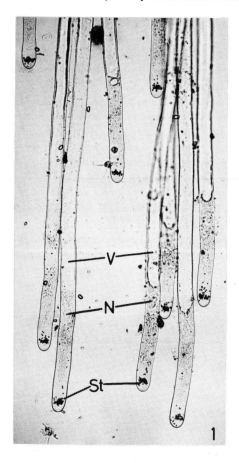

Fig. 1
Living normal growing rhizoids of *Chara foetida*; N = nucleus, St = statoliths (*Glanzkörper*), V = vacuole. Mag. 135X.

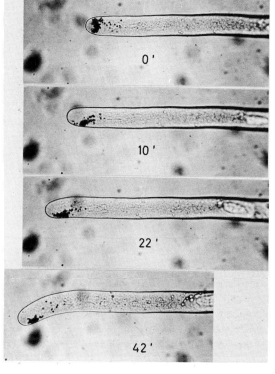

Fig. 2
Micrographs of the outline and the position of statoliths in a Chara rhizoid after 0, 10, 22, and 42 min horizontal exposure. Mag. 190X.

stronger the geotropic response. Buder concluded with good reason that the "Glanzkörper" in the Chara rhizoid do indeed act as statoliths. In the plant kingdom this is the single instance in which the statolith nature of cell particles has been proved.

Buder (1961) could not explain how these statoliths influenced the normal cell functions by their gravity-induced unilateral displacement, that is, how they influenced the direction of cell growth. The perception of the gravity stimulus remained unknown.

I have investigated Chara rhizoids by means of electron microscopy (Sievers 1965b, 1967a, 1967b, 1967c). Primarily I wished to study the normal tip growth of the cell, secondly the ultrastructure of the statoliths, and third, that of the two opposite flanks of the curving rhizoid apex.

In the normally growing rhizoid tip the Golgi apparatus participates in cell wall growth, as it does in other cells with tip growth. A longitudinal section through the tip of a growing root hair (fig. 3), which is nongeotropic, demonstrates that Golgi vesicles are concentrated in the cell apex. They are pinched off from dictyosomes in the more basal cell part and are incorporated in the growing cell wall. In tips of pollen tubes and in many other instances (for example, in the formation of the cell plate) it has been shown that the primary cell wall grows in a similar manner, by incorporation of material from the Golgi vesicles (see Sievers 1965a).

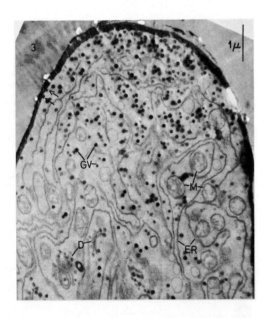

Fig. 3
Electron micrograph of a longitudinal section through the growing tip of a maize root hair, which is nongeotropic. In the tip region darkly contrasting Golgi vesicles are concentrated and incorporated in the cell wall (arrows); D = dictyosomes, ER = endoplasmic reticulum, GV = Golgi vesicles, M = mitochondria. Fixation: $KMnO_4$.

The outer tip of the growing Chara rhizoid is also filled with Golgi vesicles. In a transverse section through the outer tip zone (fig. 4) numerous Golgi vesicles are to be seen. Their contents are not so darkly contrasting as in the root hair (fig. 3); their membranes often show contact with the plasma membrane. In higher magnifications (fig. 5, a, b, and c) the plasma membrane and the vesicle membrane have the same unit membrane structure. After incorporation of the Golgi vesicles in the cell wall, both membranes are united (fig. 5, a and b). The Golgi vesicles here contain weakly contrasted contents similar to those in the inner part of the cell wall (fig. 5, c).

A section through a rhizoid in the apical region of the statoliths (fig. 6) demonstrates that more Golgi vesicles are located in the peripheral zone than in the cell center. Some statoliths with different inner structures are randomly distributed. Greater enlargements (figs. 7 and 8) show the membrane of specific vacuoles separating the statolith from the cytoplasmic ground substance. The statolith itself has a fibrillar structure, which seems to be more dense in the middle than in the outer region (fig. 7). In addition to these structures the second statolith (fig. 8) shows contrasting black particles in a typically radial arrangement. The statoliths are inclusion bodies within special vacuoles. The dictyosomes producing Golgi vesicles (fig. 9) occur only above the statolith zone of the cell.

Figure 10 is a schematic drawing of the ultrastructure of the normal growing rhizoid tip. Above the statoliths there lie plastids, mitochondria, cisterns of the endoplasmic reticulum, ribosomes, microvesicles, dictyosomes, and Golgi vesicles. Below the statoliths one only finds endoplasmic reticulum, ribosomes, Golgi vesicles, and microvesicles. The Golgi vesicles, which supply the main cell wall substances, are transported from the more basal part to the growing tip, as in the case of

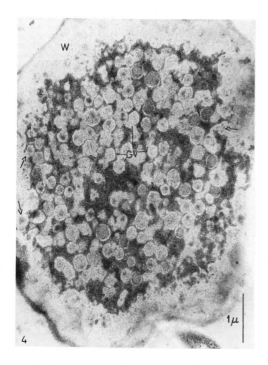

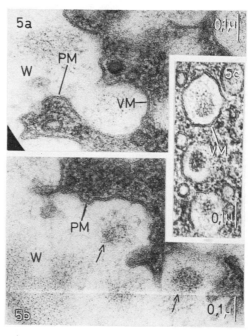

Fig. 4. Electron micrograph of a cross-section through the outer tip of a growing rhizoid of *Chara foetida*. The cell apex is filled with Golgi vesicles of differently contrasting contents. The arrows mark points at which Golgi vesicles are incorporated in the cell wall; GV = Golgi vesicles, W = cell wall. Fixation: OsO_4.

Fig. 5. Greater enlargements of plasma and Golgi vesicle membranes after (5a and 5b) and before (5c) incorporation of vesicles in the cell wall. Note the identical unit membrane structure of both membranes. The arrows in 5b mark contrasting portions, similar to the content of Golgi vesicles (5c), in the invaginations of the cell wall; PM = plasma membrane, VM = Golgi vesicle membrane, W = cell wall. Fixation: OsO_4.

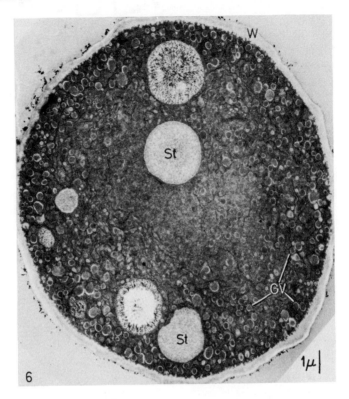

Fig. 6
Electron micrograph of a section through the apical part of the statolith region of a normal growing Chara rhizoid. There are more Golgi vesicles in the peripheral zone of this cell region than in the center; GV = Golgi vesicles, St = statoliths, W = cell wall. Fixation: OsO_4.

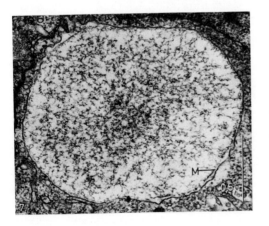
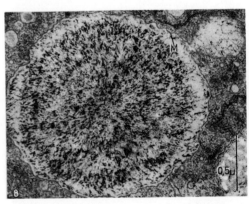

Figs. 7 and 8. Larger magnifications of two statoliths lying as inclusion bodies in specific vacuoles. The statolith in figure 7 has a fibrillar structure, and that in figure 8 in addition has black particles in a typically radial arrangement; M = membrane of the statolith vacuole. Fixation OsO_4.

Gravity Receptors in Lower Plants

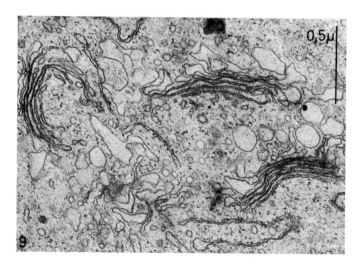

Fig. 9. Dictyosomes of a Chara rhizoid, which occur only in the cytoplasm lying above the statoliths (see next figure). Fixation: OsO_4.

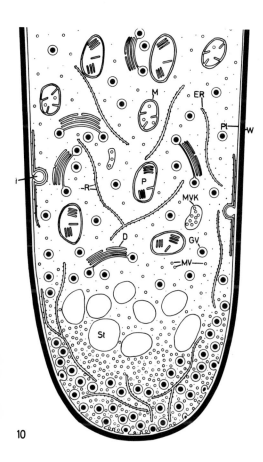

Fig. 10
A schematic drawing of the distribution of plasmatic components in a normal growing Chara rhizoid demonstrates a polar organization of the cell tip. The dictyosomes producing Golgi vesicles lie only above the statoliths. Golgi vesicles are incorporated in the apical cell wall mainly below the statolith zone; D = dictyosomes, ER = endoplasmic reticulum, GV = Golgi vesicles, i = invaginations of the plasma membrane, M = mitochondria, MV = microvesicles, MVK = multivesicular bodies, P = plastids, PL = plasma membrane, R = ribosomes, St = statoliths, W = cell wall.

the root hair (fig. 3). Their membranes are incorporated in the plasma membrane and their contents in the cell wall. The prevailing peripheral location of the Golgi vesicles in the zone of statoliths leads to the conclusion that their transport in the middle region is blocked by the statoliths.

The normally growing rhizoid has an axial symmetry (fig. 10). Only in the small region of the statoliths are the cell structures displaced by gravity. Figure 11 shows a section through the apical region of the statoliths similar to that in figure 6, but from a rhizoid grown for 10 min in horizontal exposure. In this initial phase, curvature is not yet observable, but most of the statoliths have moved into the lower half of the cell (fig. 2). Most of the Golgi vesicles are to be found in the upper part (fig. 11). The difference in number of vesicles between the upper and the lower part is obvious, particularly in comparison to the vertically growing cell (fig. 6), where the Golgi vesicles are equally distributed in the peripheral part of the whole section. There are more points of vesicle incorporation in the upper cell wall than in the lower one. A more obliquely oriented section through the basal region of statoliths (fig. 12) demonstrates the layer of statoliths in the lower part of the cell; they are relatively closely packed but do not touch the plasma membrane. In the upper part there are many active dictyosomes. The unique concentration of Golgi vesicles seen in figure 11 is absent.

In the first 10 min of horizontal exposure, before curvature is observable, the original axial symmetry of the normal growing cell tip is disturbed. A new transient, bilateral symmetry, is now established. In the vertically growing rhizoid (fig. 13, 0′) the Golgi vesicles are distributed throughout the peripheral part of the statolith zone and incorporated in the cell wall at the same level. In the horizontal growing cell (fig. 13, 10′) we can observe in the region of the statoliths a unilateral accumulation of Golgi vesicles in the upper half of the cell. We have to assume therefore that they are transported to the extreme cell tip preponderantly in this upper part of the cell. But many vesicles are also incorporated in the upper flank of the cell wall. This last observation explains the increased growth of the upper cell wall, and at the same time the difference in growth between the upper and the lower wall parts. This is the explanation I propose for the ensuing geotropic curvature.

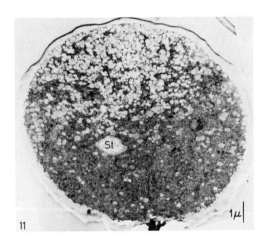

Fig. 11
Electron micrograph of a cross-section through the apical part of the statolith region similar to that in figure 6, but from a Chara rhizoid grown for 10 min in horizontal exposure. The Golgi vesicles occur mainly in the upper half of the cell where they are often incorporated in the cell wall; St = statolith. Fixation: OsO_4.

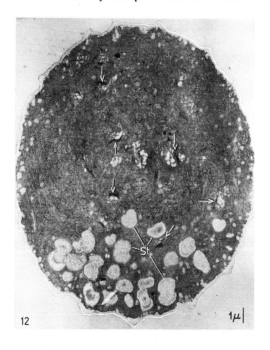

Fig. 12
A more oblique section through the basal statolith region of a Chara rhizoid after 10 min horizontal exposure. Above the statoliths lying in the lower part of the cell are some dictyosomes (arrows). A peculiar concentration of Golgi vesicles is not observable; St = statoliths. Fixation: OsO_4.

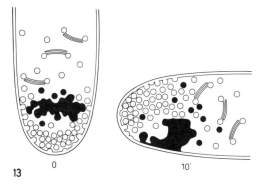

Fig. 13
Schematic drawing of the distribution of dictyosomes, Golgi vesicles, and statoliths in a vertical (0') and in a horizontal (10') growing Chara rhizoid tip.

The typical upward displacement of the Golgi vesicles is surely brought about by the gravity-induced downward movement of the statoliths. During horizontal exposure they block the acropetal transport of Golgi vesicles within the lower part of the cell and divert them to the upper part. The lower flank of the cell wall, therefore, gets considerably less wall material than the upper flank.

This blockage is only gradually removed as the cell tip curves. The axial symmetry of the cell tip regenerates by the slow redistribution of the statoliths to their normal central position. This regeneration is complete when the cell tip again grows in the vertical direction.

The physical phase of this reaction chain (susception) is the gravity-induced displacement of the statoliths. The physiological phase (perception) starts with the diversion of the acropetal transport of the Golgi vesicles to the upper half of the cell by the block of statoliths in the lower part of the cell. This differential distribution of vesicles bearing cell wall material and the consequent difference in their incorporation in the upper and lower flanks cause the curving movement (reaction). Therefore we can say that in the Chara rhizoid the Golgi and the statolith apparatus act together as a self-regulating cellular system.

Acknowledgments

The author wishes to acknowledge the excellent technical assistance of Miss Dagmar Biesang. This work was supported by grants from the Deutsche Forschungsgemeinschaft.

Summary

Gravity responses of the lower plants are rather diverse. In only two instances have the gravity receptors been analyzed in detail.

1. The negatively orthogeotropic sporangiophores of *Phycomyces blakesleeanus* possess two different kinds of georeceptors. Their specific nature is still unknown.

2. Rhizoids of *Chara foetida* are positively orthogeotropic. In their tips they contain statoliths whose chemical composition (yet to be identified) deviates from normal cell particles. As inclusion bodies they are located in specific vacuoles, and their ultrastructure shows typical arrangements of fibrils and, in addition, particles of high electron density. If the statoliths are removed from the cell tip by centrifugation, the rhizoids continue growing, but horizontal exposure does not elicit normal geotropic curvature. They regain the ability to curve only upon the renewed appearance of statoliths. Growth of Chara rhizoids is restricted to their tips, like root hairs and pollen tubes. Cell wall material is furnished by Golgi vesicles which are transported from the dictyosomes in the more basal cell part to the growing apex. In horizontal exposure the statoliths settle to the "lower" cell half. In that location they block the acropetal transport of the Golgi vesicles and divert them to the upper half of the cell. As a result the "lower" cell wall receives considerably less wall material than the "upper" one. These observations explain the increasing length of the upper side and thus the curvature of the cell tip. When the cell tip again grows in the vertical direction, the statoliths are redistributed in their normal position; the unilateral blocking of the transport of Golgi vesicles by the statoliths is thereupon eliminated. In the Chara rhizoid the Golgi apparatus and the statoliths function as a self-regulating cell system.

References

Banbury, G. H. 1962. In *Encyclopedia of Plant Physiology*, vol. 17 (2), ed. W. Ruhland, pp. 344-77. Berlin, Göttingen, and Heidelberg: Springer-Verlag.

Buder, J. 1961. *Ber. Deut. Botan. Ges.* 74: (14)-(23).

Dennison, D. S. 1961. *J. Gen. Physiol.* 45: 23-38.

Halbsguth, W. 1965. In *Encyclopedia of Plant Physiology*, vol. 15 (1), ed. W. Ruhland, pp. 331-82. Berlin, Heidelberg, and New York: Springer-Verlag.

Haupt, W. 1962. In *Encyclopedia of Plant Physiology*, vol. 17 (2), ed. W. Ruhland. Berlin, Göttingen, and Heidelberg: Springer-Verlag.

Sievers, A. 1965a. In *Funktionelle und morphologische Organisation der Zelle: Sekretion und Exkretion*, ed. K. E. Wohlfarth-Bottermann, pp. 89-118. Berlin, Heidelberg, and New York: Springer-Verlag.

———. 1965b. *Z. Pflanzenphysiol.* 53: 193-213.

———. 1967a. *Naturwissenschaften* 54: 252-53.

———. 1967b. *Protoplasma* 64: 225-53.

———. 1967c. *Z. Pflanzenphysiol.* 57: 462-73.

Discussion

WEIS-FOGH: Do you have any electrical records, or any microelectrode or microelectrophoretic experiments, to show whether these statolith vacuoles have a surface charge and if so, of what kind? Electrical forces clearly must be involved in the distribution of the two types of inclusion.

SIEVERS: We have made no such experiments, and I am unable to say anything about the electrical charges or forces associated with these cellular membranes.

AUDUS: It seems to me that there is a slight possibility that the Golgi vesicles may themselves be buoyant, and float to the upper surface. I wonder whether you examined this possibility.

SIEVERS: We should repeat Buder's experiments and determine the distribution of cell organelles after centrifugation of the rhizoids. But you are right, this is the *experimemtum crucis* for our conclusions: to see if we find not a unilateral but a normal distribution of Golgi vesicles in the subapical zone of centrifuged, horizontally growing rhizoids.

ETHERTON: Concerning the question about the Golgi vesicles being buoyant, I think that's probably answered very nicely by your normal, vertically oriented rhizoids. You didn't see Golgi vesicles all rising to the top. There they were randomly distributed.

AUDUS: Couldn't one invert the rhizoids, put them upside down? Wouldn't the statoliths then sink under their own weight?

SIEVERS: I have done this, and so has Dr. Hagen. In inverse position the statoliths fall "down" slightly, about 10 to 15 μ (1/2 cell diameter), and there they tend to rest. I should state that the statoliths have a relatively stable position axially within the cell and a quite labile one radially. During this time before the rhizoid begins to bend, it grows straight upward, at a rate faster than in normal vertical exposure. After this time you can see that a few statoliths may move—they are always moving a little bit—from the middle position to one flank. Then the curvature movement begins: the cell tip bends in a semicircular fashion within several hours. It is interesting that at first the cell grows more rapidly. Perhaps this is because the statoliths, in normal position for a vertically oriented rhizoid, block the acropetal transport of Golgi vesicles. If the rhizoids are inverted, this blockage is lost for some 10 min. The Golgi bodies by themselves do not move. It takes a displacement or motion of the statolith to cause this displacement of the Golgi vesicles.

LARSEN: This is a rather attractive idea, and relatively simple compared with what would happen in higher plants. I should like to return to a point Galston raised. Imagine a tissue composed of parallel, fused Chara rhizoids placed horizontally. If we apply Sievers' principle to two neighboring cells, the upper wall of the lower cell will be stimulated, and the lower wall of the cell above will become inhibited. If the cells are unable to slide on each other, the combined effect in even a large number of cells will be no greater than that of one cell, and will probably not lead to any curvature on account of the rigidity of the tissue. So in multicellular tissues a different principle must be at work.

SHEN-MILLER: There are quite active and relatively inactive Golgi apparatus in higher plant cells. The "inactive" ones produce very few vesicles. Could it be that there are more hypertrophied dictyosomes in the upper portion of the horizontal rhizoid, that there is really an increase in vesicle production rather than a displacement of the vesicles?

SIEVERS: I think it is a displacement. We find only dictyosomes producing vesicles in the growing Chara rhizoid, no inactive ones.

GALSTON: I would like to return to the question that Dr. Audus asked earlier. You stated several times very categorically that the Golgi vesicles are not buoyant of themselves but have to be displaced as a result of the tumbling down of the statoliths. From some of the pictures that you show, one might interpret the situation differently. I wonder if you could tell us why you are sure of a displacement? I don't accept as definitive the point that Dr. Etherton made about the vertically oriented rhizoids, because the situation in tipping would be entirely different with regard to stratification. I would like to hear this answered from the point of view of a horizontally placed rhizoid.

SIEVERS: The rhizoids grow at normal rate if the statoliths are centrifuged to the basal part.

GUALTIEROTTI: It doesn't answer the question. If the Golgi vesicles are buoyant they would go the opposite way during centrifugation. The problem of whether the main mechanism is the buoyancy of the Golgi vesicles or the higher density of the supposed statoliths is not resolved.

SIEVERS: The statoliths have a higher density than the surrounding cytoplasm, for they fall downward in horizontal exposure.

There are what I consider good reasons for my conclusion that in horizontal exposure the statoliths block the acropetal transport of Golgi vesicles in the lower part of the cell and divert them to the upper part—that it is only the sinking of the statoliths that causes displacement of the Golgi vesicles.

1. The typical unilateral distribution of Golgi vesicles occurs only in the statolith zone. You see no gravity-induced displacement of vesicles or other cell organelles in the extreme cell tip or in the more basal zones. I find it difficult to believe that only those Golgi vesicles in the statolith zone are buoyant when we turn the plant from the vertical exposure to the horizontal.

2. We must interpret the fixed electron micrographs in the light of the transport phenomena in the living, quickly growing and reacting rhizoid. Under good conditions the rhizoid grows at a rate of about $180\,\mu$/hr, that is, the diameter of the cell in 10 min. Within the first 10 min after the beginning of horizontal exposure the cell tip still grows straightforward, and the statoliths move to the lower half of the cell.

a) As is true of other cells with tip growth, the rhizoid tip grows by incorporation of Golgi vesicles bearing cell wall substances. The Golgi vesicles are produced by dictyosomes in the basal zone.

They are transported from their sources—the dictyosomes—to their sink, the plasma membrane in the cell apex. This transport is acropetally directed, and is surely an active transport. We also know from other cells that Golgi vesicles are actively transported. In the Chara rhizoid the Golgi vesicles have to pass the statolith zone. In a cross-section through this zone of a normally growing rhizoid we can see that the statoliths are located mainly at the cell center and the Golgi vesicles only near the cell wall. We conclude that in this zone the vesicles are transported chiefly in this small peripheral part of the cell. Perhaps the mass of statoliths in the center blocks the transport of vesicles elsewhere.

b) The second phenomenon regarding transport is that of the statoliths. If we turn the rhizoid from the normal vertical exposure to the horizontal, the statoliths sink to the lower cell flank. This downward shift of statoliths must interfere with the tipward transport of vesicles. In cross-sections through the statolith zone of a 5- to 10-min horizontally grown rhizoid we see this typical arrangement of the cell structures: Golgi vesicles in the upper cell half and statoliths in the lower one. I think that in the lower part of the cell the pathway of the vesicles is blocked by the statoliths, whereas in the upper part it is more open. It is necessary to remember that vesicles do not rest in the upper part of the cell but must be transported to the cell tip. There must be a continuous flow of vesicles, for cell growth does not stop. If we assume these motions of the cell organelles, we can explain the first electron micrographs of the apical statolith zone where only a few statoliths occur in the lower cell part and many Golgi vesicles in the upper part. On horizontal exposure the statoliths occupy a cell space at the lower flank of about 20 to 30 μ in length. This large block may force the vesicles to the upper pathway so that more of them reach the upper cell flank than the lower.

WESTING: How can you explain on the basis of statolithic action the downward curvature of a Chara rhizoid that occurs after it is inverted to a point straight upward?

SIEVERS: We didn't investigate inverted rhizoids by means of electron microscopy. But I'll try to give an explanation. You know that the statoliths fall somewhat in the basal direction if we invert the plant. You can see in a horizontal microscope that the statoliths are always moving a little. If some statoliths accidentally move nearer the cell wall than others, they may begin to block the vesicle transport. This may be the initiation of the following curvature movement. We know from Buder's experiments that the presence of only a few statoliths can cause the bending. If the bending is started you can observe in the microscope that more and more statoliths move to the concave flank. Thus the block may gradually become larger and cause the whole curvature movement.

6
Gravity Receptors in Phycomyces

David S. Dennison, *Dartmouth College*

The study of gravity receptors in multicellular plant organs is complicated by the possibility that the cells participating directly in the geotropic response are not themselves the principal gravity receptors; thus some kind of stimulus transmission between cells must be postulated in certain cases (see chap. 13). In unicellular geotropic plant organs this complication does not arise, and the problem of gravity perception may be pursued at the subcellular level. A very promising approach is the examination of the fine structure of these cells with the electron microscope. Sievers has shown that in the positively geotropic rhizoids of Chara there is a much greater concentration of Golgi vesicles in the upper half of a horizontal cell than in the lower half (chap. 6, this volume; Sievers 1967*a*, 1967*b*). This asymmetry is nicely correlated with a higher concentration of statolith-like particles in the lower half of the cell, although the causal relationship between the distribution of particles and that of the Golgi vesicles is not yet understood. In the negatively geotropic sporangiophores of Phycomyces, the situation is less clear, since no sedimenting particles have been observed in this cell (Banbury 1962). New techniques are being developed that are more successful in fixing this cell for electron microscopy (Peat and Banbury 1967), and it may soon be possible to correlate gravity perception with changes in ultrastructure in Phycomyces as well as Chara. This paper is concerned with physiological studies of gravity perception in Phycomyces, which may, I hope, complement the ultrastructural data.

Phycomyces Sporangiophores

In plants the application of acceleration (by gravity or centrifuge) may act along two different pathways. Acceleration acting on particles or liquid phases whose density differs from that of the surrounding cytoplasm may lead to their displacement. The acceleration will also cause stresses to be set up in the organ as a whole, which may lead to its physical distortion. In principle, either of these effects can act as the primary gravity stimulus; in the sporangiophores of Phycomyces, *both* are effective but elicit different responses.

Phycomyces sporangiophores are fast-growing, unicellular structures which emerge from the fungal mycelium about 4 days after inoculating an agar medium with vegetative spores. The cell is about $100\,\mu$ in diameter and reaches a length of about 20 cm. At its upper end, the cell bears a relatively massive sporangium, a spherical structure with a diameter of about $500\,\mu$ and containing about 10^4 vegetative spores. Mature sporangiophores attain a linear growth rate of 3 mm/hr, which is maintained for many hours. Growth is confined to a 3 mm region just beneath the sporangium, and bending responses to unilateral stimuli are likewise confined to this region.

Geotropic Responses

Since sporangiophores show a rather slow geotropic bending rate when subject to the acceleration of natural gravity, such studies are more conveniently carried out on a centrifuge (fig. 1). With a vertical axis of rotation and an acceleration of 4.1 g at the location of the sporangiophore, there is a steady resultant acceleration of 4.2 g acting on the cell in a direction 13.5° below horizontal. Observations of the angle of the growing portion of the cell can be made stroboscopically while the centrifuge is turning.

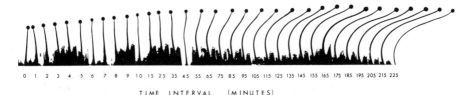

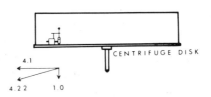

Fig. 1. A typical centrifuge experiment. Ten sporangiophores are mounted in balanced pairs on the disk, but only one is shown. The addition of acceleration vectors acting on this sporangiophore is also indicated. Photographs are taken by stroboscope while the disk is turning. All photographs and measurements reported in this paper are made using red light, which is inert for phototropism. A series of these shows a single sporangiophore before the centrifuge is turned on (0 min) and at various times after the centrifuge is brought up to a steady speed (4.1 g).

A series of stroboscope photographs made during a typical experiment is shown in figure 1. At time 0 the centrifuge is revolving very slowly, giving essentially zero acceleration. During the 30 sec prior to the photograph at 1 min, the centrifuge is brought up to a steady speed, corresponding to 4.1 g.

The clearly noticeable difference in sporangiophore angle between time 0 and 1 min is due to the passive bending of the cell. The action of the centrifuge on the massive sporangium displaces the upper end of the cell outwards from the axis of rotation, while the lower end of the cell is fixed. Measurements have indicated (Dennison 1961) that the curvature of this passive bend is constant within the growing zone and drops to a much lower level in the mature, nongrowing region. During the next 6 min, a bending response occurs in the opposite direction, namely, toward the axis of rotation. As may be seen in figure 2, this response is relatively rapid, but does not persist beyond the first 5 to 7 min of the centrifuge run. This rapid, short-lived response may be termed the *transient* geotropic response. There is a second response, the *long-term* geotropic response, which is evident about 20 min after the start of the centrifuge run (figs. 1 and 2). This response has a bending speed only about one-tenth that of the transient response and seems to be located somewhat lower in the growing region. The long-term response is not transient but continues steadily until the growing region of the sporangiophore is oriented parallel to the resultant acceleration vector.

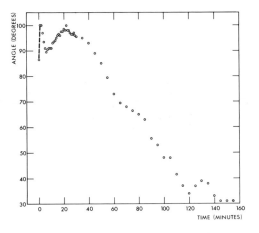

Fig. 2
The same experiment as that pictured in figure 1. The angle plotted is that between the upper 1 mm of the sporangiophore and horizontal. The dashed line indicates the passive bending that occurs when the centrifuge is brought from rest to its steady speed.

Role of Stresses in Geotropism

Since the long, slender cell carries a relatively massive sporangium at one end, the lateral acceleration causes considerable physical distortion in the cell surface. This distortion accompanies whatever internal rearrangement may also be caused by the acceleration, and these two physical results of acceleration must be separated if one is to identify the primary gravity receptor. The cell surface distortion could be largely removed if the sporangiophore were centrifuged while immersed in a fluid whose density is equal to that of the sporangium; the buoyant force on the sporangium would then exactly compensate for the force due to acceleration. Actually, a fluid much denser than the sporangium was used (Dennison 1961). This liquid, perfluorotributylamine (Fluorochemical FC-43 of the Minnesota Mining and Manufacturing Co.), has a density of 1.87 and does not appreciably affect the growth of sporangiophores except for the suppression of phototropism (Delbrück and Spropshire 1960). Because FC-43 is much denser than the sporangium, the buoyant force dominates, and as a result the stress is applied to the cell surface in a direction opposite to that in air.

Figure 3 shows the results of a centrifuge run in which the sporangiophore is mounted in FC-43. As expected, the direction of passive bending is reversed by immersion in FC-43. Furthermore, the transient response is likewise reversed in direction. The long-term geotropic response (not shown) is substantially unaffected by immersion in FC-43, however, and continues at the usual rate until the direction of sporangiophore growth is opposite to that of the resultant acceleration vector, as in air.

The reversal of the transient response in FC-43 suggests that it is purely a response to the physical distortion of the cell surface and is not a gravity response in the narrow sense. This view is supported by the fact that the transient tropic response can also be obtained by applying a lateral force of 0.5 mg to the sporangium of a vertical sporangiophore by means of a calibrated fine glass fiber (Dennison 1961). As before, the direction of bending in the transient response is opposite to the direction in which the cell is bent passively, either by mechanical means or by centrifuging.

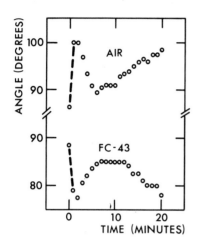

Fig. 3
The effect of FC-43 on the transient response. Both centrifuge runs are at 4.1 g, but in one case the container around the sporangiophore is filled with FC-43, an inert fluid of density 1.87. At 0 min the sporangiophore is upright (sporangium up) in both cases, and the angle is that between the upper 1 mm of the sporangiophore and horizontal. The dashed lines indicate passive bending.

Thus the sporangiophore responds to gravity (or centrifugation) in two distinct ways. In the transient response, the physical distortion of the cell causes a brief but rapid bending in the opposite direction. In the long-term response, some intracellular mechanism, not related to mechanical distortion of the cell surface, responds to lateral acceleration and triggers a steady but slow bending that continues until the cell axis is lined up with the resultant acceleration vector.

Linear Growth Response to Stress

The transient tropic response is a differential growth response to the asymmetric deformation of the cell; the side of the cell that was compressed grows faster and the side that was stretched grows more slowly. This response to deformation has been studied in greater detail under conditions where the deformation is distributed around the cell (Dennison and Roth 1967).

This type of deformation is produced by hanging weights on an inverted cell by means of a small hook attached to the sporangium (fig. 4). The weight hangs from the hook by a silk fiber, and the load on the cell can conveniently be removed by raising a pan to support the weight; lowering the pan causes the load to be reapplied.

The cell deformations are remarkably elastic (reversible) with loads in the range of 8 mg, in spite of the fact that the amount of deformation is not a linear function of load. Under a load of 8 mg, the sporangiophore stretches passively about 1% of its length; about half of this extension is in the 3-mm growing zone, which may elongate 10% or more. Removal of this load causes a passive contraction of the cell by the same amount.

The response to adding a 5-mg load (fig. 5) is a transient drop in growth rate which begins about 1 min later and lasts for about 5 min. Removal of the load causes a transient increase in growth rate, also following a 1-min latency and lasting for about 5 min. This symmetrical dependence of the growth response on the deformation stimulus permits an understanding of the transient geotropic response. If one assumes that a lateral force on the sporangium (applied mechanically or through acceleration) causes a compression along one flank of the cell (concave) and a stretching along the other flank (convex), then the above results would lead one to expect an increase in growth rate on the compressed side and a decrease in growth rate on the stretched side, resulting in a bend against the applied force (as observed).

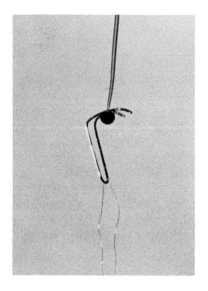

Fig. 4
A sporangiophore with a wire hook attached to its sporangium. The sporangium diameter is 0.5 mm. Weights are hung on the lower end of a 5-cm silk fiber, whose upper end is looped through the hook as shown.

Fig. 5
Growth responses to changing the load on a sporangiophore. A positive growth response is seen following the removal of a 5-mg weight ("OFF 5 mg"), and a negative growth response is seen following the application of a 5-mg load ("ON 5 mg").

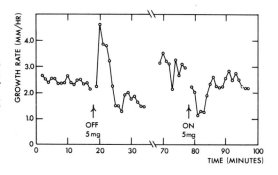

Although the growth rate deviates from normal for 5 min following the change in load, it subsequently returns to normal even though the load remains constant. This point was checked carefully by measuring growth rates 50 min after the application of loads, to see if there is any systematic difference in growth rate due to the steady action of the load. No significant effect of load on growth rate was found. This suggests that the primary stimulus for the stretch response is not a static structural deformation caused by a steady load, but rather a *change* in some structure or membrane system lying just beneath the plasma membrane.

Receptor Mechanism of the Long-term Geotropic Response

A few speculations may be made about the receptor mechanism of the long-term geotropic response. Since the long-term response is the same in FC-43 as in air, its mechanism cannot be associated with deformation of the cell surface. Rather, its mechanism must be wholly internal, having to do with relative displacements of particles or liquid phases within the cell.

Phycomyces sporangiophores have no conspicuous particles that seem to have a statolith function. The cytoplasm is very granular, with many particles about 0.5 μ in diameter and some yellow lipid droplets about 1 μ in diameter. These are in constant movement in longitudinal channels of protoplasmic streaming. It would be possible for these particles or droplets to form a vertical concentration gradient due to a tendency to either float or sink in the surrounding cytoplasm. However, no such aggregations have been noted in living cells.

Sporangiophores have a large axial vacuole, which might be of lower density than the cytoplasm. If this were so, the transverse acceleration might cause a lateral displacement of the vacuole, making the layer of cytoplasm thicker on the lower side of the cell than on the upper. This differential thickness of cytoplasm might then cause differential rates of growth on the two sides and thus a geotropic reaction. Unfortunately, observations of such a vacuole shift have not been attempted. In addition, systematic studies have not yet been made of possible gravity-induced alterations in cell fine structure.

Summary

In sporangiophores of Phycomyces, there are two responses to gravity or centrifugation. One response is a transient but rapid bending, which is triggered by the physical deformation of the cell, due to the action of the acceleration on the massive sporangium. The transient response can also be elicited by the application of a purely mechanical force to the sporangium and is reversed by centrifuging in FC-43. The mechanism of cell growth is sensitive to mechanical deformation, as shown by the negative and positive growth responses to the application and removal of a longitudinal force on the sporangiophore. The long-term response is a slow but steady bending which continues until the direction of sporangiophore growth is opposite to the net acceleration vector. This response is not reversed in FC-43 and hence is associated with an intracellular effect of acceleration, such as the sedimentation of lipid droplets or perhaps the entire central vacuole.

Acknowledgments

I thank Mr. William D. Graham, Mrs. Elsa Braestrup, Miss Muriel Cole, and Mrs. Carolyn Roth for their technical assistance in various parts of this work. This research was supported by grants from the National Science Foundation (G-8719 and G-18889).

References

Banbury, G. H. 1962. In *Encyclopedia of Plant Physiology*, vol. 17 (2) ed. W. Ruhland, pp. 344-77. Berlin: Springer-Verlag.

Delbrück, M., and Shropshire, W., Jr. 1960. *Plant Physiol.* 35: 194-204.

Dennison, D. S. 1961. *J. Gen. Physiol.* 45: 23-38.

Dennison, D. S., and Roth, C. C. 1967. *Science* 156: 1386-88.

Peat, A., and Banbury, G. H. 1967. *New Phytologist* 66: 475-84.

Sievers, A. 1967a. *Naturwiss.* 54: 252-53.

———. 1967b. *Z. Pflanzenphysiol.* 57: 462-73.

Discussion

BROWN: Have you examined the streaming rate, say by analysis of film? You see no difference in streaming patterns. Would a 50% or 100% difference really be detected?

DENNISON: I think it would, but I don't have the equipment to observe this under high magnification while the sporangiophore is being centrifuged. It is a great pity that centrifuge microscopes of good resolution are not commercially available.

HERTEL: The optical properties of the lens may vary depending on the location of the vacuole. Are there any investigations which use the lens effect in phototropism to study a possible displacement of the vacuole?

DENNISON: That's a very interesting suggestion. The problem is that the refractive index is not very different between the vacuole and the cytoplasm.

WESTING: In those strain experiments in which you hung a weight from an inverted Phycomyces sporangiophore, the inversion adds an unnecessary confounding factor to the system. Thus, a minor refinement would be to grow the sporangiophore in its normal, upright position and have the weight attachment go up and over a pulley.

DENNSION: If it weren't more difficult I would have done it that way.

JOHNSON: If, in the slow geotropic response, you stop the centrifuge and allow it to remain static, do the sporangiophores recover or do they stay in the same position?

DENNISON: They would remain static for a long time and then very slowly bend back to the vertical. The reason is that the bending speed in the slow response is a function of the intensity of the acceleration, and the response rate I showed for $4\,g$ is about four times the rate seen at $1\,g$ which means it would take a longer time to bend back up to vertical.

WEIS-FOGH: Have you or anybody else ever taken into consideration what may happen to the sporangium itself when the stimulus is applied between the sporangium and the tip?

DENNSION: We don't know what is going on in the sporangium; it is packed with about 10,000 spores that propagate this fungus vegetatively. It doesn't contain the same apparatus that a growing cell contains. We know that if we remove it the cell will stop growing in about 10 min. You cannot get a growth response by shining light on a sporangium, so its response, if any, to a stimulus is not intimately associated with the growth of the cell.

LEOPOLD: As I understand it, the growth of this sporangiophore has a helical pattern. If this is true, you could get bending by changing the orientation of the helical angle without any change in the rate of the laying down of cell walls. If the micelles on one side were laid down at a wider angle than those on the opposite side, this could give you a relative foreshortening. You could change the helical pattern and get a tropistic turning without alteration of growth rates on either side.

DENNISON: True. But the helix is basically symmetrical around the axis; you still have to establish some kind of asymmetry in order to get a bending reaction.

7
The Susception of Gravity by Higher Plants

Poul Larsen, *University of Bergen*

General Survey

Statolith Hypotheses

We shall concern ourselves with the initial, physical process, or *susception*, in which a gravitational stimulus is received by the higher plant. It is apparent that the pertinent physical action of gravity on a plant can consist only in an interaction between the mass of the earth and the masses of some constituents of the plant, a plant cell, or the contents of the cells. That this is so has been confirmed by the substitution of centrifugal forces for gravity in producing "geotropic" responses.

The action might consist in changes in the distribution of pressure on sensitive loci, exerted either by the entire cell contents or by particles (statoliths) heavier or lighter than the surrounding medium. The action might also consist in movements or reorientation of such particles. Most hypotheses concerning the susception of gravity in plants assume that certain particles in the cells function as statoliths. Early in this century it was suggested that starch grains, particularly starch grains which are mobile under the influence of gravity, acted as statoliths. Precursors of the idea can be found in the writings of Nägeli, Berthold and Noll. In the year 1900 Haberlandt and Němec, independently, reported experiments and observations that showed a high degree of correspondence between the occurrence and motility of starch grains on the one hand, and the presence of geotropic sensitivity and the required minimum stimulation times (presentation times) on the other.

Figure 1 shows an example of the occurrence of movable starch grains in cells of the root cap of garden cress (*Lepidium sativum*). When the root is in its normal position, the starch grains in the cells, with the exception of the oldest (that is the most distal) stories, are assembled on the cytoplasm close to the lower cell wall, the "floor" of the cell. If the root has been kept horizontal for 15 min, the starch grains in the corresponding 4 stories have accumulated on the lowermost tangential walls of each cell (fig. 2). If the root is turned from its normal to the inverse position, the starch grains start falling through the cell toward its opposite end. In our experiments, however, even after 20 min (fig. 3), the starch grains in most of the cells do not get as close to the ceiling as they were to the floor before the inversion.

Other workers have made similar observations of movable starch grains in the tips of coleoptiles (fig. 4) and in specific layers in young stems.

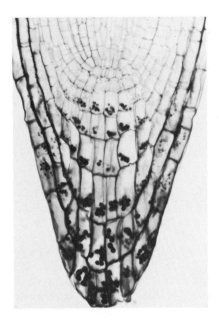

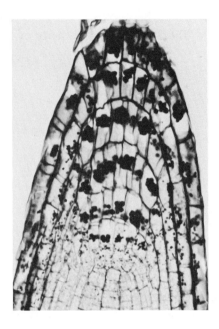

Fig. 1. Longitudinal section of the root tip of the garden cress *(Lepidium sativum)* in normal position. Stained by the PAS method (periodic acid and Schiff's reagent; see Jensen 1962). The amyloplasts in the youngest (most proximal) root cap cells are close to the floor of each cell. Mag. 300X.

Fig. 3. As figure 1, except that the root was inverted for 20 min before fixation. In most of the cells the amyloplasts have moved toward the ceiling of the cell but are close to the ceiling only in two of the youngest stories. Mag. 300X.

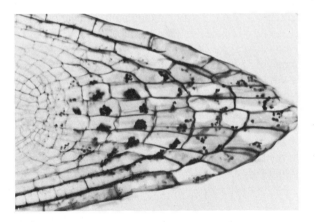

Fig. 2. As figure 1, except that the root was kept horizontal for 15 min before fixation. The amyloplasts in the youngest root cap cells are close to the lower, longitudinal cell wall, mostly in the floor end of each cell. Mag. 300X.

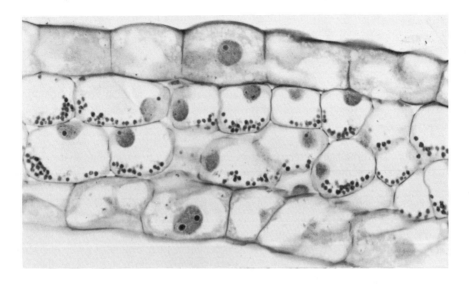

Fig. 4. Longitudinal section of a subapical portion of a wheat coleoptile which has been kept horizontal for 1 hr. Stained by the PAS method. The amyloplasts are close to the lower side walls. From Pickard and Thimann 1966; reproduced by permission of the authors and the publisher, the Rockefeller University Press, from the *Journal of General Physiology* 49 (1966): 1080.

It has been suggested that other particles might function as statoliths, for instance, mitochondria, microsomes, or protein molecules. Audus (1962) and Gordon (1963) made calculations to see if such particles, under the influence of gravity and within known minimum stimulation times, would travel a distance large enough to create what could be accepted as the necessary minimum reorientation of the cell contents. They both concluded that starch grains could certainly so travel and mitochondria could do so in certain instances. The cytoplasmic medium in which the movements are to take place is assumed to have a density of 1 and a viscosity between 5 and 20 centipoises. The density of starch grains is about 1.5 and that of mitochondria and microsomes about 1.2. Fully developed starch grains vary considerably in size, from about 1 μ to about 100 μ in the largest dimension. They are formed within special organelles, amyloplasts. An amyloplast may contain a single starch grain but usually contains several. The presumed starch statoliths are actually amyloplasts, small packages of starch grains. Mitochondria range from about 1 to a few μ in length and from 0.5 to 0.8 μ in width.

Microsomes of about 30 nm diameter and spherical protein molecules of 5 nm diameter would, according to the calculations by Audus and by Gordon, require considerably longer than the observed minimum stimulation times to move a distance that could have physiological consequences. Observed minimum stimulation times for geotropic stimulation vary between a fraction of a minute and several minutes. The directional movements and the orientation of the smaller particles would also be greatly disturbed by thermal agitation (Brownian movements). Still, it does not seem improbable to me that a large number of small rod-shaped particles, heavier at one end, could *statistically* be oriented under the effect of gravity, even though each individual particle need not be parallel or at the same angle to the direction of gravity at all times. I felt forced to suggest such a pattern some time ago in an attempt to reconcile the seemingly controversial results reported by Zimmermann and von Ubisch, but after resolving this controversy by other means (as reviewed below), I no longer see a need for a model more complicated than the simple amyloplast system.

In the calculations by Audus and by Gordon, the required minimum distances of travel were chosen arbitrarily. Actually we have no good guide on this point. The cell may be able to respond to very small changes in the arrangement of its organelles, and small changes in orientation may be sufficient to alter the polar properties of the tissues.

The operation of statoliths lighter than the surrounding cytoplasmic medium, such as oil droplets, has also been suggested, but this idea has not received much attention. The starch statolith hypothesis, on the other hand, has been the object of extensive experimentation and speculation. The status of this hypothesis has been reviewed, by, among others, Rawitscher (1932), Brauner (1962), Audus (1962), and more recently by Wilkins (1966). Movable starch grains occur in root caps, in coleoptile tips, and in the endodermis (the innermost layer of the cortex) of certain stems, all being parts of the plant which exhibit a high degree of geotropic sensitivity. Strong support for the statolith function of this starch comes from plants such as the onion and iris, which do not manufacture starch in their chloroplasts and do not employ starch as the storage form of carbohydrate. These plants do, however, possess movable starch in geotropically sensitive organs, such as the root cap and the cotyledons. Staining tests with iodine indicate that this starch is chemically somewhat different from starch in most other plants.

The literature contains a few examples of the presence of movable starch in geotropically insensitive organs. These cases, of course, do not disprove the starch statolith hypothesis, since other links in the reaction chain may be missing. Audus cites two examples of geotropic sensitivity in starch-free organs. These may be explained by assuming that other bodies function as statoliths. As a whole, the correlation between the occurrence of movable starch and the presence of geotropic sensitivity is excellent and very suggestive of the statolith function of the movable starch. This applies even to the quantitative relationship between the amount of statolith tissue and the degree of geotropic sensitivity, as stressed by Audus (1962), mainly in reference to investigations by Hawker (1932).

Experimental Approaches to the Starch Statolith Problem

Time Relationships. Several authors, including Haberlandt, have claimed that the time required for a number of the amyloplasts to move from their initial position to the lower sides of the cells after a plant organ has been placed horizontal, agrees with known minimum stimulation times (from about 1 min to about 20 min in most instances). The methods used to determine the rate of fall of the amyloplasts, however, are often poorly documented, but in the studies by Hawker (1932, 1933) the procedures are described in detail. She found a very close agreement between rate of falling and minimum stimulation time at temperatures varying from $10°$ to $40°C$. At $30°$, the rate of falling is highest and the minimum stimulation time shortest.

Experimental Removal of Statolith Starch. Many attempts have been made to remove the statolith starch from various plant organs and study the effect of gravitational stimulation on the starch-free organs. Němec (1902) embedded roots of peas and broad beans in plaster of paris. The tips of such roots became starch-free and lost their geotropic sensitivity. Prolonged chilling, in experiments by

Haberlandt (1903), had the same two effects on stems. Similar results have been obtained by heat treatment or starvation (in darkness), and by treatment with chemicals such as potassium alum or sulfur dioxide. Recovery of geotropic sensitivity, usually accompanied by the re-formation of starch, by returning the plants to normal conditions was taken as evidence for the absence of permanent damage. In most of these experiments, however, the reactions were not uniform in the entire material, and quantitative data on growth rates and degrees of curvature are often scarce. In experiments by Went (1909) in collaboration with Pekelharing, for instance, it was reported that some starch-free Lepidium roots were still able to respond to gravity. The results of the older experiments on the removal of starch from geotropically sensitive plant organs have been variously interpreted either for or against the starch statolith theory by various reviewers.

Somewhat more clear-cut results were obtained by von Bismarck (1959), who found that sphagnum shoots which had lost their starch after chilling were still able to respond to gravity. These results may, however, have no bearing on geotropism in seed plants.

More recently Pickard and Thimann (1966) have succeeded in making excised wheat coleoptiles completely starch-free by a relatively mild treatment with gibberellic acid plus kinetin at 30°C for 34 hr. Such coleoptiles were still able to carry out geotropic curvatures. They curved more slowly than excised coleoptiles that had been incubated in water or 0.03% sucrose solution. They also elongated more slowly, but the rate of curvature and the rate of elongation were reduced in about the same proportion, so that the curvature per mm elongation was nearly the same, namely about 50° per mm, after pretreatment in water, in sucrose, and in gibberellin plus kinetin. Pickard and Thimann conclude that starch grains are not necessary for graviperception in wheat coleoptiles, but that perhaps the pressure distribution of the entire cell contents may be decisive.

In Bergen, Iversen has attempted to duplicate Pickard and Thimann's experiments using roots of cress. He found that the roots of intact seedlings, raised in the dark at 21°C, could be made completely starch-free simply by transferring the seedlings to 35°C for more than 55 hr. By the application of gibberellic acid plus kinetin, the disappearance of starch could be accelerated, becoming complete in 29 hr. When returned to 21°C and allowed 3 hr for temperature adaptation, the starch-depleted roots elongated at a rate of 0.48 mm/hr, but proved unable to respond to gravity. Under the same conditions control roots pretreated in plain water at 21°C and at 35°C elongated at 0.64 and 0.33 mm/hr, respectively (Iversen 1969). The absence of geotropic responsiveness in the gibberellin-treated, but still elongating, roots is the result to be expected if the statolith function is performed by the starch grains. Gibberellin-treated roots did not form starch unless the seedlings were illuminated. With the re-formation of starch, after 20 to 24 hr, geotropic responsiveness was restored.

Figure 5 shows a section of a root treated with gibberellin and kinetin. Its root cap cells contain no visible trace of starch (compare figs. 1-3). This section was stained by the PAS method, but the absence of starch in treated roots was confirmed by other staining methods and by electron micrography.

Removal of the statoliths has also been achieved by removing the tissue containing them, most frequently in roots, where the sensitive and the reacting parts are clearly separated in space. Although decapitated roots are usually unable to respond to gravitational stimulation, results of experiments involving resectioning of the root cap or part of it have often been deemed inconclusive on technical grounds. Juniper et al. (1966), however, made use of the discovery that the root caps of maize and barley can be snapped off cleanly with tiny tweezers without damage to the remaining tissues. Such decapped roots showed normal elongation, but no geotropic response. With regeneration of the root cap (see also Cercek 1970), the geotropic sensitivity was gradually restored. Such results are in complete agreement with the starch statolith hypothesis.

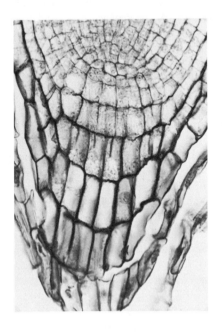

Fig. 5
As figure 1, except that the root has been treated with a mixture of 10^{-4} M gibberellic acid and 10^{-4} M kinetin at 35° for 30 hr. The starch grains have disappeared from the amyloplasts. Mag. 300X.

Manipulation of Amyloplasts. It has also been attempted by suitably chosen manipulations of the plants to direct the amyloplasts to certain specified regions of the cell before or after stimulation (or both). The ensuing geotropic responses are compared with the expected or observed movements of the amyloplasts and the results evaluated in relation to the starch statolith hypothesis.

The results of some of these experiments (Jost 1902, Fitting 1905) were interpreted as not supporting the starch statolith hypothesis; others (Richter 1914, Buder 1908, Zimmermann 1927) were taken as supporting it. In 1928, Gerda von Ubisch published a paper in which she claimed to have performed an *experimentum crucis* against the starch statolith hypothesis. Her work extended experiments reported by Zimmermann in 1924 and 1927.

Zimmermann determined the minimum geotropic stimulation time for roots of Lepidium. After stimulation, the seedlings were placed for 5 min in a vertical position, either normal or inverted. Thereafter they were rotated parallel to the horizontal axis of a clinostat for 45 min (to compensate unilateral gravitational stimulation during the development of the curvatures), and the response was recorded by counting the number of reacting roots. Zimmermann found the minimum stimulation time much shorter when the aftertreatment was inversion that in the alternative case. These basic observations were confirmed by von Ubisch (1928) and, in part with other plant materials, by Reiss (1934), Dolk (1936) and Larsen (1953, 1965).

Zimmermann interpreted his results in terms of the starch statolith hypothesis. When a root is turned from the normal to the horizontal position (fig. 6, *A2*) and left there for a few minutes, the starch grains, starting from the floor end of a cell, will slide down and gradually move a certain distance into the cytoplasm along the lower horizontal wall. If the root is now returned to the normal position (*A3n*), the starch grains will slide back the short distance to the floor of the cell. If, on the other hand, the root is inverted (*A3i*), the starch grains will move toward the ceiling of the cell, traversing a longer distance within the presumably sensitive cytoplasm than in the former case and "rubbing" this for a longer time. If such "rubbing" constitutes the stimulation, inversion should either increase the response, or, if only a minimum response is desired, permit a shortening

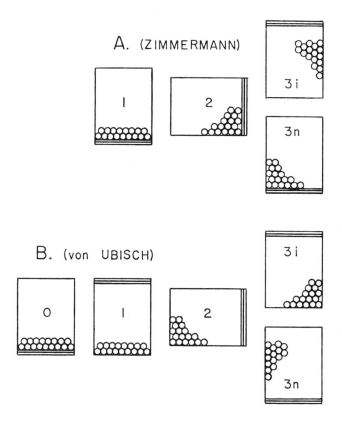

Fig. 6. Displacement of amyloplasts in root cap cells in successive stages of geotropic experiments shown schematically (modified from Zimmermann [1927] and von Ubisch [1928]). Triple line indicates the floor of the cell (the normally lower wall). Series A = Zimmermann's experiment: 1 normal; 2 horizontal; 3*i* inverted; 3*n* normal. In 3*i* the amyloplasts can slide a longer way along the previously lower, horizontal, longitudinal wall than in alternative 3*n*. Series B = von Ubisch's design: 0 normal, 1, 2, 3*i*, and 3*n* as in A. In this case position 3*n* offers a longer path for the sliding of amyloplasts than alternative 3*i*. With Lepidium and Artemisia roots, treatment 3*i* gives rise to larger curvatures than 3*n* in both A and B. Reproduced by permission from Larsen 1965).

of the time of horizontal exposure. The starch statolith hypothesis assumes that the cytoplasm along the outer tangential walls is more sensitive than that lining the inner tangential walls, so the discussion can be restricted to a "statocyst" located near the vertical median plane in the lower half of the horizontal root.

Von Ubisch (1928) designed a variant of this experiment. Lepidium seedlings were first to be inverted for 20 min, a time found sufficient for the starch grains in the root cap to settle on the ceiling of the cells (fig. 6, *B1*). The inverted plants were then to be stimulated in the horizontal position for a certain time (*B2*), and then turned to a vertical position, either normal (root tip down, *B3n*) or inverted (*B3i*). In the latter case the starch grains would slide only a short distance back to the ceiling end of the cell, and the stimulus should be less than that produced by turning the seedlings to the normal position. Still, von Ubisch claimed that in this case inversion after the horizontal stimulation produces the largest curvatures or requires the shortest minimum stimulation time.

Von Ubish used this model experiment in her discussion of the so-called "rotation curvatures" described by Zimmermann, but her paper contained no report to show that the experiment had actually been done with plants. Her prediction was based on other experiments in which the starch grains before horizontal stimulation were evenly distributed over the entire cytoplasm, a condition achieved by a 2-hr rotation of the seedlings parallel to the horizontal axis of a clinostat.

I have repeated the experiments described by Zimmermann and by von Ubisch (Larsen 1965) with improved methods and quantitative recording of the degrees of curvature. The results, and the predictions by the two authors were confirmed fully. Von Ubisch has been criticized for insufficient documentation of her results, but nobody has disputed the strength of her argument, provided the proposed experiment gave the predicted results. Having verified von Ubisch's predictions, I originally agreed with reviewers who regarded the outcome of her experiment as a serious blow to the starch statolith hypothesis and felt forced to abandon any hypothesis involving free, unattached statoliths. Instead I proposed a model containing several pendulums, the statistical behavior of which under gravity would accord with von Ubisch's and my own observations. The pendulums were further assumed to be electrically charged particles. However, after Grahm and Hertz (1962, 1964) showed that the geoelectric effect develops only after a latent time, and depends on metabolism and the presence of auxin, assignment of the geoelectric effect as a primary physical phenomenon in geotropic stimulation lost its foundation. Some relatively large, rod-shaped bodies present in the geosensitive regions of roots of Lepidium were studied by electron microscopy at the Anatomical Institute, University of Bergen (Iversen and Flood 1969). These bodies carry some of the characteristics of the otherwise hypothetical pendulums, but they move easily in centrifugal fields. Their role in geotropism, if any, would thus be the same as that of the amyloplasts: free, unattached statoliths. So other means are needed to reconcile the results of von Ubisch and Zimmermann.

As mentioned before, Zimmermann took his results as strong support for the starch statolith hypothesis. A critical examination of his reasoning, however, shows that his results may have no bearing whatsoever on the statolith hypothesis so far as geotropic stimulation is concerned. In its original and uncomplicated form, the hypothesis assumes that it is the pressure of the starch grains on the sensitive cytoplasm along the outer tangential walls that constitutes the stimulation. If this is accepted, the sliding of starch grains along these walls in a vertical direction should not involve any stimulation. Neither Zimmermann's nor von Ubisch's results would thus have any bearing on the statolith hypothesis for geotropic stimulation. The clear-cut effects of the manipulations of the plants by these two authors must be regarded as "tonic," that is, related to the conditions of sensitivity or reactivity, not to the susception proper. Zimmermann clearly assumed that the tonic and the tropic effects were based on the same mechanism, namely the movement of starch grains. Abandoning this assumption may furnish a fresh approach to the problem.

The experiments on starch grain manipulation should be redesigned to allow the starch grains to exert some pressure on the presumably sensitive sites. This can be achieved by letting them slide, not in a vertical direction, but along the cell walls at some angle to the plumb line. In Bergen we have carried out some experiments along these lines, and our results, so far are in good accord with the starch statolith hypothesis.

A Reinvestigation of the Relationship between Amyloplast Movements and Geotropic Response

The curves in figure 7 show the relationship between geotropic response and stimulation of cress roots at various angles. In all instances the roots were stimulated for 3.75 min and then rotated parallel to the horizontal axis of a clinostat. Curvatures were recorded after 30 min of rotation. Curve A represents the response when roots are taken from the normal position and stimulated at the angles indicated on the abscissa. The optimum angle of stimulation under these conditions is 120° to 135° (Larsen 1962, pp. 169-70; Larsen 1969). This can readily be explained by the starch statolith hypothesis. Figure 8 (A) shows the expected behavior of the starch grains in the experiment. When the roots are stimulated at 45°, the starch grains slide a short distance along the floor of the cell and exert pressure only on the lower part of the lower longitudinal wall. When stimulated at 135°, the starch grains will slide over a longer portion of the lower longitudinal wall.

If there is a casual relationship between statolith behavior and geotropic response in this experiment, we should expect the optimum angle to be 45° to 60° when using roots which before

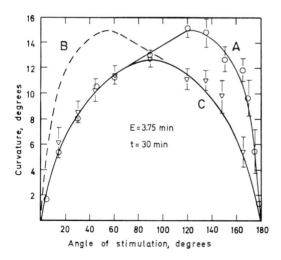

Fig. 7
Relationship between geotropic response and 3.75-min stimulation of cress roots at various angles, 90° meaning horizontal, 135° meaning root tips pointing obliquely upward. Responses recorded after 30-min rotation parallel to the horizontal axis of a clinostat at 0.25 rpm. Curve A: roots were in the normal position before stimulation; optimum 120 to 135°. Curve B: hypothetical curve for roots having their statoliths accumulated in the ceiling end of the cell after a certain time of inversion before stimulation. Curve C: experimental curve for roots which were kept inverted for 8 min before stimulation; optimum 90°. Reproduced by permission from Larsen 1969.

the stimulation have had their statoliths accumulated in the ceiling end of the cells. The expected behavior of the starch grains under these conditions is illustrated in figure 8 (*B*). Stimulation at 45° now offers the starch grains a longer path for sliding along the lower longitudinal wall than stimulation at 135°. The responses should follow curve *B* in figure 7.

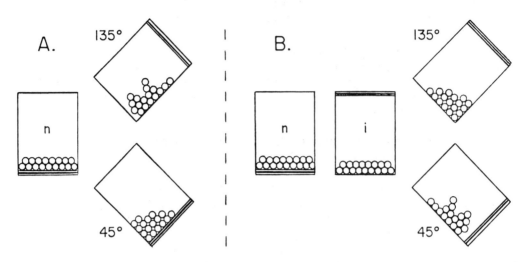

Fig. 8. Displacement of amyloplasts in root cap cells in successive stages of geotropic experiments shown schematically. Triple line indicates the floor of the cell (the normally lower wall). Series *A*: roots coming from the normal position (*n*). When stimulated at 45°, the amyloplasts slide a short distance along the floor and exert pressure only on the lower part of the lower longitudinal wall. When stimulated at 135°, the amyloplasts will slide over a longer portion of the lower longitudinal wall. Series *B*: roots coming from the normal position (*n*) are inverted (*i*) long enough for the amyloplasts to accumulate on the ceiling of the cell. In this case stimulation at 45° offers the amyloplasts a longer path for sliding along the lower longitudinal wall than stimulation at 135°. Compare figure 7. Reproduced by permission from Larsen 1965.

In order to check this idea experimentally one must know how much time is needed for the starch grains to accumulate on the ceiling of the cells in inverted roots. The root cap of cress has usually 7 stories of what may be called statocysts, but in only 3 or 4 of these do the starch grains move appreciably within the few minutes usually employed for stimulation. In preliminary microscopic studies on fixed root tip sections we found that the starch grains in these stories were packed on the ceiling of the cell after 6 min of inversion. We had made sure that fixation was complete in 45 seconds. Unnecessary prolongation of the inversion time should be avoided to prevent the roots from carrying out too large spontaneous curvatures, which might expose them to slight geotropical stimulation during this stage. For these reasons 8 min was chosen as the time in the inverted position before stimulation under various angles.

The responses of roots stimulated at various angles after being kept inverted for 8 min are shown in figure 7, curve *C*. The optimum at 120° to 135° had disappeared, but the optimum was now 90°, not 45° to 60°. This led to various speculations and to a more rigorous study of the time relationships of starch movement during the period of inversion, using more refined methods and thinner sections (Iversen, Pedersen, and Larsen 1968).

Figure 1 shows the position of amyloplasts in cells of a root in the normal position. The amyloplasts occupy the same positions in sections of roots which are fixed immediately after inversion. Other roots were fixed after 2, 6, 8, 10, 16, and 20 min of inversion. Figures 9 and 3 show sections of roots fixed after 8 and 20 min of inversion, respectively.

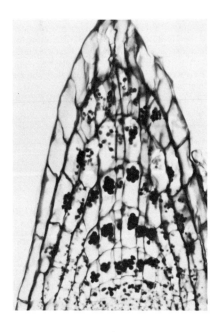

Fig. 9
As figures 1 and 3, except that the root was kept inverted for 8 min before fixation. The most advanced amyloplasts are only a little over halfway through the cell. Mag. 300X.

Using a series of photomicrographs, a scoring procedure was applied to the cells near the median line of the root cap. The scorers estimated the distance from the floor (distal cell wall) to the position of the center of density of the group of amyloplasts in each cell. The results were expressed as percentages of the total cell length. The stories were numbered from 1 to 7, number 7 being the most distal one containing amyloplasts.

The means of several scores in stories 2, 3, and 4 are plotted against inversion time in figure 10. There was no directional movement of the amyloplasts in the other stories. The curves illustrate the comparatively rapid movement in stories 2, 3, and 4 during the first 5 min. If the percentages are converted to μ on basis of the average lengths of the respective cells, it turns out that the amyloplasts fall about $6\,\mu$ in the first 5 min (at 21°C), in reasonable agreement with Hawker's results (1933) on cells from the endodermis of epicotyls of *Lathyrus odoratus*: 54 μ/hr at 20°C. Similar experiments were made with barley roots by Cercek (1969). She also reported that pretreatment of the roots with ionizing radiation in air slowed down the movement of the amyloplasts considerably, whereas the same pretreatment in the absence of oxygen increased the rate of sedimentation.

Figure 10 further shows that after a certain time, shortest for cells in the oldest stories, the movement of the amyploplasts comes to a stop. In stories 4, 3, and 2, the amyloplasts stop at 45, 70 and 80%, respectively, of the total cell length, after about 6, 12, and 16-18 min.

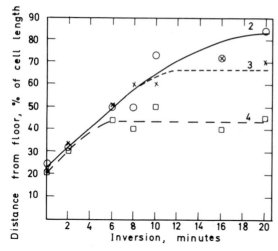

Fig. 10
Estimated distance (percentage of total cell length) of amyloplasts from floor of cell plotted against time of inversion of the root. Numbers refer to the stories of cells in which amyloplast movement can be observed, number 2 being the youngest and most proximal of these.

Amyloplasts in stories 2, 3, and 4 in roots which are placed in the horizontal position get close to the (longitudinal) cell wall in 15 min or less (figure 2). This is also the case with amyloplasts in roots which are first inverted for 8 min and then turned to the horizontal position for 15 min.

Even though the amyloplasts have not reached the opposite end of all the cells in 8 min of inversion, it is understandable on the basis of the starch statolith hypothesis that stimulation at 135° is smaller in previously inverted roots than in roots coming from the normal position. In the previously inverted roots the futher movement toward the ceiling of the cell is too slow to increase the stimulation. The absence of enhancement of the stimulation by exposure at 45°, on the other hand, requires a different explanation.

The experiments were continued after the conference. By variation of the time of inversion and the duration of stimulation it proved possible to find conditions which yielded response curves of the type of curve B in figure 7. In one series (Iversen and Larsen, in preparation), the inversion time was extended to 16 min, whereby the amyloplasts in the younger stories do get closer to the ceiling of the cells. When Lepidium seedlings were stimulated for 10 min at various angles after 16 min of inversion, the curvatures of the roots, measured after 10 and 20 min of clinostat rotation, were actually larger at 45° than at 135°. This was also true in some cases after 30 min of clinostat rotation, but at about this time the roots stimulated at 45° stopped curving, whereas roots stimulated at 135° continued increasing their curvature for at least another 30 min. Before 30 min of rotation, the response curves are thus skewed to the left, as curve B in figure 7, and after 30 min they are skewed to the right, as the curves for roots that had not been previously inverted. The behavior of the roots during the first period is in agreement with predictions based on the starch statolith hypothesis.

The cessation, and later reversal, of the development of the curvatures at a certain time (30-40 min) after the end of stimulation at 45° or smaller angles takes place both in normal and previously inverted roots. This effect is interpreted as tonic (see p. 81).

The behavior of both normal and previously inverted roots can be interpreted as the result of at least two effects: (1) a stimulation due to the movement of amyloplasts, which is enhanced if these are allowed to slide along the cytoplasm lining the cell walls and (2) a modification of the development of the resulting curvatures by tonic effects, which are inhibitory between the angles of stimulation 0° and 90°, and absent or enhancing between 90° and 180°.

In the experiments the inhibitory, tonic effects are aftereffects, induced during the stimulation. In nature, on the other hand, under continuous geotropic stimulation, the roots will approach their normal position, the inhibitory, tonic effects getting stronger and stronger and serving to counteract or prevent "overshoot" of the curvature.

Summary and Conclusions

Observations on the behavior of amyloplasts in geotropically sensitive organs, as well as theoretical considerations, point with considerable strength toward the movable amyloplasts (plastids with starch grains) in certain cells as the particles whose interaction with gravity initiates the geotropic reaction chain. In other words, the amyloplasts may be regarded as statoliths. Although exceptions exist, most of the evidence seems to support the starch statolith hypothesis. On theoretical grounds, however, the argument by von Ubisch (1928), when supported by experimental results (Larsen 1965), has been regarded as crucial against the hypothesis. A closer scrutiny, however, reveals that the design of the experiments by Zimmermann (1927) and von Ubisch (1928) is inadequate as far as proving or disproving the hypothesis is concerned. During the critical part of the experiment, the amyloplasts, falling in a vertical direction, do not fulfill the requirements of the hypothesis of exerting a pressure against the cytoplasm along the outer tangential walls of the statocysts. The Zimmermann-von Ubisch controversy over Lepidium roots can thus be "explained away."

New experiments according to the Zimmermann-von Ubisch design, but allowing the amyloplasts to slide at an angle to the vertical and thus exerting a pressure on the wall-lining cytoplasm, were carried out with Lepidium roots in the author's laboratory in Bergen. They furnished a strong support for the starch statolith hypothesis.

Whereas normal roots produce a response curve which has its peak at about 135°, roots previously inverted for 16 min yield a curve with a peak at about 45°, as predicted by the hypothesis. This is true, however, only during a certain time (about 30 min) of clinostat rotation. After that time the curvatures at 45° develop no further, but roots stimulated at 135° continue to curve, and the peak thereby moves to larger angles of stimulation. These changes can be ascribed to tonic effects which in contrast to Zimmermann's suggestion (1927), cannot be explained by statolith movements.

Dismissing von Ubisch's argument as irrelevant, most of the existing evidence can be reconciled with the view that amyloplasts do function as statoliths. The susception of gravity, however, can evidently also take place in the absence of statolith starch, at least in certain organs, as evidenced by the results of Pickard and Thimann on wheat coleoptiles. Starch-free, but still elongating, Lepidium roots, on the other hand, did not respond even to several hours of gravitational stimulation (Iversen 1969).

Acknowledgment

Research by the author and his associates reported in this paper was carried out in the Botanical Laboratory, University of Bergen, Norway. The construction of instruments was supported in part by Norges Almenvitenskapelige Forskningsraad. This support and the conscientious and patient cooperation on several aspects of this work by Knut Pedersen, university lecturer, Tor-Henning Iversen, cand.real., and Mrs. Ellen Larsen are gratefully acknowledged.

References

Audus, L. J. 1962. In *Symp. Soc. Exptl. Biol.*, vol. 16, pp. 197-226. London.

Bismarck, R. von. 1959. *Flora (Jena)* 148: 23-83.

Brauner, Leo. 1962. In *Encyclopedia of Plant Physiology*, vol. 17 (2), ed. W. Ruhland and E. Bünning, pp. 74-102. Berlin: Springer-Verlag.

Buder, J. 1908. *Ber. Deut. Botan. Ges.* 26: 162-93.

Cercek, L. 1969. *Intern. J. Radiation Biol.* 16: 419-29.

———. 1970. *Intern. J. Radiation Biol.* 17: 187-94.

Dolk, H. E. 1936. Geotropie en groeistof. Thesis, Utrecht, 1930. English translation: Geotropism and the Growth Substance. *Rec. Trav. botan. néerl.* 33: 509-85.

Fitting, H. 1905. *Jahrb. Wiss. Botan.* 41: 221-398.

Gordon, S. A. 1963. In *Space Biology*, pp. 75-105. Corvallis: Oregon State University Press.

Grahm, L. 1964. *Physiol. Plantarum* 17: 231-61.

Grahm, L., and Hertz, C. H. 1962. *Physiol. Plantarum* 15: 96-114.

———. 1964. *Physiol. Plantarum* 17: 186-201.

Haberlandt, G. 1900. *Ber. Deut. Botan. Ges.* 18: 261-72.

———. 1903. *Jahrb. Wiss. Botan.* 38: 447-500.

Hawker, Lilian E. 1932. *Ann. Botany (London)* 46: 121-57.

———. 1933. *Ann. Botany (London)* 47: 503-15.

Iversen, T-H. 1969. *Physiol. Plantarum* 22: 1251-62.

Iversen, T-H., and Flood, P. R. 1969. *Planta* 86: 295-98.

Iversen, T- H.; Pedersen, K.; and Larsen, P. 1968. *Physiol. Plantarum* 21: 811-19.

Jensen, W. A. 1962. *Botanical Histochemistry*. San Francisco and London: W. H. Freeman & Co.

Jost, L. 1902. *Biol. Cbl.* 22: 161-79.

Juniper, B. E.; Groves, S.; Landau-Schachar, B.; and Audus, L. J. 1966. *Nature* 209: 93-94.

Larsen, P. 1953. *Physiol. Plantarum* 6: 735-74.

———. 1962. In *Encyclopedia of Plant Physiology*, vol. 17 (2), ed. W. Ruhland and E. Bünning, pp. 153-99. Berlin: Springer-Verlag.

———. 1965. *Physiol. Plantarum* 18: 747-65.

———. 1969. *Physiol. Plantarum* 22: 469-88.

Němec, B. 1900. *Ber. Deut. Botan. Ges.* 18: 241-45.

———. 1902. *Ber. Deut. Botan. Ges.* 20: 339-54.

Pickard, Barbara Gillespie, and Thimann, K. V. 1966. *J. Gen. Physiol.* 49: 1065-86.

Rawitscher, F. 1932. *Der Geotropismus der Pflanzen*. Jena: Gustav Fischer.

Reiss, Elisabeth. 1934. *Planta* 22: 543-66.

Richter, E. 1914. *Ber. Deut. Botan. Ges.* 32: 302-8.

Ubisch, Gerda von. 1928. *Biol. Zb.* 48: 172-90.

Went, F. A. F. C. 1909. *Koninkl. Akad. Wetenschap., Amsterdam. Proc., Sect. Sci. 12*, pt. 1, 343-45.

Wilkins, Malcolm B. 1966. *Ann. Rev. Plant Physiol.* 17: 379-408.

Zimmermann, W. 1924. *Ber. Deut. Botan. Ges.* 42: (39)-(52).

———. 1927. *Jahrb. Wiss. Botan.* 66: 631-77.

8
The Susception of Gravity by Higher Plants: Analysis of Geotonic Data for Theories of Georeception

Barbara Gillespie Pickard, *Washington University*

It is my opinion that no modern worker has contributed more to our knowledge of geotropism than has Dr. Larsen; therefore it is a singular honor to me to be asked to discuss his very interesting, scholarly presentation.

Experimental Removal of Statolith Starch

The more we learn about the possible participation of statoliths in geotropism, the more complex the situation seems. Without going into the extensive early literature on the subject, which has already been reviewed at this symposium and which still seems somewhat controversial to many workers, I wish to stress that in 1961 Buder published a paper which demonstrated to everyone's satisfaction that statoliths (though not amyloplasts) are indeed necessary for the geotropic response of rhizoids of Chara. On the other hand, the work of von Bismarck (1959) on Sphagnum made it difficult to believe that starch grains, the obvious candidates for the role of statolith in that plant, are in any way essential to its geotropism. Thimann and I (1966) undermined the hypothesis that starch grains obligatorily participate in gravity reception by the wheat coleoptile, but in contrast Iversen's painstaking and closely comparable studies (1969) on the root of Lepidium emphatically support the hypothesis that starch grains are critically involved in that organ. It thus appears that there may be different mechanisms of georeception in different organs and in different species.

Even in several of these cases a considerable degree of uncertainty about statoliths still prevails. Filner and Hertel (1970), for example, present evidence that in corn coleoptiles the activity of the lateral auxin transport system which mediates the bending response is related to the size and abundance of movable starch (see also chap. 4, this volume). This support for the starch statolith hypothesis cannot be ignored; it does not constitute proof, but in the absence of contradictory experiments it would certainly seem convincing. Also, there remains one caviling objection to the experiments of Iversen. When Thimann and I carried out preliminary experiments with wheat coleoptiles, we found that geotropism indeed disappeared when starch disappeared, returning when starch grains were resynthesized (Pickard and Thimann 1966). Reasonable growth of the coleoptiles occurred even during the period of geotropic failure. Because we were soon able to find experimental conditions under which starch-free coleoptile tips did consistently respond to gravity, the negative preliminary results were judged misleading. Unfortunately, we did not pursue the cause of the specific geotropic inhibition, which seems to be assuming some importance. Thus, although the experiments of Iversen are well designed and executed and are very convincing, it seems worthwhile to keep the possibility of this last small loophole in mind.

Time Relationships

Dr. Larsen has reviewed studies, such as those of Hawker, in which the fall time of starch grains and the minimum geotropic stimulation time are in close agreement. Elsewhere (Larsen 1957) he has pointed out that the standard method of evaluating the minimum stimulation time is not a sensitive one. He has found for roots that geoinduction is approximately logarithmic with exposure time, and he has extrapolated his experimental response curve to show that the true minimum stimulation time for roots is much shorter than was once believed—under 10 sec (Larsen 1957, 1969). It might at first seem inconsistent to argue in one case that geotropism begins when the starch grains have finished falling and in another that geotropic induction is actually well under way before the starch grains are observed to settle. However, the inconsistency might disappear if one adopts the appealing notion that only a few grains need move a small distance in order to stimulate a small amount. The notion is unfortunately difficult to reconcile with an induction curve that is concave downward; if the plasmalemma is involved in reception, one would have expected the curve to be concave upward until the statoliths complete their fall. If the plasmalemma is not involved, but rather some other organelle forms a particular relationship with starch grains, the settling time of statoliths should not be relevant to their action.

One can only conclude that, although it is clear that there is sometimes correlation between the fall time of starch grains and geotropic sensitivity, it is not clear that time course data available to date consistently favor a causal role for statoliths.

Manipulation of Amyloplasts: A General Difficulty

It is not unlikely that georeception by both roots and coleoptiles is a function of pressure exerted within individual cells. This makes it difficult to design an experiment which would discriminate between a role for pressure exerted by statoliths and a role for the pressure exerted differentially by the entire contents of the cell on the upper and lower membranes. The differential pressure due to the cell's contents must be proportional to acceleration, and must begin the moment a cell is displaced from a position of geotropic equilibrium. On the other hand, both the falling rate of an individual statolith and the net fall time for a population of statoliths are probably proportional to acceleration, and the pressure exerted by falling and fallen statoliths against a membrane must be proportional to that component of statolith acceleration which is normal to the membrane. This consideration is among the strongest and most general of those which tend to render arguments based on manipulation of amyloplasts rather indecisive.

Larsen, in refining the ideas and experiments of Zimmermann and of von Ubisch, has provided some information which is obviously important to our understanding of georeceptive processes. He has designed experiments in which amyloplasts are manipulated, and shows how the results are partially in agreement with predictions based on certain variations of the starch statolith hypothesis but must in part be interpreted as tonic and unrelated to statoliths. Since Iversen (1969) has provided support for a role for statolith starch in these roots, this view makes sense. However, strong though the argument for statoliths in Lepidium roots may be, I do not believe the experiments on the dependence on stimulation angle and influence of inversion make it significantly stronger. The results are very complex, and the separation of effects due to statoliths and those due to other causes has not been experimentally justified. It is still possible to interpret the experiments by postulating that georeception results from the development of differential pressure—exerted either by statoliths or by the total contents of the cells.

Angle of Stimulation

By way of example I would like to demonstrate how the difference between response to stimulation at supplementary angles in the first and second quadrants may be compatible with a pressure mechanism. In figure 1 are replotted Larsen's data on the dependence of geotropic response on the angle of stimulation for an exposure of 3.75 min. In order to develop an expression which fits these points, let us assume first that response to first-quadrant stimulation, $M(\theta)$, is purely geotropic in nature, and second, that there exists an axial or longitudinal effect of gravity negligible when the stimulus angle θ is no greater than 90° and proportional, perhaps in a time-dependent manner, to $\cos \theta$ when θ is between 90° and 180°. Third, assume that the longitudinal effect modifies the purely geotropic effect by multiplicative interaction so that the resultant curvature is given by $M(180° - \theta) [1 + k(t)|\cos \theta|]$. As shown in figure 1, the fit of Larsen's second-quadrant curvature data points by the curve determined by this expression, using $k = 0.7$, encourages one to consider the possibility that there does exist a kind of georeception proportional to the pressures on the end wall of the cell—whether exerted by statoliths or by the cellular contents as a whole. Similar analysis of the more extensive data from figure 3 in Larsen's 1969 publication is also in accord with this possibility.

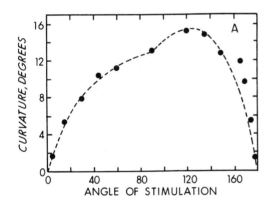

Fig. 1
All points represent experimental data and are transferred from figure 7 in Larsen's 1969 publication. The first-quadrant branch of the curve $M(\theta)$ was drawn by eye to fit the points. The second-quadrant branch was determined by the equation $M(\theta) = M(180°-\theta) [1 + k |\cos \theta|]$, with $k = 0.7$.

If there is a longitudinal effect closely proportional to the cosine of the stimulation angle, one might at first expect geotropism in its absence to be proportional to the sine. It was shown by Rutten-Pekelharing (1910) that a threshold response of Lepidium roots always results if the product of exposure duration and the sine of the angle of stimulation equals a certain value. Thus, reception depends in some undefined manner on the sine of the stimulation angle. For greater stimuli, Rutten-Pekelharing found marked deviations from reciprocity. There are several reasons why this might be so, but three are of particular interest.

The first is trivial: in designing the experiment to test the reciprocity relation, Rutten-Pekelharing chose a product of time and acceleration and then systematically decreased the stimulus angle while increasing the exposure duration compensatorily. Clearly, in this experiment, small stimulus angles require long time intervals which can greatly exceed the time required for curvature response to the stimulus received during the first few minutes of exposure. Thus, the true limitation on Rutten-Pekelharing's test might be the maximum reasonable exposure time and not a response value above which reciprocity does not hold.

A second reason is that response might be a nonlinear function of the product of stimulus angle and duration. Many nonlinear functions can be approximated as linear over a small range, particularly near the origin. This approach becomes more interesting when one examines figure 3 in Larsen's 1969 paper, because it is apparent that in the first quadrant geotropism is, very roughly, logarithmic with time, and also, very roughly, logarithmic with the sine of the stimulus angle. Therefore I tested a variety of functions in order to fit the experimental points of the figure and found, as illustrated in figure 2, that the points all fall on the line determined by the equation

$$M(t,\theta) = k_1 \ln(1 + k_2 t \sin\theta), \quad 0° < \theta < 90°$$

with t representing stimulus duration and with $k_1 = 6.0$ and $k_2 = 1.5$. It must be emphasized that no physical or physiological significance is placed on this arbitrarily chosen function. It simply shows that ultimate response may be regarded as a function of the product of stimulus duration and the sine of the angle of stimulation when a proper experiment is performed.

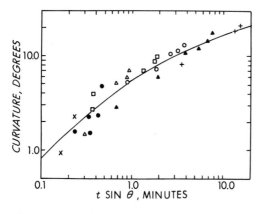

Fig. 2
All points represent experimental data and are transferred from the first quadrant of figure 3 in Larsen's 1969 publication. The curve is $M(\theta) = k_1 \ln(1 + t \sin\theta)$, with $k_1 = 6.0$, $k_2 = 1.5$.

The equation above can clearly be modified for the second quadrant by including a term to account for the cosinusoidal component of acceleration, in much the way described for the second quadrant equation shown in figure 1. Indeed, several different functions of the cosine could be multiplied with the first-quadrant equation to give plots which fit the second-quadrant data points (both those of fig. 1 and those of fig. 3, Larsen 1969) tolerably well. Since more extensive data would be required to choose between alternative equations, and since the first-quadrant equation has been selected because its fit is acceptable rather than because physical or biological significance is postulated for the function which relates $t \sin\theta$ to M, it seems premature to engage in further mathematical manipulations.

Although it appears that the second explanation of Rutten-Pekelharing's nonconforming results may provide an adequate basis on which to interpret Larsen's first- and second-quadrant data, the last of the three explanations also works in the same direction. Günther-Massias (1928) found with Lepidium roots that intermittently administered stimuli may be twice as effective as the same total stimulus administered continuously. If any such enhancement occurs, stimulation at small angles might well be relatively more effective than expected on the basis of any simple sine relation. Moreover, the deviations from a sine relation would be more conspicuous when the product of stimulus duration and angle is large enough to produce a big curvature, because it is hard to assess small deviations from expectation when the curvature itself is barely detectable. It is a simple matter to normalize Larsen's data points in accord with this outlook.

Assuming that the measured curvature response $M\theta$ deviates from a simple mathematical law only because of the simultaneous occurrence of secondary reactions and the primary georeceptive reaction, define

$$M(\theta) = D(\theta) M_p(\theta)$$
$$= D(\theta) K M_n(\theta)$$

where $D(\theta)$ is the deviation coefficient due to which $M(\theta)$ differs from $M_p(\theta)$, the response which would be measured if the receptor step were not modified by secondary processes. $M_n(\theta)$, the normalized value of $M_p(\theta)$, differs from $M_p(\theta)$ by the factor K. $M_n(90°) = 1$. Assume that for first-quadrant stimulation, axial effects are negligible, and that $M_n = \sin\theta$. Then

$$KD(\theta) = \frac{M(\theta)}{\sin\theta}. \qquad 0° < \theta < 90°$$

For the second quadrant,

$$M_n(\theta) = \frac{M(\theta)}{KD(\theta)}. \qquad 90° < \theta < 180°$$

Assume that secondary reactions are symmetric about $90°$; that is, with or without an axial effect

$$D(\theta) = D(180° - \theta). \qquad 0° < \theta < 180°$$

Then

$$M_n(\theta) = \frac{M(\theta)}{KD(180° - \theta)}. \qquad 90° < \theta < 180°$$

But

$$KD(180° - \theta) = \frac{M(180° - \theta)}{\sin(180° - \theta)}. \qquad 90° < \theta < 180°$$

Thus the formula for normalization is

$$M_n(\theta) = \frac{M(\theta)}{M(180° - \theta)} \sin\theta. \qquad 90° < \theta < 180°$$

Normalized data points are plotted in figure 3. The curve of the same figure is obtained in the first quadrant from the expression

$$\sin \theta \qquad\qquad 0° < \theta < 90°$$

and in the second quadrant from the expression

$$\sin \theta \, [1 + k |\cos \theta |], \qquad\qquad 90° < \theta < 180°$$

with $k = 0.7$. The equation for the curve, incidentally, is equivalent to that of Metzner (1929) taken for the second quadrant only. The curve and normalized data points match each other satisfactorily.

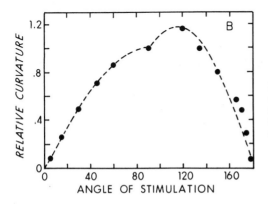

Fig. 3
Points representing Larsen's data are normalized as described in the text and expressed as curvature relative to that measured at 90°. The first-quadrant branch of the curve is simply $\sin \theta$; the second-quadrant branch is $\sin \theta \, [1 + k|\cos \theta|]$, with $k = 0.7$.

It will be shown elsewhere that for the coleoptile, under conditions in which enhancement by intermittent stimulation does not occur, first-quadrant curvature is a linear function of the sine of the angle of stimulation and second-quadrant curvature is given by $\sin \theta \, [1 + k|\cos \theta|]$. Comparative interpretation is, of course, complicated by the fact that statoliths seem not to be required for coleoptile geotropism. For simplicity the two different types of analysis represented in figures 1, 2 and 3 have been treated individually, but since both work in the same direction they could be combined readily (with adjustment of parameters) to give the same outcome.

In sum, the purpose of these analyses is to formally separate georeception into reasonable functions of the perpendicular and longitudinal components of gravity. If the separation is valid, it supports a physical model invoking pressures exerted within cells. Since numerous interpretations of these kinetic data are possible, the data themselves do not persuade one to believe in any particular model, but should serve to help design further experiments.

The separation of georeception into perpendicular and longitudinal components in no way suggests the nature of the mechanisms which mediate the ultimate response. However, a physiological interpretation of the modifying axial response might be based on the long-known fact that gravity can act longitudinally to influence growth (see in particular Larsen 1953). Any such change in growth rate would be expected to influence the differential growth which brings about tropistic curvature of seedling roots and shoots. Such an effect on the differential growth, which would show up as the deviation from symmetric orthogeotropic response about the horizontal position of stimulation, and as influences of prior or subsequent inversion, is by definition geotonic.

The geogrowth reaction is particularly hard to understand in the root, in which the nature of hormone transport and action is still so controversial. However, the behavior of the coleoptile is better understood, and may provide at least a very crude analogue to serve as a starting point for thoughts about roots. In 1962 and 1963, Hertel and Leopold provided an explanation for the geogrowth reaction. Comparing long-term transport of radioactive auxin in upright and inverted coleoptile sections, they demonstrated a difference in the amount of auxin emerging from the basal cut surface. More recent work by Ouitrakul and Hertel (1969) has shown that a decrease in the transport of the synthetic auxin naphthaleneacetic acid occurs rapidly after a Zea coleoptile section is inverted. Such changes in auxin transport would of course necessarily bring about changes in growth rate.

Preinverted Roots

The time courses of response following inversion described by Dr. Larsen are complex, and it is here that he attributes early response stages to statoliths and ultimate response to unidentified factors. It is difficult to make sense out of the data at this stage, but this is not surprising in view of the uncertainties surrounding the role and behavior of growth factors in roots. Without suggesting that mechanisms in roots and shoots are the same, it is worth noting that transients in the hormone system seem to occur even in the simpler coleoptile, for the data of Ouitrakul and Hertel (1969) indicate that auxin transport may overshoot that of controls when inverted organs are turned back to the upright position. A thorough, systematic description of all the systems interacting in root geotropism may be required for a fully satisfying explanation of the complex influences of pre- and postinversions.

Summary

Current evidence suggests that different georeceptive tissues may transduce acceleration in different ways. Iversen has presented a strong case, based on starch depletion, that amyloplasts function as statoliths in roots of Lepidium. Nevertheless, the kinetics of the Lepidium root response can probably be explained without reference to statolith behavior, though they seem also to be compatible with it. Current kinetic evidence must thus be viewed as contributing a great deal to our knowledge of geotropism without providing proof for a functional role for statoliths in roots.

Acknowledgment

Preparation of this discussion was supported by the Center for the Biology of Natural Systems through Public Health Service grant no. ES-00139 from the Division of Environmental Health Sciences.

References

Bismarck, R. von. 1959. *Flora* 148: 23-83.

Buder, J. 1961. *Ber. Deut. Botan. Ges.* 74: 14-23.

Filner, B., and Hertel, R. 1970. Submitted to *Planta*.

Günther-Massias, M. 1928. *Z. Botan.* 21: 129-72.

Hertel, R., and Leopold, A. C. 1962. *Naturwissenschaften* 16: 377-78.

———. 1963. *Planta* 59: 535-62.

Iversen, T.-H. 1969. *Physiol. Plantarum* 22: 1251-62.

Larsen, P. 1953. *Physiol. Plantarum* 6: 735-74.

———. 1957. *Physiol. Plantarum* 10: 127-63.

———. 1969. *Physiol. Plantarum* 22: 469-88.

Metzner, P. 1929. *Jahrb. Wiss. Botan.* 71: 325-85.

Ouitrakul, R., and Hertel, R. 1969. *Planta* 88: 233-43.

Pickard, B. G., and Thimann, K. V. 1966. *J. Gen. Physiol.* 49: 1065-86.

Rutten-Pekelharing, C. J. 1910. *Rec. Trav. Botan. Neerl.* 7: 241-346.

9
A Case against Statoliths

Arthur H. Westing, *Windham College*

The geotropism of plants is an *internal* growth response to an *external* stimulus (Larsen 1962a). Moreover, the stimulus—gravity—has an orientation in space, as does the response (Westing 1964). In the case of certain tropistic responses, for example, phototropism, one can explain the lateral asymmetry of the growth response by a gradient of stimulus across the organ in question. Since the force of gravity between objects (masses) is inversely related to the distance between them, there is in geotropism also a gradient of stimulus across an organ. However, this gradient can, I believe, be dismissed as too small to be recognized by the plant. One is left, therefore, as Professor Larsen has pointed out, with seeking a mass displacement within the plant as one likely alternative.

Larsen has presented a persuasive argument in favor of a starch statolith mechanism of gravitational perception, as have others before him, including Haberlandt (1909), Zimmermann (1926-27), Rawitscher (1932), Brauner (1962), and Audus (1962, 1964). Indeed, some of the best experimental evidence to support such a hypothesis has been provided us by Hawker (1932, 1933) using (among others) a variety of coniferous species, plants for which I have a particular fondness. For my commentary, however, I shall play the devil's advocate.

First of all, in my opinion, Larsen dismissed all too briefly the reports in the literature of geotropic responses (either normal or virtually so) that have been observed to occur apparently without the benefit of movable starch grains. I believe that he alluded to only two such cases. In fact, the list, although not large, spans four major plant taxa and encompasses stems, leaves, flowers, and roots. A somewhat cursory search of the literature provides the following examples:

Clivia (the Kafir lily): the perianth, the starch grains of which do not sink (Gius 1905);

Laelia (a cattleya-like orchid): the naturally starch-free aerial roots (Tischler 1905);

Caulerpa (a marine alga): the stems, the starch grains of which are immobile (Haberlandt 1906);

Lepidium (the garden cress): artificially starch-free roots (Went 1909);

Pisum (the garden pea): artificially starch-free roots (Guttenberg 1933);

Zea (corn): artificially starch-free roots (Guttenberg 1933);

Vicia (the broad bean): artificially starch-free roots (Syre 1938-39; Younis 1954);

Spahagnum (a moss): artificially starch-free gametophytic stems (Bismarck 1959-60);

Triticum (wheat): artificially starch-free coleoptiles (Pickard and Thimann 1966); and

The geosensitive fungi: the naturally starch-free sporangiophores of Phycomycetes and sporocarps of Basidiomycetes (Banbury 1962).

Of these, Dr. Pickard has just discussed her researches with *Triticum* seedlings, providing us with one of the most incisive, best documented, and least ambiguous examples in the list (Pickard and Thimann 1966).

Apropos some of the difficulties mentioned by both Larsen and Pickard, one way to avoid energy depletion simultaneously with artificial starch depletion might be to immerse the organs in a sugar solution in the manner of some work I once described (Westing 1962). There is also the possibility that organs could be rendered starch-free by applying extracts of *Allium* (Hart et al. 1960-61). Finally, artificial starch depletion might be accomplished more easily and be a less traumatic experience for the plant if species or organs of plants naturally low in starch (such as *Allium* or *Iris*) were employed.

If not starch grains (or amyloplasts), can some other cell organelles or inclusions function as statoliths? Considerations of mass, specific gravity (or density), size, Brownian movement (thermal agitation), cyclosis, occurrence, distribution, and demonstrability have led Audus (1962), Gordon (1963), Griffiths (Griffiths and Audus [1964]), and perhaps others virtually to preclude such a possibility. Pollard (1965; and chap. 2, this volume) has suggested nucleoli, but the nucleolus is an unlikely candidate since the nuclei in a multicellular organ are located at random within their various cells. The suggestion by Larsen that nonamylaceous statoliths might exist under certain circumstances is based, I believe, purely on the negative, circumstantial evidence that when starch grains are absent it stands to reason that some other organelle must be serving the statolithic function.

The first major thorn in the side of the statolithophiles was introduced in 1928 by von Ubisch. She reported that experimental manipulations which should intensify statolithic stimulation had no influence on subsequent geotropic response. Through the years she has been explained away repeatedly (for example, by Rawitscher [1932], Audus [1962], and Wilkins [1966]) on the basis of poor documentation. Unfortunately, Larsen (1965) himself has fully confirmed von Ubisch's data. As we have heard, Larsen reconciled himself several years ago with these data by hypothesizing that statoliths could still be the basis of the sensing mechanism if they behaved as pendulums rather than as free-falling bodies (1962b). More recently (1965) he advanced an alternate hypothesis that the geotropic response is confounded by a superimposed geotonic response (quite correctly so, in my view); but he suggests that while the former is perceived by free-falling statoliths, the latter is not.

I find it rather disconcerting to think that we now have to seek out two primary mechanisms of gravitational perception when we are having so much trouble with one. Actually it is, of course, a valid speculation that the plant has two or more independent mechanisms of inertial perception, but I am aware of no concrete evidence that points in this direction.

Along another tangent, Hertel and Leopold (1962-63) have presented some evidence to suggest that a statolith mechanism is required for the phenomenon of auxin polarity. Recently, however, Gordon and his group (Dedolph, Naqvi, and Gordon 1966) have presented other evidence which seems to dispute this.

Some time ago Günther-Massias (1928-29) found that periods of intermittent stimulation of only a few seconds each gave a more than additive response when the total stimulation time equaled the presentation time; and her results have been confirmed more recently in similar experiments by Bünning and Glatzle (1948-49). Pickard has already pointed out that such results are difficult to reconcile with a statolith mechanism (Pickard and Thimann 1966).

Pickard has further suggested that we take a second look at Czapek's suggestion of eight decades ago (1898) that the stimulus might well be perceived via the weight of the cytoplasm pressing upon the cell membrane beneath it. This force, although minute (of the order of 1 dyne/cm^2), apparently exceeds by a large margin the physical force that would be exerted by the amyloplasts. She has pointed out that stimuli of this magnitude have been shown in other systems to be amplified and translated into chemical action (Pickard and Thimann 1966). On the other hand, I should add that it has been suggested (validly or not) that the osmotic force of the cytoplasm would seem capable of swamping the downward force exerted by its own weight (Audus 1962).

I must also mention the rather far-fetched model presented in the literature a few years ago (Westing 1965) in which statoliths were likened to doubly hinged pendulums (thus doing Professor Larsen's model one better). The phenomena that were being accounted for were plagiogeotropism, topophysis, and training in gymnospermous stems and branches. Owing to the inability of anyone to come to the rescue by discovering such a statolith, one can instead speculate that chronic geotropic stimulation (of whatever sort) leads more or less rapidly to geotropic desensitization of whichever wall in the cell is lowermost and, contrariwise, that chronic freedom from static irritation leads to sensitization (Westing 1968).

To conclude, I wish to simply stress that we must not close our minds (and thus our researches and their interpretation) to the possibility that statoliths are *not* involved in gravitational perception—or that if they are, that they are not obligately involved. Moreover, the means may not be the same in all plants. Larsen opened a recent article on this subject (1965) with the statement, "Geotropically sensitive plant organs *must* contain statoliths of some kind or other" (italics mine). It is obvious that I feel much less confident about this than he does.

Summary

Since Professor Larsen interpreted his research findings (and those of others) to support a statolith theory of gravitational perception, I have felt that the most useful commentary might be to raise some objections to such an interpretation.

It was pointed out that of the various cell organelles and inclusions only starch grains (or amyloplasts) seem capable of serving the statolithic function. A list of cases was presented in which plant organs perceive gravity apparently without the benefit of mobile starch. Certain additional lines of evidence that speak against statoliths were also mentioned. The question was raised of whether it is likely that more than one means of gravitational perception exists within a single plant. It was suggested that the mechanism may not be the same in all plants.

Added Note

Recently I have been reading Ludwig Jost's "Vorlesungen über Pflanzenphysiologie" (3d ed., Jena: Gustav Fischer 1913) and find that he mentions (p. 594) two cases of geotropically responsive, naturally starch-free root hairs that had been described by Knoll (1909, Sitzungsberichte d. Kaiserlichen Akad. d. Wissenschaften, Math.-Naturwiss. Cl., Wien, Pt. 1, 118: 576 and Bischoff (1912, Beihefte z. botanischen Centralblatt, Leipzig 28 [Pt. 1]: 94).

Jost also (pp. 594-95) alludes to experiments in which the geotropic response can be observed despite manipulations (intermittent stimulation or rotation) which prevent the settling of statoliths.

References

Audus, L. J. 1962. *Symp. Soc. Exptl. Biol.* 16: 197-226 + 2 pl.

———. 1964. *Physiol. Plantarum* 17: 737-45.

Banbury, G. H. 1962. In *Encyclopedia of Plant Physiology*, vol. 17 (2), ed. W. Ruhland, pp. 344-77. Berlin: Springer-Verlag.

Bismarck, R. von. 1959-60. *Flora* 148: 23-83.

Brauner, L. 1962. In *Encyclopedia of Plant Physiology*, vol. 17 (2), ed. W. Ruhland, pp. 74-102. Berlin: Springer-Verlag.

Bünning, E., and Glatzle, D. 1948-49. *Planta* 36: 199-202.

Czapek, F. 1898. *Jahrb. Wiss. Botan.* 32: 175-308.

Dedolph, R. R.; Naqvi, S. M.; and Gordon, S. A. 1966. *Plant Physiol.* 41: 897-902.

Gius, L. 1905. *Z. Österr. Botan.* 55: 92-97.

Gordon, S. A. 1963. In *Space Biology: Proceedings of the Twenty-fourth Biology Colloquium*, ed. F. A. Gilfillan, pp. 75-105. Corvallis: Ore. State Univ. Press.

Griffiths, H. J., and Audus, L. J. 1964. *New Phytologist* 63: 319-33 + pl. 24-26.

Günther-Massias, M. 1928-29. *Z. Botan.* 21: 129-72.

Guttenberg, H. von. 1933. *Planta* 20: 230-32.

Haberlandt, G. 1906. *Sitzber. Akad. Wiss., Math.—Natur. Cl., Wien,* pt. 1, 115: 577-98 + 1 pl.

———. 1909. *Physiologische Pflanzenanatomie,* 4th ed. Leipzig: Wilhelm Englemann.

Hart, R. D.; Haldiman, R.; Gries, G. A.; and Rogers, B. J. 1960-61. *Bot. Gaz.* 122: 148-50.

Hawker, L. E. 1932. *Ann. Botany* 46: 121-57.

———. 1933. *Ann. Botany* 47: 503-15.

Hertel, R., and Leopold, A. C. 1962-63. *Planta* 59: 535-62.

Larsen, P. 1962a. In *Encyclopedia of Plant Physiology,* vol. 17 (2), ed. W. Ruhland, pp. 34-73. Berlin: Springer-Verlag.

———. 1962b. In *Encyclopedia of Plant Physiology,* vol. 17 (2), ed. W. Ruhland, pp. 153-99. Berlin: Springer-Verlag.

Larsen, P. 1965. *Physiol. Plantarum* 18: 747-65.

Pickard, B. G., and Thimann, K. V. 1966. *J. Gen. Physiol.* 49: 1065-86.

Pollard, E. C. 1965. *J. Theoret. Biol.* 8: 113-23.

Rawitscher, F. 1932. *Geotropismus der Pflanzen.* Jena: Gustav Fischer.

Syre, H. 1938-39. *Z. Botan.* 33: 129-82.

Tischler, G. 1905. *Flora* 94: 1-67.

Ubisch, G. von. 1928. *Biol. Zb.* 48: 172-90.

Went, F. A. F. C. 1909. *Proc. Koninkl. Ned. Akad. Wetenschap.* (pt. 1) 12: 343-45.

Westing, A. H. 1962. *Proc. Indiana Acad. Sci.* 72: 115-17.

———. 1964. *Science* 144: 1342-44.

———. 1965. *Botan. Rev.* 31: 381-480.

———. 1968. *Botan. Rev.* 34: 51-78.

Wilkins, M. B. 1966. *Ann. Rev. Plant Physiol.* 17: 379-408.

Younis, A. F. 1954. *J. Exp. Botany* 5: 357-72.

Zimmermann, W. 1926-27. *Jahrb. Wiss. Botan.* 66: 631-95.

Discussion (of the Three Preceding Papers)

AUDUS: The majority of the quoted exceptions to the correlation between occurrence of starch statoliths and possession of sensitivity to gravity have practically no numerical data to substantiate their case. I put this forward as a plea for more work on the so-called exceptions. I think we will find when they are looked at more carefully that perhaps they aren't exceptions at all. I suspect that in many cases the curvatures recorded were not curvatures in response to gravity but responses to some other field which the early worker didn't attempt to eliminate.

WILKINS: In discussing the validity of the von Ubisch objection to the statolith theory, I did, in my review of 1966, go rather carefully into the literature of the very precise work of Hertz and his collaborators in using the geoelectric effect as a measure of gravity perception. After all, if you want to discuss geoperception mechanisms, the geoelectric effect is as good as the geotropic response as an observable process. Hertz was able to show that he could repeat, with coleoptiles, the Zimmermann procedure. He was also able to get the predicted results when using the von Ubisch type of procedure, which involves initial inversion. But, he couldn't get the von Ubisch procedure to work unless he inverted the plants for about 5 hr before he started changing their orientation again. If he inverted them for only 20 min or so he couldn't get the differences which one might predict. It makes a big difference how long the organs are inverted beforehand. This is obvious from the work of Hertz, and I think we have to bear it in mind when we take a plant that is normally vertical and invert it. The plant is stimulated to a great extent by inversion, and we have to wait for it to recover before meaningful experiments can be done. I think this is what happened in the case of all experiments of this type. The material is merely inverted for 20 min, then put on its side, and so on. In other words, it is still probably responding to the initial inversion when it is put on its side. I believe that this type of work needs to be gone into a good deal more thoroughly with particular emphasis on the initial inversion.

PICKARD: Grahm and Hertz believe their data to be consistent with that version of the statolith theory which supposes stimulation to be due to contact or rubbing rather than pressure of the statoliths. I view the data as persuasive though by no means conclusive evidence against a critical role for statoliths. The importance of the Grahm and Hertz experiments to me is that they suggest that certain results of inversion build up fairly rapidly toward a maximum, such that further uprighting or reinversion is but slightly stimulatory if it is of brief duration. If inversion lasts long enough (as 5 hr), plants achieve a steady state comparable, in at least some respects, with that of upright controls, a steady state such that returning plants to their normal position has an influence similar, in at least some respects, to that of the original inversion. If, on the other hand, one wishes to attribute the effects of a 10- or 20-min inversion prior to geotropic stimulation to a peculiarly irrelevant geostimulation, then one must also view inversion for 10 or 20 min immediately following geotropic stimulation as having an irrelevant upsetting effect on the plant's behavior. Grahm and Hertz have an exceptionally elegant system for studying responses to the longitudinal component of gravity. However, I could, if asked, produce a simple nonstatolith receptor hypothesis which yields predictions that overlap those of at least several versions of the sliding and falling statolith hypothesis. I therefore think that at present it would be a mistake to base a belief for or against statoliths on experiments of the Zimmermann and von Ubisch style.

This is why I have tried to point out that a slightly more abstract approach yields a more general and secure frame of reference within which one can design future experiments. The important question to be answered is whether the tonic effects are direct or indirect. I see no firm evidence that they are direct, and many clues that changes in axial transport of auxin could indirectly influence tropistic response, even though the clues do not fit together well in my mind. I especially want to note that Grahm and Hertz came close to adopting my view as they examined the possibility that the inversion effects might be explained in terms of influences controlling the axial growth rate. Perhaps because they, unfortunately, compared their own electrical data with Zimmermann's root curvature experiments rather than with legitimate controls, and perhaps because they had not worked out under defined conditions the relation of the magnitude of auxin asymmetry and of curvature to the geoelectric potential, they rejected axial effects as of secondary significance. They instead supported the statolith hypothesis with cautious enthusiasm. Also, they did not emphasize the marked difference between results of the 20- and 60-versus 300-min preinversions, because they thought they could attribute this difference to their use of decapitated sections (with cells postulated to be rectangular in axial section) for the 300-min preinversion and intact tips (with apical cells postulated triangular in axial section) for the briefer inversions. Inspection reveals that certain cells in the apex do taper a bit; however, this tapering is in the epidermal cells (which have amyloplasts) and is in the wrong direction for the argument.

LARSEN: Returning to the von Ubisch studies, I have known since 1953 that she was right. I repeated her experiments with improved methods and reported this at a conference in England, but there were no published proceedings. I did, however, include my results in an article in the *Encyclopedia of Plant Physiology* in 1962, still regarding them as strong evidence against the starch statolith hypothesis. This is why I felt forced to propose something else. However, as stated in my paper, the results of Zimmermann's and von Ubisch's experiments may have no bearing on the starch statolith hypothesis, and I am now inclined to accept it. Still, there are several points that need to be cleared up, such as, for example, the response of starch-free organs and of certain decapped roots. Younis removed the cap from roots of *Vicia faba*, but in such roots the curvature proceeded at the same rate as in control roots and to the same final angle. Geotropic responses by decapitated Vicia roots were also mentioned in some of the work by Haberlandt, who referred to other cells containing movable starch as being responsible for geoperception. But he regarded the root cap as the most effective georeceptor of the whole plant. So, removal of the cap should at least reduce the response, but in Younis's experiments it did not. Since Younis worked in London, perhaps our British colleagues might furnish additional information on his experiments.

AUDUS: I don't think your interpretation of the results is quite correct. As I read these results I discover that Younis couldn't in fact separate precisely growth and geotropic response. This was very difficult, because his decapitation methods were surgical, and you cannot remove the whole of the cap without removing the quiescent center—in other words, without reducing the growth. I hope to say a few words about this particular kind of effect later in my own contribution.

LARSEN: Then the Juniper method of removing the root cap might resolve the problem of the presence of a secondary statolith apparatus, for instance, in the cortex of the root. But perhaps Juniper's method cannot be successfully applied to the roots of *Vicia faba*?

AUDUS: Juniper and his coworker discovered the decapping technique. In certain varieties of maize, the root cap can be removed completely without damaging the quiescent center or the meristem in any way. When you do this the roots become completely insensitive to gravity, and will continue to grow horizontally at the same rate as undecapped roots. There is *no* effect on growth rate, and the roots are completely insensitive to gravity. They grow horizontally for 36 hr, after which time the root cap has regenerated. At the same time they acquire geotropic sensitivity, which coincides with the appearance of sedimented starch. More studies need to be done to establish this unequivocally. This seems to be an absolute correlation, and I am quite convinced that the root cap is the only sensor in the maize root. Whether it is the starch statolith that is responsible for gravity perception is another matter, but I claim that we have located the sensitive *region* in the root cap of maize.

LAVERACK: Is there in the cytoplasm a basal body of a cilium or a kinetosome? In animals it seems obvious at the moment that these are likely to be mechanically sensitive, certainly at the surfaces of mechanoreceptive cells. If somewhere in the cytoplasm of your plant you have a similar structure, you might have a similar answer.

GENERAL REPLY: Not as far as we are aware.

WILKINS: These amyloplasts have a membrane around them. There are very few structures with a membrane around them that do not carry a charge. We do not necessarily look for a whisker that is banged by a calcium carbonate granule as in animals, but maybe this amyloplast is charged; when it reaches the vicinity of the plasmadesmata, which may also be charged, one can ask whether this alters the charge in the tube and perhaps alters the movements of ions through the tube. One may have to look at the electrophysiology of the phenomenon as well as at its structural correlates.

III Correlation in Plant Geotropism

10
Hormone Movement in Geotropism

Malcolm B. Wilkins, *University of East Anglia*

A geotropic response is, in general terms, the curvature developed by a plant organ when its normal orientation with respect to gravity is disturbed. In a geotropic response, however, the plane and direction of curvature are determined entirely by the orientation of the organ in the gravitational field and by the physiological state of the tissues. A geotropic response therefore differs from a geonastic response in that the direction of bending is not restricted or predetermined by the structural characteristics of the organ. This distinction highlights the central problem for elucidation in studies of geotropism, the mechanism controlling the direction of curvature of an organ.

Geotropic responses are limited to organs having extending cells or cells capable of extension. The responses fall into three broad mechanistic categories depending upon whether curvature is brought about by: (1) the development of different rates of growth of the upper and lower halves of an organ which is already growing by irreversible expansion of cells in its subapical region (root and shoot); (2) the initiation of growth on one side of an essentially nongrowing region of an organ (grass node); (3) the differential increase in turgidity of cells in the upper and lower halves of an organ which is nongrowing and has no potential for further growth (pulvinus of Phaseolus).

Space will not allow detailed consideration of the mechanism of the response in all these categories; indeed, the third has scarcely been investigated from the mechanistic standpoint. In this paper discussion will be confined to the relationship between auxin transport and the geotropic response of roots and shoots.

Geotropic Responses of Roots and Shoots

The geotropic curvature of roots and shoots takes place in the subapical regions of the organs where cells are actively growing. As early as 1882 Sachs had established that the downward curvature of horizontal roots was due to the lower half growing more slowly than the upper half. In horizontal shoots Weber (1931) using Hordeum coleoptiles, and Navez and Robinson (1932) using Avena coleoptiles found that upward curvature was due to the growth rate of the lower half being increased and that of the upper half decreased. The problem of explaining the geotropic response of both roots and shoots thus resolves itself into establishing the mechanism for the induction of different rates of growth of the upper and lower halves of the organ.

The early work of Boysen-Jensen (1910, 1911, 1913) and of Paál (1914, 1919) led to the concept of growth-regulating substances being produced at the apex of roots and shoots and of these substances then moving in a basipetal direction toward the growing zone of the organ. It was Paál

(1914, 1919) who first suggested that the curvature responses might be due to an unequal distribution of the apically produced growth-regulating substances in the two halves of the organ. There are, of course, a number of ways in which such asymmetry of growth substance concentration could arise, for example, by differential synthesis, degradation, or immobilization. However, simultaneous investigations by Cholodny and Went led to independent papers in 1926 in which both authors suggested that the asymmetry of growth substances in roots and shoots arises as the result of a lateral movement of hormone from the upper half to the lower half of the horizontal organ. The opposite responses of roots and shoots were attributed to their opposite response to the concentration gradient of growth substance established across the organ. This general concept became known as the Cholodny-Went hypothesis and it seemed to find experimental support in the studies made during the period immediately after Went (1928) had developed the first method for the quantitative assay of plant growth substances. Recent investigations appear to have established the validity of the Cholodny-Went hypothesis for geotropism in shoots but have cast very considerable doubt upon its validity for the responses of roots. For this reason the evidence for the lateral movement of growth hormones in geotropically stimulated roots and shoots must be examined separately.

Shoots

The first experimental investigation to test the validity of the Cholodny-Went hypothesis for shoot tissues was carried out by Dolk (1929), using the general experimental procedure outlined in figure 1 (*A* and *B*). Apices of coleoptiles of Avena and Zea, and segments of Avena coleoptiles were placed on bisected blocks of agar. In the case of the coleoptile segment (fig. 1*B*), an agar block containing the natural growth-regulating diffusate from coleoptile tips was applied to the apical cut surface. The tissues were then oriented vertically or horizontally, and after about 90 min the amount of growth-promoting activity in the agar receiving blocks was determined by the Went Avena curvature test. The general distribution of growth-promoting activity is illustrated in the figures 1*A* and 1*B*. With horizontal tissue more activity was found in the lower receiving block than in the upper one. No difference between the total amount of growth-promoting activity diffusing out of vertical and horizontal tips could be detected. Navez and Robinson (1932) and, much more recently, Gillespie and Briggs (1961) used essentially the same experimental procedure (fig. 1*A*) with tips of Avena and Zea coleoptiles respectively and obtained results closely similar to those published by Dolk (1929). Dijkman (1934), using hypocotyls of Lupinus and the procedure shown in figure 1*B*, also found more growth-promoting activity in the lower receiving block than in the upper one in the case of horizontal segments.

A somewhat different experimental approach in which the growth substances have been chemically extracted from the upper and lower halves of the organ by various solvents has also shown there to be more growth-promoting activity in the lower half than in the upper half of horizontal tissue (Boysen-Jensen 1936; van Overbeek, Olivo and de Vazquez 1945).

These studies show that an asymmetry in net growth-promoting activity is established between the upper and lower halves of horizontal shoots and between blocks of agar in contact with the basal ends of the upper and lower halves of shoot segments. They do not, however, provide evidence to establish how this gradient arises. While the gradient could undoubtedly have been set up by the lateral movement of growth-promoting substances from the upper to the lower half of the tissues, it could also have arisen in a number of other ways. For example, in the experiments with tips of coleoptiles a changed pattern of auxin synthesis involving an increase in the lower half and a decrease in the upper half could alone give rise to the observed gradient. In the experiment with segments of tissue, different rates of uptake and longitudinal movement of auxin by the lower and upper halves of the tissue could lead to an asymmetry of growth substances both within the tissues and in the receiving blocks. A further difficulty in interpreting the results of experiments in which the asymmetry of growth substances is estimated by the means of their diffusion into agar blocks is that the data refer only to the net growth-promoting activity diffusing out of the tissues. This net activity could be determined by the relative amounts of a number of different substances entering the agar block. The asymmetry of net growth-promoting activity need not reflect the redistribution of growth-promoting substances such as auxin at all, but perhaps the upward movement or changed pattern of synthesis of a growth-inhibiting substance.

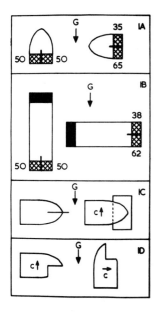

Fig. 1
The various experimental techniques used in the study of geotropism of plant organs. Donor blocks of agar are shown in black, and receiver blocks are cross-hatched. The plant tissue, either coleoptile tip (1A, C, and D) or subapical segment (1B) is unshaded. The figures beside the receiving blocks show the percentage distribution of the net growth activity emerging from the basal end of the tissue. Direction of gravity is shown by the arrow (G) and the direction of curvature by the arrow (C). For explanation see text.

There are two investigations which suggest strongly that lateral movement of a growth-regulating substance is involved in the geotropic responses of shoots. Brauner and Appel (1960) inserted in the apex of Avena coleoptiles small plates of mica. The coleoptiles were then oriented horizontally, in some the mica plates being horizontal and in others vertical as shown in figure 1C. The coleoptiles in which the plates were vertical bent upward to a much greater extent than those in which the plates were horizontal. The existence of a mechanical barrier to the lateral movement of a substance within the coleoptile thus reduces the geotropic response. This clearly implies that lateral movement of a substance is involved in the geotropic response. The fact that coleoptiles having the horizontal plate eventually bent upward to a significant extent does not invalidate this conclusion since the lateral movement of growth substances could have occurred in subapical regions of the coleoptiles where the mechanical barrier did not exist. The experiments of Brauner and Appel (1960) do not, of course, indicate whether substances move from the upper to the lower half of the organ or vice versa. The experiments of Koch (1934) also point strongly to the lateral movement of a substance during geotropic responses in shoots and further suggest that the movement takes place from the upper to the lower half of the organ. Koch (1934) excised one half of the tip of an Avena coleoptile and then oriented the segments in the ways shown in figure 1D. Because growth of coleoptiles is inhibited by total decapitation, the fact that the direction of curvature in relation to the intact half of the tip was reversed in the horizontal coleoptile shows that a substance must have moved from the upper to the lower half. Since the growth rate of the lower half has been enhanced relative to that of the upper half, it appears that the substance which has moved laterally is a promoter of growth or a precursor of a growth promoter.

By 1955 it had been established with a reasonable degree of certainty that indole-3-acetic acid (IAA) is the principal growth hormone produced at the apex of the shoot, and that it moves basipetally and controls the growth rate of the cells in the subapical region. In addition, by this time radioactive IAA had become available. It was possible therefore to test directly the validity of the Cholodny-Went hypothesis by determining whether or not radioactive IAA moved laterally from the upper half to the lower half of geotropically stimulated shoots.

The first attempts to demonstrate the occurrence of a lateral asymmetry of IAA in horizontal shoot tissues were unsuccessful. Reisener (1957) and Reisener and Simon (1960) applied labeled IAA to the apices of intact Avena coleoptiles and found no difference between the amount of radioactivity in the upper and lower halves of the organ after geotropic stimulation. Similarly no asymmetry of radioactivity could be detected after geotropic stimulation of roots and shoots of Pisum, Phaseolus and Zea which had previously been immersed vertically in a solution of radioactive IAA (Ching and Fang 1958).

A number of subsequent investigations have shown, however, that an asymmetry of radioactivity develops both in geotropically stimulated shoot segments and in the receiving blocks at their basal ends. Gillespie and Thimann (1961) used the technique shown in figure 1B with Avena coleoptile segments. The symmetrical donor block contained labeled IAA. More radioactivity was found in the lower receiving block than in the upper one, but the amounts of radioactivity in the upper and lower halves of the segment were not significantly different. In similar experiments with Zea coleoptiles (Gillespie and Thimann 1963), more radioactivity was found in the lower half than in the upper half of the horizontal segment, in addition to the asymmetry found between the upper and lower receiving blocks. These results are in close agreement with those of the earlier experiments in which asymmetry in the distribution of the natural growth-promoting diffusate from coleoptile tips was established in coleoptile segments (Dolk 1929). Furthermore, they strongly suggest that it is the growth-regulating hormone indole-acetic acid which is asymmetrically distributed.

Another experimental procedure, used both by Hertel and Leopold (1963a) and by Gillespie and Thimann (1963), has been to place agar blocks in contact with the lateral cut surfaces of half segments of Zea coleoptiles as shown in figure 2A. The amount of IAA moving into the lateral receiving block from an apical donor block was then determined as a function of orientation in the gravitational field. In both investigations more radioactivity was found in the lateral receiving block when this was beneath the segment than in any other orientation.

None of these experiments provides conclusive evidence that the asymmetry of radioactivity arises from a lateral transport of IAA within the tissues. The movement of labeled IAA from an apical donor block to a basal or lateral receiving block involves at least three processes: (1) uptake, (2) movement, and (3) secretion. Because of the complexity of the overall process the interpretation of experimental results based on amounts of radioactive IAA in receiving blocks is extremely difficult. The appearance of more radioactivity in a lateral receiving block when the lateral cut surface of the coleoptile is directed downward could be due to enhancement of secretion of IAA and not necessarily to lateral transport. The extent to which uptake, movement, and secretion of IAA are interrelated is not known at the present time, but secretion might be affected by gravity independently of movement.

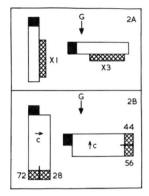

Fig. 2
Further techniques used in the study of geotropism in plant tissues. Tissue unshaded, donor blocks black and receiving blocks cross-hatched. Figures beside the receiving blocks show in *A* the relative amounts of radioactivity present when the donor contains labeled IAA, and in *B* the percentage distribution of the total radioactivity emerging from the base of the segment. Direction of gravity is indicated by the arrow (*G*) and of curvature by the arrow (*C*). For explanation see text.

Similarly, experiments in which sections are supplied with symmetrical apical donors and two receiving blocks at the basal end are open to several possible interpretations. The amount of radioactive IAA found in the upper and lower receiving blocks at the basal end of the segment has very little significance in attempting to establish whether lateral transport has occurred since, as has already been pointed out, changes in the rates of uptake, longitudinal transport and immobilization of IAA by the upper and lower halves of the sections could give rise to asymmetry of IAA in the receiver without the occurrence of lateral movement of IAA. In addition, such changes could also give rise to an asymmetry in the IAA present in the upper and lower halves of the tissues. How then can unequivocal evidence be obtained for or against the occurrence of lateral transport of auxin *within* the plant tissues?

The method adopted by Goldsmith and Wilkins (1962, 1964) was to apply radioactive IAA asymmetrically to the apical end of coleoptile segments of *Zea mays* (fig. 2*B*). The IAA will then be taken up on one side of the coleoptile only, and any radioactive IAA found in the opposite half of the segment *must* have moved laterally within the tissue. It is therefore possible to determine whether the amount of IAA which moves laterally is affected by the orientation of the segment in the gravitational field.

Goldsmith and Wilkins (1964) used 15-mm Zea coleoptile segments and found that uptake of IAA from the asymmetrical donor was not significantly affected by orientation in the gravitational field, at least during the 2-hr experimental period. The amount of radioactivity reaching the opposite half of the section was always greatest when the section was horizontal with the asymmetrical donor in contact with the upper half of the segment. This increase in lateral movement of IAA seemed to occur in all regions of the coleoptile segment since in the basal third of the segment the gradient of radioactivity could be reversed in horizontal sections as compared with that in vertical sections. The distribution of radioactivity between the two receiving blocks also showed a reversal of the gradient when the segments were horizontal. Even though radioactive

IAA was being supplied only to the upper half of the horizontal segment, more radioactivity was recovered in the lower receiving block than in the upper one (fig. 2B). Since both Hertel and Leopold (1963b) and Parkes (personal communication) have shown that when carboxyl-labeled IAA is supplied to the apical end of Zea coleoptile segments, all the radioactivity to emerge into receiving blocks at the basal end is confined to the IAA molecule, the radioactivity data of Goldsmith and Wilkins (1964) clearly establish that radioactive IAA is transported laterally to a greater extent in horizontal Zea coleoptile segments than in vertical ones. This conclusion is supported by the curvature of segments supplied with asymmetric donors. The horizontal segments with upper donors exhibited upward curvature, that is, curvature towards the donor block. Vertical sections developed much greater curvatures in the opposite direction, that is, away from the donor block.

There has not yet been an unequivocal demonstration of lateral movement of auxin in Avena coleoptiles, but the data of Koch (1934) using coleoptiles from which one half of the tip had been removed strongly suggest that such a lateral movement occurs.

While there is no doubt that a lateral movement of auxin occurs during geotropic stimulation in Zea coleoptiles, it is not known whether other substances move laterally as well, especially molecules and ions that might act as cofactors for the promotion of growth by auxin. The experiments of Brauner and Hager (1958) show that auxin depleted, decapitated Helianthus shoots could be placed horizontally for up to 14 hr without curvature taking place. If the segments were restored to the vertical position and supplied with a symmetrical source of IAA at their apical end, they developed a curvature consistent with the original geotropic stimulation. The segments appeared to have been polarized laterally by the geotropic stimulation in the absence of auxin. This polarization persists for some hours after the segments are restored to the vertical position since curvature will still develop even if 12 hr elapse between righting the segment and applying the IAA. Rather similar results have been reported for Avena and Zea coleoptiles by Hahne (1961).

The lateral polarity induced by geotropic stimulation in auxin-free hypocotyls has been attributed by Brauner and Hager (1958) to the lateral movement of an auxin cofactor. This would render the former lower side of the righted hypocotyl more responsive to auxin than the former upper side, and hence result in the development of curvature when an apical source was applied. An alternative explanation for the induction of a lateral polarity in auxin-depleted hypocotyls has been offered by Hertel and Leopold (1963a). They suggest that there may be residual auxin in the starved hypocotyl and that this may undergo lateral displacement. While the concentration of auxin might be too low to induce growth, the gradient set up may be sufficient to maintain a higher rate of metabolic activity on the lower side of the organ. This higher rate of metabolism could maintain a higher rate of longitudinal transport on the former lower side when the segment is righted and supplied with a symmetrical source of auxin. Hertel and Leopold (1963a) have indeed shown that in Zea coleoptile segments the ability to transport auxin declines with time after excision, but this decline can be prevented by maintaining a supply of auxin to the tissue. A further possibility, suggested by Wilkins (1966), is that the auxin lateral transport system is induced in starved tissues and that the ability to transport auxin laterally is maintained for some hours after the tissue is righted. This suggestion could easily be tested experimentally using the technique of Goldsmith and Wilkins (1964) on the righted tissues. Furthermore, the hypotheses of Brauner and Hager (1958) and of Hertel and Leopold (1963a) could also be tested by placing a mechanical barrier in the auxin-depleted segments during geotropic stimulation in much the same way as in the experiments of Brauner and Appel (1960). This would establish whether or not anything was transported laterally during geotropic stimulation. It is rather important that the interesting question posed by the work of Brauner and Hager (1958) on Helianthus hypocotyls and by that of Hahne (1961) on auxin-depleted coleoptile segments of Avena and Zea should be answered, since it raises the important point of whether substances other than indole-3-acetic acid are transported laterally as the result of geotropic stimulation.

Both the normal longitudinal, and gravity-induced lateral, polar movement of IAA in coleoptile tissues are entirely dependent upon energy derived from metabolism. Although Naqvi, Dedolph, and Gordon (1965) reported that longitudinal basipetal transport of IAA is unaffected by depriving Zea coleoptiles of oxygen, these findings have not been substantiated by several other investigations. Goldsmith (1967), Wilkins and Martin (1967), and Wilkins and Whyte (1968) have shown basipetal movement of IAA to be greatly reduced under strictly anaerobic conditions, but never to the level of acropetal movement. Even under anaerobic conditions, therefore, a slight longitudinal polarity of IAA movement exists in Zea coleoptile segments. This slight polarity under anaerobic conditions alone. Naqvi, Dedolph, and Gordon (1965) have also reported that the the presence of the metabolic inhibitors sodium fluoride and iodoacetic acid (Wilkins and Whyte 1968). In Avena longitudinal polarity of auxin movement appears to be totally abolished under anaerobic conditions along. Naqvi, Dedolph, and Gordon (1965) have also reported that the asymmetry of radioactivity between the upper and lower receiving blocks at the basal end of a Zea coleoptile segment supplied with a symmetrical donor, is even greater in anaerobic conditions than in air. This would suggest that the lateral polar movement of IAA known to be induced in geotropically stimulated coleoptiles is largely independent of energy derived from aerobic metabolism.

Wilkins and Whyte (unpublished) have made an extensive investigation of the effects of anaerobic conditions on lateral movement of IAA in Zea coleoptile segments using the asymmetric donor techniques of Goldsmith and Wilkins (1964). The results are rather similar to those obtained for longitudinal transport. Deprivation of oxygen leads to a large decrease in the amount of lateral movement of IAA in geotropically stimulated coleoptiles but it is still greater than that found in vertical segments. When the segments are treated with sodium fluoride as well as being deprived of oxygen, the lateral movement of IAA in vertical and horizontal segments is identical. The report of Naqvi, Dedolph, and Gordon (1965) that the asymmetric distribution of IAA in horizontal coleoptiles, and the basipetal movement of IAA in vertical coleoptiles are independent of aerobic metabolism have been attributed by Goldsmith (1967) and Wilkins and Martin (1967) to the use of imperfect anaerobic conditions.

Despite a report to the contrary (Dedolph, Breen, and Gordon 1965), coleoptiles of Avena and Zea do not show a geotropic response under anaerobic conditions even after 15 hr in the horizontal position (Wilkins and Shaw 1967). When oxygen is readmitted to the tissues, however, a geotropic curvature develops at approximately the same rate as in freshly harvested and stimulated coleoptiles in the case of Avena, and slightly more rapidly in the case of Zea. The occurrence of a small amount of lateral movement of IAA in Zea coleoptile segments geotropically stimulated under anaerobic conditions (Wilkins and Whyte unpublished) could account for their more rapid upward curvature on the readmission of oxygen to the tissues as observed by Wilkins and Shaw (1967). A gradient of auxin concentration would already exist in the segments when oxygen was restored and this could lead to the more rapid onset of differential growth and curvature than is found in freshly geotropically stimulated coleoptiles.

Lateral movement of IAA in geotropically stimulated shoots has therefore been firmly established. This movement is polarized downward, and the polarity is entirely dependent upon the availability of metabolic energy.

At the present time it is not known whether the lateral polar movement of auxin is the only process which gives rise to the asymmetric distribution of auxin in horizontal coleoptiles. There is now a considerable amount of evidence to suggest that gravity can influence the longitudinal basipetal movement of auxin in Zea coleoptile segments. Naqvi and Gordon (1966) found that less IAA reached the basal receiving block of segments inverted for the 5 hr previous to the application of the donor. Similarly, Little and Goldsmith (1967) found less IAA in the basal receiving block when the segment was inverted during the 8-hr transport period. Goldsmith and Wilkins (1964) could find no difference in the amounts of IAA reaching the receiving blocks of horizontal and upright segments for the first 2 hr after segments were placed in the horizontal position. After this time, however, less IAA appeared to reach the receiver of the horizontal segments. On the other hand, Hertel and Leopold (1963b) found a more rapid effect of inversion in Zea segments; even after only 50 min of inversion, less IAA appeared to reach the receiver of the inverted segments. While there is little doubt that gravity does affect the basipetal movement of auxin in Zea coleoptile segments, in general the effect seems to be rather too delayed to play a major role in geotropism. Of very great interest, however, is the finding of Naqvi and Gordon (1966) that while placing a segment in the horizontal or inverted position for 2 hr has no effect on the overall basipetal movement of auxin as judged by the total amount of IAA reaching the receivers, the upper and lower halves of the horizontal segments transported different amounts of IAA. The lower half transported more IAA and the upper half less IAA than a vertical half segment. This differential change in the longitudinal basipetal movement of IAA by the upper and lower halves of the horizontal segment could contribute to the asymmetry of auxin distribution in the segment. It must play a relatively minor role in the establishment of the asymmetry of auxin in comparison with the lateral transport system, since Naqvi and Gordon (1966) have shown that the effect of longitudinal transport alone gives rise to a ratio of 1.2 for the amounts of auxin in the lower and upper receiving blocks at the basal end of a totally split segment. In whole segments, where lateral movement occurs, this ratio is 3.4.

Future research must be aimed at elucidating the mechanism of the lateral transport system induced in geotropically stimulated coleoptiles and establishing whether it transports only IAA and molecules of similar configuration. On the basis of present evidence the validity of the Cholodny-Went hypothesis for shoots seems to be well established. It must be remembered, however, that the hypothesis may be an oversimplification and that processes other than lateral transport of auxin may contribute to the establishment of an asymmetric distribution of auxin and the induction of differential growth in horizontal shoots.

Roots

Audus and Brownbridge (1957) have recently confirmed the report by Sachs (1882) that the downward geotropic curvature of roots is due to the growth rate of the lower half being decreased to a greater extent than that of the upper half (fig. 3A). The Cholodny-Went hypothesis attributes this differential growth to the establishment of an auxin concentration gradient across the organ by the lateral transport of auxin from the upper to the lower half. The application of this hypothesis to root responses demands an explanation of how the increased concentration of auxin in the lower half gives rise to the decreased growth rate.

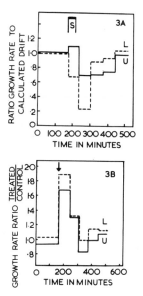

Fig. 3
A, the changes in the growth rate of the upper (U) and lower (L) halves of roots as a function of time after the onset of geotropic stimulation. Period of geostimulation is indicated by the lines (S). B, the promotion of growth of the upper (U) and lower (L) halves of horizontal roots by immersion in 10^{-11} M IAA at the time indicated by the arrow. Both diagrams from Audus and Brownbridge 1957 by permission.

A number of independent studies have shown that the growth rate of a root is increased when the apex is cut off (Wiesner 1884; Cholodny 1926; Bünning 1928). Cholodny (1926, 1931), and Keeble, Nelson, and Snow (1931) have shown that the lower growth rate could be restored by replacing the severed apex or by applying the apex of a coleoptile. Cholodny (1924) also found that while decapitated roots did not respond to geotropic stimulation, their responsiveness can be restored by placing on the cut surface the apex of a root or a coleoptile. On the basis of his results Cholodny (1926) concluded that the apices of roots and shoots secrete a substance which retards root growth and promotes shoot growth, and that the lateral redistribution of this substance in horizontal organs would account for the upward curvature of shoots and the downward curvature of roots.

There are several ways in which a substance might act as an inhibitor of root growth and a promoter of shoot growth. First, the substance might act specifically in inhibiting the growth of roots and promoting the growth of shoots. Second, whether the substance inhibits or promotes the growth of either roots or shoots may depend entirely upon its concentration. The early work of Kögl, Haagen-Smit, and Erxleben (1934) indicated that low concentrations of IAA which promote shoot growth inhibit root growth, and later Thimann (1937) showed that there was an optimum type of relationship between concentration of auxin and growth rate for both roots and shoots. The concentration of auxin which gave the maximum growth rate was about 10^{-4} ppm in roots and 5-10 ppm in shoots. In each case the growth rate progressively decreased as the concentration was increased above these optimum values. At some concentrations, therefore, shoot growth is promoted and root growth is inhibited.

The results of the decapitation experiments with roots could be explained on the basis of the concentration of auxin normally being supraoptimal. Furthermore, the effectiveness of a coleoptile tip in reducing the growth rate of, and restoring the geotropic responsiveness to, decapitated roots could be ascribed to secretion of auxin at the cut surface of the tip at concentrations inhibitory to root growth. The general formulation of the Cholodny-Went hypothesis for roots was thus as follows: the concentration of auxin is normally supraoptimal, and geotropic stimulation leads to the lateral transport of auxin so that the concentration of auxin in the lower half is further increased, thereby reducing the growth rate and inducing downward curvature.

The occurrence of an asymmetry of net growth substance activity in horizontal roots was established by experiments similar to those carried out with shoots. Hawker (1932) geotropically stimulated *Zea mays* roots for 3 hr and then removed the apices. The apices were sliced into upper and lower halves and applied asymmetrically to decapitated vertical roots. The lower half-tip gave nearly three times the curvature toward the side of application (inhibition of growth) than that given by the upper half. Boysen-Jensen (1933) placed root tips of Vicia horizontally with their basal cut surfaces in contact with upper and lower agar receiving blocks. The growth activity of the blocks was then assayed on detipped Avena coleoptiles and more growth-promoting activity was found in the lower block than in the upper. Similar results were later obtained by Boysen-Jensen (1936) using an auxin extraction technique. While these experiments show that asymmetric distribution of one or more substances which inhibited root growth and promoted shoot growth is established during geotropic stimulation, they cannot, for the reasons outlined in the discussion of shoot geotropism, allow any conclusions to be drawn about the mechanism involved in the establishment of the asymmetry. Whatever this mechanism is in roots, it does not appear to be restricted to the tip, since Cholodny (1924) has shown that differential growth could be induced in geotropically stimulated, decapitated roots after they had been returned to the vertical position and then supplied with an unstimulated tip.

The validity of the Cholodny-Went hypothesis as an explanation of root geotropism clearly rests on establishing that auxin is (1) present at supraoptimal concentrations and (2) transported laterally as the result of geotropic stimulation. These two conditions are interdependent, since, if the concentration is suboptimal and lateral transport of auxin occurs, the root will then bend upward—a response which does not normally occur. If the concentration of auxin is suboptimal, the downward curvature could be brought about by the upward transport of auxin or the downward movement of an inhibiting substance. Both of these possibilities, however, are rather different from the original ideas expressed by Cholodny and Went.

A number of recent investigations have given results which suggest that in roots auxin is neither present at supraoptimal concentrations nor transported laterally during geotropic stimulation. A further difficulty in accepting the Cholodny-Went hypothesis for roots is that the changes in the growth rate of the upper and lower halves of the horizontal organ are not consistent merely with the redistribution of growth substances.

The existence of a supraoptimal concentration of auxin in normal roots was originally proposed to explain the enhancement of the growth rate after the apex was removed. These earlier findings have not been substantiated by some more recent investigations (Gorter 1932; Younis 1954; Vardar and Tözün 1958). One reason for the conflicting results from this type of experiment may be that different amounts of tip were removed in the different investigations. A much more important objection to the existence in the roots of supraoptimal concentrations of auxin is that their growth is markedly promoted by immersion in low concentrations of IAA. If the concentration of auxin were supraoptimal, a further increase in concentration resulting from the application of external auxin could not possibly enhance the growth rate. Larsen (1955) has collated these data, but of particular importance are the elegant measurements of Audus and Brownbridge (1957) shown in figure 3B. They found that during downward geotropic curvature of pea roots when, according to the Cholodny-Went hypothesis, the supraoptimal concentration of auxin is being still further increased in the lower half, the growth rate of both the upper *and* the lower halves of the root is promoted by low concentrations of IAA and inhibited by higher concentrations.

Several investigations have shown by the usual methods of chromatographic analysis that IAA is present in roots (Audus and Thresh 1956; Britton, Housley, and Bentley 1956; Thurman and Street 1960). It is not present in sufficient quantities, however, to allow unequivocal chemical identification. Audus (1959) has pointed out that if the concentration is assumed to be uniform throughout the root, the amount of IAA found to be present in several independent investigations would support the idea that the concentration of auxin is supraoptimal. Audus also points out that if the calculated concentrations were present, other auxin responses, such as lateral extension of root cells would occur and these are not found in normal roots. Almost nothing is known about the distribution of total assayable auxin within the intact root between (1) the free and bound forms and (2) the various intracellular locations. The actual concentration of free auxin might be suboptimal with most of the extractable auxin in a bound and inactive state.

The so-called antagonists of auxin action have been found in several cases to promote the growth of roots (Åberg 1950; Burström 1950). If the auxin concentration in roots is supraoptimal, this finding could be explained on the basis of the auxin antagonists reducing the effective level of auxin and thereby enhancing the growth rate. Unfortunately it is not known what direct effects these auxin antagonists have on metabolism and cell extension apart from their possible interference with the action of auxin. Until their own physiological activity in root cells has been established, the argument for the existence of supraoptimal concentrations of auxin in roots based on the promotion of growth by auxin antagonists cannot carry much weight.

The changes in the growth rate of the upper and lower halves of a geotropically stimulated root (fig. 3*A*) are not consistent merely with a redistribution of auxin. If the auxin concentration were supraoptimal, then a lateral redistribution in the horizontal organ should decrease the growth rate of the lower half and enhance the growth rate of the upper half, and there should be little or no overall change in the growth rate of the root. Audus and Brownbridge (1957) and Bennet-Clark, Younis, and Esnault (1959) have shown that there is a marked overall decrease in the growth rate of the root following geotropic stimulation. The data of Audus and Brownbridge (1957) show that there is a slight transient increase in the growth rate of the upper half but this is soon replaced by a marked decrease, so that the growth rate of both the upper and lower halves is decreased for a considerable time.

Taking a broad view of the somewhat conflicting evidence presently available, I feel that the existence of supraoptimal concentrations of auxin in the active form in normal roots has not yet been established.

The task of establishing whether or not lateral movement of auxin occurs during the geotropic response of roots has been attempted in several laboratories. Evidently no indication of lateral movement has been obtained since no data have been published. This lack of publication is regrettable since negative evidence is of much value in a discussion such as this. The whole question of auxin movement in root tissues needs to be thoroughly reinvestigated, since, even for the longitudinal movement, conflicting reports have been published (Yeomans and Audus 1964; Pilet 1964). Könings (1965) has concluded that the asymmetry of radioactivity he found in roots supplied with labeled IAA was not caused by the lateral movement of this substance.

The changes in the growth rate of geotropically stimulated roots, together with the absence of convincing evidence for the occurrence of either supraoptimal concentrations of auxin in roots or lateral transport of IAA during geotropic stimulation, make the validity of the Cholodny-Went hypothesis to explain root geotropism extremely doubtful. What alternative mechanisms could be responsible for the occurrence of differential growth of the upper and lower halves of horizontal roots? In view of the overall depression in the growth rate of geotropically stimulated roots found by Audus and Brownbridge (1957) and Bennet-Clark, Younis, and Esnault (1959) and the complex pattern of growth-promoting and growth-inhibiting substance present in roots (Audus and Thresh 1956; Audus and Gunning 1958; Lahiri and Audus 1960; Thurman and Street 1960; Bennet-Clark, Younis, and Esnault 1959), the two most attractive possibilities are (1) the release of an inhibitor of root cell extension from a bound inactive form or from a particular intracellular location separated from the site of the growth-controlling processes and (2) the *de novo* synthesis of an inhibitory substance. If either of these processes occurred preferentially or only in the lowermost half of the horizontal organ, it would give rise to the differential growth rate.

The analysis of Audus and Thresh (1956) has shown that roots contain a number of ether-soluble growth-active substances. At least 4 inhibitors and 1 promoter were found, and it is of interest that the inhibiting substance present in the largest amount, as judged from its activity in the pea root segment test, had the same Rf value as IAA. Three or 4 water-soluble growth-active substances have also been found in roots (Audus and Gunning 1958). These water-soluble substances appear to be interconvertable, since when any one is eluted from the paper it yields the same 3 or 4 substances on further chromatographic separation. Several growth-regulating substances are present in *Vicia faba* root tips, but none appears to be IAA (Bennet-Clark, Younis, and Esnault 1959). The part played by each of these regulators in controlling root growth at any one time is not known, but the growth rate of a root at any instant is most probably controlled by a delicate balance between the concentrations of a number of promoting and inhibiting substances.

A key investigation in elucidating the geotropic response in roots is to establish whether or not geotropic stimulation induces changes in the relative amounts of the various growth regulators in the organ. Audus and Lahiri (1961) have made such an analysis in stimulated and unstimulated roots of *Vicia faba* and found that an active root growth inhibitor begins to increase in amount about 30 min after the onset of geotropic stimulation. The amount of this substance reaches a maximum at 40 min and then decreases. This substance (APii) promoted the growth of Avena mesocotyls, inhibited the growth of pea root segments and had an Rf value closely similar to that of IAA (fig. 4). The increase in the amount of this root inhibitor seems to be associated with a decrease in the amount of another substance (APiii) which also promotes Avena mesocotyl growth and inhibits pea root growth. This finding suggests that the inhibitor (APii) arises from the substance (APiii). The new substance (APii) must be more inhibitory, molecule for molecule, than the precursor (APiii). The analyses of Audus and Lahiri (1961) show that the decrease in (APiii) is observed only in the Avena mesocotyl assay and not in the pea root assay in which only a small amount of inhibition appears at this Rf value in stimulated and unstimulated roots. The substance (APiii) may therefore be a relatively mild inhibitor of root growth, but has the capability of giving rise to a more potent inhibitor (APii) on geotropic stimulation. The substance (APii) has strong promoting action in shoot tissues and strong inhibiting action in root tissues.

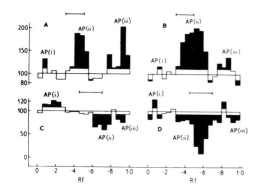

Fig. 4
Changes in the relative amounts of growth regulators present in unstimulated and stimulated root tips of *Vicia faba*. A and C, extracts of vertical tips; B and D extracts of root tips placed horizontal for 1 hr; A and B, Avena mesocotyl segment assay; C and D, Pisum root segment assay. The black portions of the histograms represent responses significant at the 5% level. The horizontal lines above the histograms indicate the Rf of IAA. From Audus and Lahiri 1961 by permission.

The involvement of an inhibitory substance in the geotropic response of roots is also suggested by the experiments of Bennet-Clark et al. (1959). They found that roots placed at various angles to the vertical underwent rapid curvature for 3-10 hr, but then, despite the fact that they were still far from having regained the vertical position, the rate of curvature became extremely low. The overall growth rate of the roots was markedly reduced during the development of rapid curvature. When the rate of curvature had decreased to a low value, rapid curvature could once again be induced by turning the root through 90°. Using starch column chromatography Bennet-Clark et al. (1959) found no difference between the amounts of the various growth regulators present in stimulated and unstimulated roots. However, the roots were assayed 5, 12 and 18 hr after geotropic stimulation began, and it is highly likely that any differences that occur as the result of reorientation of the root would have almost disappeared after 5 hr. The geotropic response begins within approximately 30 min of stimulation, and it is at about this time that differences in growth substance content should be most marked. This has indeed been found to be so by Audus and Lahiri (1961) who were able to demonstrate that changes in growth substance content occurred 30 min after the onset of geotropic stimulation but began to disappear shortly afterwards. This probably explains why no differences were found by Bennet-Clark et al. (1959). Because of their inability to establish differences in the amount of growth substances present in stimulated and unstimulated roots after 5 hr, Bennet-Clark et al. (1959) favored the idea that the inhibition of growth which occurs during geotropic stimulation is the result of the release of an inhibitor from a bound form, or perhaps from the vacuole into the cytoplasm, rather than a *de novo* synthesis. However, the validity of this conclusion depends upon whether their assays were made at the time after stimulation began when a difference in growth substance concentration would have been most likely to occur.

The growth studies of Audus and Brownbridge (1957) suggest that the inhibitor found by Audus and Lahiri (1961) is synthesized preferentially or solely in the lower half of the horizontal root and then diffuses throughout the root. This obviously requires direct experimental confirmation by assaying the growth substance content of the upper and lower halves of geotropically stimulated roots. The technical difficulties of this task however, are, very considerable, and it has not yet been successfully undertaken.

A gradient of net growth-regulating activity is undoubtedly set up across horizontal roots, but at the present time there is no substantial evidence to suggest that this asymmetry arises from the lateral movement of growth hormones. Rather the asymmetry appears to be the result of *de novo* synthesis of a growth-inhibiting substance in the lower half of the organ. The Cholodny-Went hypothesis cannot be applicable to roots unless auxin is present at supraoptimal concentrations, and this does not appear to be the case. The capability of coleoptile tips to restore the geotropic responsiveness to detipped roots could be due to the lateral transport of auxin in the tip, giving rise to a more growth-inhibiting level of auxin in the lower half of the root than in the upper half. This situation, however, has little to do with the response of the normal intact root.

The literature on root geotropism contains so many conflicting facts that it is virtually impossible to make an acceptable statement about how differential growth is induced. For example, it has been reported that curvature can be induced in geotropically stimulated decapitated roots after they have been righted and supplied with an unstimulated apex (Cholodny 1924). The decapitated root can therefore detect the gravitational stimulus. Despite this fact the removal of the root tip allegedly enhances the growth rate (Cholodny 1926; Bünning 1928) and removes the geotropic sensitivity (Cholodny 1924). If the auxin concentration is supraoptimal, and lateral movement of auxin is involved in the induction of differential growth, some curvature would be expected in decapitated roots, perhaps positive at first and then negative as the concentration of auxin becomes suboptimal. On the other hand, if the auxin concentration is normally suboptimal and lateral movement is not involved in inducing differential growth, then these conflicting facts might be explained on the basis that geotropic stimulation differentially changes the sensitivity of the cells in the upper and lower halves of the growing zone of the roots to an inhibiting substance which is produced at the apex or in the root cap. In the absence of the tip no differential growth would be possible because of the absence of the specific inhibitor. This differential change in sensitivity of the cells to the inhibitor could occur in addition to geoinduced changes in the amount of inhibitor synthesized in the upper and lower halves of the tip itself.

Speculative schemes to explain the conflicting facts serve little purpose in advancing our understanding of the geotropic responses of roots. It is much more important at the present time to ascertain with certainty whether the growth rate is enhanced, and geotropic sensitivity lost, when roots are decapitated. Juniper et al. (1966) recently made the important discovery that removal of the root cap from Zea roots did not change the growth rate but eliminated geotropic responsiveness. This observation opens up some exciting new possibilities of being able to determine whether growth-regulating substances (especially the inhibitors) are synthesized in the cap itself, and whether they are differentially synthesized or laterally transported in the cap or the root during geotropic stimulation.

Obviously much remains to be done in elucidating the mechanism of the geotropic response in roots and the parts played in this response by the various growth-regulating substances known to be present. Priority must be given to the chemical identification of the growth-promoting and growth-inhibiting substances so that the whole question of growth regulator synthesis and transport in root tissues can be thoroughly examined.

Acknowledgments

The author wishes to thank the Space Science Board of the National Academy of Sciences for inviting him to participate in this symposium and for financing his visit to the United States. The author's unpublished work, discussed in this paper, was supported by grant no. 83/6 of the United Kingdom Agricultural Research Council.

References

Åberg, A. 1950. *Physiol. Plantarum* 3: 447-61.

Audus, L. J. 1959. In *Proceedings of the Symposium on Plant Growth Regulators*, pp. 9-22. Poland: Torun.

Audus, L. J., and Brownbridge, M. E. 1957. *J. Exptl. Botany* 8: 105-24.

Audus, L. J., and Gunning, B. E. 1958. *Physiol. Plantarum* 11: 685-97.

Audus, L. J., and Lahiri, A. N. 1961. *J. Exptl. Botany* 12: 75-84.

Audus, L. J., and Thresh, R. 1956. *Ann. Botany (London)* n.s. 20 (79): 439-59.

Bennet-Clark, T. A.; Younis, A. F.; and Esnault, R. 1959. *J. Exptl. Botany* 10: 69-86.

Boysen-Jensen, P. 1910. *Ber. Deut. Botan. Ges.* 28: 118-20.

———. 1911. *Bull. Acad. Roy. Danmark* 1: 3-24.

———. 1913. *Ber. Deut. Botan. Ges.* 31: 559-66.

———. 1933. *Planta* 20: 688-98.

———. 1936. *Kgl. Danske Videnskab. Selskab., Biol. Med.* 15: 321-31.

Brauner, L., and Appel, E. 1960. *Planta* 55: 226-34.

Brauner, L., and Hager, A. 1958. *Planta* 51: 115-47.

Britton, G.; Housley, S.; and Bentley, J. A. 1956. *J. Exptl. Botany* 7: 239-51.

Bünning, E. 1928. *Planta* 5: 635-59.

Burström, H. 1950. *Physiol. Plantarum* 3: 277-92.

Ching, T. M., and Fang, S. C. 1958. *Physiol. Plantarum* 11: 722-27.

Cholodny, N. 1924. *Ber. Deut. Botan. Ges.* 42: 356-62.

———. 1926. *Jahrb. Wiss. Botan.* 65: 447-59.

———. 1931. *Planta* 14: 207-16.

Dedolph, R. R.; Breen, J. J.; and Gordon, S. A. 1965. *Science* 148: 231-61.

Dijkman, M. J. 1934. *Rec. Trav. Botan. Neerl.* 31: 391-450.

Dolk, H. E. 1929. *Rec. Trav. Botan. Neerl.* 33: 509-85.

Gillespie, B., and Briggs, W. R. 1961. *Plant Physiol.* 36: 364-68.

Gillespie, B., and Thimann, K. V. 1961. *Experimentia* 17: 126-29.

———. 1963. *Plant Physiol.* 38: 214-25.

Goldsmith, M. H. M. 1967. *Plant Physiol.* 42: 258-63.

Goldsmith, M. H. M., and Wilkins, M. B. 1962. *Plant Physiol.* 37: xvii.

———. 1964. *Plant Physiol.* 39: 151-62.

Gorter, C. J. 1932. Doctoral thesis, University of Utrecht.

Hahne, I. 1961. *Planta* 57: 557-82.

Hawker, L. E. 1932. *New Phytologist* 31: 321-28.

Hertel, R., and Leopold, A. C. 1963a. *Naturwissenschaften* 22: 695-96.

———. 1963b. *Planta* 59: 535-62.

Juniper, B. E.; Groves, S.; Landau-Schachar, B.; and Audus, L. J. 1966. *Nature* 209: 93-4.

Keeble, F.; Nelson, M. G.; and Snow, R. 1929. *Proc. Roy. Soc. (London), Ser. B.* 105: 493-98.

———. 1931. *Proc. Roy. Soc. (London), Ser. B.* 108: 537-45.

Koch, K. 1934. *Planta* 22: 190-220.

Kögl, F.; Haagen-Smith, A. J.; and Erxleben, H. 1934. *Z. Physiol. Chem.* 228: 104-12.

Konings, H. 1965. *Plant Physiol. Suppl.* 40: xxxii.

Lahiri, A. N., and Audus, L. J. 1960. *J. Exptl. Botany* 11: 341-50.

Larsen, P. 1955. In *Modern Methods of Plant Analysis,* vol. 3, pp. 565-625. Berlin: Springer-Verlag.

Little, C. H. A., and Goldsmith, M. H. M. 1967. *Plant Physiol.* 42: 1239-45.

Naqvi, S. M.; Dedolph, R. R.; and Gordon, S. A. 1965. *Plant Physiol.* 40: 966-68.

Naqvi, S. M., and Gordon, S. A. 1966. *Plant Physiol.* 41: 1113-18.

Navez, A. E., and Robinson, T. W. 1932. *J. Gen. Physiol.* 16: 133-45.

Overbeek, J. van; Olivo, G. D.; and Vazquez, E. M. S. de. 1945. *Botan. Gaz.* 106: 440-51.

Paal, A. 1914. *Ber. Deut. Botan. Ges.* 32: 499-502.

Paal, A. 1919. *Jahrb. Wiss. Botan.* 58: 406-58.

Pilet, P. E. 1964. *Nature* 204: 559-62.

Reisener, H. J. 1957. *Naturwissenschaften* 44: 120.

Reisener, H. J., and Simon, H. 1960. *Z. Botan.* 48: 66-70.

Sachs, J. 1882. *Textbook of Botany,* 2d English ed. Oxford: Clarendon Press.

Thimann, K. V. 1937. *Am. J. Botany* 24: 407-12.

Thurman, D. Z., and Street, H. E. 1960. *J. Exptl. Botany* 11: 189-96.

Vardar, Y., and Tözün, B. 1958. *Am. J. Botany* 45: 714-18.

Weber, U. 1931. *Jahrb. Wiss. Botan.* 75: 312-76.

Went, F. W. 1926. *Proc. Koninkl. Ned. Akad. Wetenschap.* 30: 10-19.

———. 1928. *Rec. Trav. Botan. Neerl.* 25: 1-116.

Wiesner, J. 1884. *Ber. Deut. Botan. Ges.* 2: 72-78.

Wilkins, M. B. 1966. *Ann. Rev. Plant Physiol.* 17: 379-405.

Wilkins, M. B., and Martin, M. 1967. *Plant Physiol.* 42: 831-39.

Wilkins, M. B., and Shaw, S. 1967. *Plant Physiol.* 42: 1111-13.

Wilkins, M. B., and Whyte, P. 1968. In *Proceedings of the Sixth International Congress on Plant Growth Substances.* Canada: Ottawa.

Yeomans, L. M., and Audus, L. J. 1964. *Nature* 204: 559-62.

Younis, A. F. 1954. *J. Exptl. Botany* 5: 357.

Discussion

BROWN: What evidence is there for active transport in roots? As far as I am aware, the lateral movement that has been discussed cannot be demonstrated to go against a concentration gradient. Is this correct?

WILKINS: It does go against a concentration gradient in coleoptile tissue; the situation in roots is not known.

BROWN: What is the evidence? Receiver blocks?

WILKINS: Receiver blocks, and if the organ is split longitudinally, all of the tissue segments on the lower side have more auxin from a symmetrically applied apical source.

BROWN: Would not increased longitudinal movement of the auxin and diffusion across with the concentration gradient be sufficient to account for the results you obtained?

WILKINS: There is always that possibility. There is evidence to suggest that the longitudinal auxin transport may be different between the upper and lower sides of the organ.

HERTEL: I am not convinced that this is an inhibition of the longitudinal transport on the upper side. I have done some experiments with longitudinally sliced sections (for a completely different purpose) and there it appears that the large cut surface will inhibit the transport. Now, imagine an "upper half" and assume lateral transport does take place: more auxin than in a vertical control will move toward the cut surface where its axial transport will be strongly inhibited. The opposite of course is true for the "lower half."

WILKINS: No, but it could occur. I am not saying that the mechanism that Dr. Gordon is proposing is the cause of active transport.

HERTEL: It could occur, but one could also explain it in a different way.

WILKINS: I think we are obsessed with putting auxin in at one end of a piece of tissue and watching it come out of the other end. What does matter is what's in the tissue, and that's why in all our recent work we've been chopping up the tissues and determining what is in them. Of course, with respect to the concentration gradient we do not know whether the auxin is uniformly distributed throughout all the cells. If we look at the upper side and the lower side, we can always show that, assuming that it is evenly distributed, there is more in the lower side than in the upper side.

POLLARD: Concerning the question of whether there is an active lateral transport—if one varies the temperature and the effect on transport follows an Arrhenius plot, the transport has got to be active.

WILKINS: You can show that it is simply by putting the tissue under anaerobic conditions. Once you stop aerobic metabolism, the amount you get in the vertical section moving across the tissue will be reduced, and the rate will fit a diffusion curve.

AUDUS: I would like to make three remarks to supplement what Dr. Wilkins said. *First of all*, we have evidence now from our laboratory, as well as from experiments in Street's laboratory, that the substance which accumulates on the lower side of roots is certainly not indoleacetic acid. It may be an indole compound and there is some evidence that it is a complex of IAA, but it is certainly not indoleacetic acid itself. *Second*, we did in fact try at one stage to see whether we could verify our theory that there was a synthesis of this substance in the lower half of the root tip. A young man has spent three very frustrating, trying years doing just this. I must confess that the results were entirely negative. There was in fact a general overall increase in auxin concentration in the root tip as shown by Lahiri and me, but so far we have not been able to demonstrate any differential distribution. *Third*, evidence has been obtained by Mrs. Arslan-Cerim in our laboratory that if you feed radioactive calcium to the roots of Helianthus seedlings, and then you turn the seedlings on their sides, there is an accumulation of calcium on the upper side. This is completely blocked under anaerobic conditions. Here again is evidence for the possible lateral transport of substances other than auxins.

PICKARD: We should mention the elegant case made by Stanley and Ellen Burg* for (*a*) the stimulation of root growth by indoleacetic acid, (*b*) the stimulation of ethylene production in roots by indoleacetic acid, and (*c*) the inhibition of both root growth and lateral transport by ethylene, with the consequent feedback system controlling geotropic response in roots. This is a mechanism which, while not proved thoroughly, serves to tie a number of phenomena together.

LARSEN: I have a few comments on the question of retardation versus stimulation of root growth with auxin. Some years ago I went through the literature and found that in all the reported cases of stimulation of intact roots by addition of auxin, the stimulation was observed only after several hours. Geotropism is something that occurs within a few minutes, less than one hour, and the explanation must be consistent with these times.

AUDUS: I have evidence to the contrary; we can get stimulations within the first hour. As far as I am aware no one else has ever really analyzed this situation thoroughly. Our work was done on intact roots and also on root segments.

GALSTON: Dr. Wilkins has summarized very convincingly the evidence for the lateral transport of labeled indoleacetic acid in coleoptiles under the influence of gravity. Yesterday we heard about many experiments in which centrifugal forces were used as a substitute for gravitational force. I would now ask whether there is any evidence at all for the lateral transport of labeled auxin under centrifugal rather than gravitational forces. And if not, I submit that this would be a very interesting experiment to do.

WESTING: Dr. Galston, are you wondering about centrifugal forces of a few g, in what might be called a physiological range? If so, I wonder whether you have any basis for supposing that the inertial force of a centrifuge differs from the inertial force of gravity?

GALSTON: We keep making the assumption that the only component of the force field to which the biological object responds when we're talking about gravity is equivalent to that which can be produced by centrifugal forces. I assume that this is correct, but I don't know that one has tested it critically in the experiments such as this.

LARSEN: You asked about the lateral transport of *labeled* auxin under the influence of a centrifugal force. I do not know if this has been studied. But there are experiments with *unlabeled*, endogenous auxin. Boysen-Jensen in 1933 stimulated *Vicia faba* roots with $2g$ for 4 hr on a centrifuge and found 37% auxin in the "upper" and 63% in the "lower" half of the root tips.

WESTING: Experiments have been reported in the literature in which stem cuttings of both woody and herbaceous plants have been centrifuged at a few hundred g and the growth substances driven to either the apical or basal end depending on which way the cutting was oriented in the centrifuge (e.g., *Ann. Botan.* 39: 359; *Sugar J.* 14(8): 11; *Fiz. Rast.* 5: 368).

GORDON: Your own work, Dr. Westing, offers indirect support for an equivalence for centrifugal fields of much lower magnitudes. You showed that the *geotropic* response of centrifuged organs, if one took care of confounding factors such as phototropism, could be predicted quite precisely using the vector sum of the gravitational and angular accelerations. It seems to me that this permits the inference that similar auxin redistributions take place in the two force fields.

---

*See Burg, S. P., and Burg, E. A. 1968. In *Biochemistry and Physiology of Plant Growth Substances.* F. Wightman and G. Setterfield, Eds. Ottawa: Runge Press. pp. 1275-94 (S.A.G.).

11
On Hormone Movement in Geotropism

Henry Rufelt, *University of Uppsala*

I shall comment on the paper of Professor Wilkins and supplement it with respect to the geotropism of roots. Wilkins has concluded that probably no transverse transport of auxin takes place in roots and further, that the Cholodny-Went theory, which implies a supraoptimal content of auxin in roots, is not valid. Finally he has referred to results that contradict the belief that indoleacetic acid (IAA) is the growth regulator functioning in roots.

I agree that transverse transport of auxin probably does not mediate the tropism of roots, but I doubt the validity of the other two points. In my experiments with wheat roots I find no reason to abandon either the concept of a supraoptimal auxin content or the presence of IAA in roots. My main argument comes from the experiments with the isobutyric acids, which are considered to be auxin antagonists, particularly with indole-isobutyric acid. This substance increases the cell elongation by more than 100% and causes wheat roots to curve negatively geotropically (Rufelt 1957a). Wilkins has omitted the results with auxin antagonists, commenting that we do not know if they have effects other than the possible interference with auxin. I think this argument is valid for all growth regulators which are supplied from outside, including IAA, and many experiments are of questionable validity on that basis. I am suspicious of conclusions about substances which have been extracted from root tissues and tested on intact roots. We never know where the extracted substance is localized in the cell and cannot be sure that it exists as such in the intact plant. In addition we have of course the difficulties of bioassay pointed out by Wilkins.

Results supporting the opinion that IAA is present in roots have also been published by Konings (1964), who found an influence of IAA-oxidase on the geotropic reaction of pea roots. On the other hand, his results do not support the concept of a supraoptimal concentration. The contradictory results may be due to species differences.

A pertinent point in the study of root geotropism is the pattern of the time course of the reaction. Many investigators have observed the pattern shown in figure 1. The straightening of the curvature

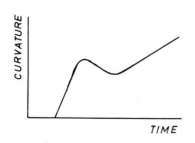

Fig. 1
Generalized time course of geotropic curvature development in roots.

illustrated by the decrease in the curve has been found to be caused by a growth inhibition on the upper side of the root. It has been suggested that this inhibition is a consequence of the diffusion of a growth inhibitor, formed after geotropic stimulation, from the lower side. There is another possible explanation—that the straightening is originated by a second growth inhibitor. Consider the following. If both inhibitions are caused by the same substance, the situation would be equivalent to the existence of a weaker geotropic reaction. Thus only the velocity of the bending could be influenced, and after a somewhat longer period we should obtain vertically growing roots. Even a weak geotropic bending tendency will cause the reaction to continue until verticality has been reached. But this is not what happens. In all instances where the end result has been recorded, the final liminal angle is not zero (for references, see Rufelt 1962). Furthermore, it is influenced by the position from which the reaction starts. Such behavior can be understood if two geoinduced processes are postulated, presumably assignable to two more or less independent georeactions. Both must go on simultaneously, however, and both must be of importance for the determination of the final length of the cells, because the final shape of the geotropic curvature is a manifestation of the final length of the cells.

Starting with this view, we can look for processes that fit the assumptions. The geotropic behavior of wheat roots is most easily understood if we assume the existence of two bending tendencies working in opposite directions (Rufelt 1957b). The negative tendency is more sensitive to such environmental factors as pH, oxygen supply, and temperature. Both are influenced in different ways by growth-regulating substances. The liminal angle is influenced by the same factors. A further argument is that different liminal angles are found in different varieties of wheat: In 400 varieties of this material the liminal angle was found to vary between 15° and 80°.

In current experiments in Uppsala I have found that the liminal angle is in some way influenced by a substance given off by the roots themselves. To determine the liminal angle, roots were placed horizontally between two agar plates and left overnight to form a geotropic curvature. It was found that the plagiotropic behavior was changed when activated carbon was mixed in one of the agar plates. The liminal angle was decreased (table 1). I have interpreted this result as an indication that some exudate from the roots influences the course of the reaction. Work is underway to identify the active substance.

Table 1. Influence of Activated Carbon on Growth and Liminal Angles of Four Wheat Varieties

Variety (number)	Growth (mm/24 hr)		Liminal Angle (degrees)	
	+C	−C	+C	−C
431	27.2 ± 2.3	24.1 ± 1.8	39 ± 6.4	54 ± 6.4
433	28.3 ± 1.9	22.5 ± 1.5	22 ± 7.3	39 ± 7.1
407	26.2 ± 2.1	23.0 ± 1.6	44 ± 8.6	56 ± 4.7
402	30.0 - 1.6	25.0 ± 1.2	38 ± 6.7	45 ± 8.0

The positive geotropic reaction of wheat roots has been found to fit the Cholodny-Went theory rather well. However, the reason for an asymmetric distribution of the growth inhibitor may not be a lateral transport (Rufelt 1957b). The mechanism behind the negative geotropic reaction is not known. Theoretically, curvatures may arise in three ways: by an earlier start of the elongation in the upper or the lower part, by differences in the rates of elongation, or by a later or earlier halt in the elongation of the upper or the lower part. It may be difficult to distinguish between the three ways experimentally, but under certain conditions negative curvatures start in the distal part of the elongation zone of wheat roots (Rufelt 1957a). From a morphological point of view, elongation starts from isodiametric cells, which begin a rapid extension in the longitudinal direction. In normal roots the nucleus is found near the extending walls during the whole period of elongation. We have found that in coumarin-treated roots the nuclei occur near the short walls, and the cells simultaneously undergo an abnormal primary growth which results in swollen roots (Rufelt 1959). It was further found that coumarin influences the cells in the meristem, but the effect cannot be observed until the affected cells reach the elongation zone. Thus, we are dealing with a process resulting in short cells, but its mechanism is quite different from what we find in studies of auxin effects. There is some reason to believe that the negative geotropic reaction in wheat roots is mediated by a reaction chain similar to the coumarin effect. If this is true, it is possible to interpret the negative reaction as a result of a geoinduced dorsiventrality in the meristem which has a delayed effect on cell elongation. It may be thought of as a geoinduced epinastic movement.

Finally, I shall mention two of our other recent experiments. Czapěk (1906) found that root tissue reduces silver from a silver-ammonia solution. Metzner (1934) published pictures showing that more silver is precipitated in the upper than in the lower part of a geotropically stimulated root. Starting from these observations, we tried to determine how soon after the beginning of stimulation the effect can be detected. We used roots of *Zea mays* about 50 mm long and confirmed Metzner's results very nicely. The effect could be demonstrated 5 min after the beginning of stimulation.

We also found that a change in pH value in the root could be observed 5 min after the start of the geotropic stimulus. The roots were placed on the freezing attachment of a microtome and cut longitudinally. The slices were placed in a solution of brom-thymol blue. We found that the lower part of the root acquired a darker color than the upper part. The color was concentrated at the cell surfaces; no change could be found within the cells. It is clear that 5 min after the roots have been placed horizontally the region at the cell surfaces in the lower tissues has a higher pH value than that in the upper part.

References

Czapěk, F. 1906. *Jahrb. Wiss. Botan.* 43: 361-467.

Konings, H. 1964. *Acta Botan. Neerl.* 13: 566-622.

Metzner, P. 1936. *Jahrb. Wiss. Botan.* 83: 781-808.

Rufelt, H. 1957a. *Physiol. Plantarum* 10: 500-520.

———. 1957b. Studies on the Geotropism of Wheat Roots. Dissertation, Lund University.

———. 1959. *Svensk. Botan. Tidskr.* 53: 312-18.

———. 1962. In *Encyclopedia of Plant Physiology*, vol. 17(2), ed. W. Ruhland and E. Bünning, pp. 322-42. Berlin: Springer-Verlag.

Discussion

COHEN: During increased protein synthesis in neurons of insects there is a shift of the nucleus to an eccentric position in the cell from a previously central location. The same thing happens in lymphocytes making antibodies. Presumably the size increase of the plant cell is accompanied by protein synthesis. One may speculate whether the shift in nuclear position is a correlate of cell elongation or volume increase or of protein synthesis. And whether the cells enlarge symmetrically when the nucleus does not shift to the periphery.

BROWN: May I make a very brief comment that may be helpful to zoologists who aren't familiar with the growth process in roots and shoots? During the stage of very rapid organ growth, many cells change from isodiametric to spectacularly elongated cells. The process involves the development of a larger vacuole, and most of the volume increase is due to increase in the size of this vacuole. Therefore, growth is largely fluid uptake. There is synthesis of new protoplast and cell wall materials of course but the principal gain is water. In a sense it is the large vacuole present in these cells which distinguishes their growth from that of animal cells.

COHEN: This indicates then that the vacuole is just pushing the nucleus off to one side. In animal cells there is a similar nuclear shift without a vacuole.

RUFELT: I should like to add something. In the experiments I described, roots on an agar plate were covered with another agar plate. When there was no covering agar plate, the roots were more positively geotropic. I can think of two possible explanations. The first is that there was a greater supply of oxygen with only one agar plate. The second is that some substances are exuded from the roots, affect their growth, and change the curvature response to gravity. To check this I covered the root with an agar plate containing activated charcoal; they curved much more. Their growth rate was also greater. Eliason also obtained more rapid root growth when he added activated carbon to nutrient solutions. I suggest that there is some substance which is exuded from the root, a substance active as a regulator during the geotropic response.

GORDON: From the work of the Burgs I would suggest that both the growth and the tropic effects could be interpreted as a removal of ethylene by the activated charcoal.

12
Hormone Movement in Geotropism: Additional Aspects

A. C. Leopold and R. K. de la Fuente, *Purdue University*

In commenting on the subject of geotropism, two items may be useful. The first is that the lateral movement of auxin in response to a geotropic stimulus may involve both active and passive components; the second is that geotropism may involve other types of stimuli besides auxin in the achievement of differential growth.

The lateral movement of auxin following geotropic stimulation has been thought to involve active transport, for it is subject to inhibition by known inhibitors of auxin transport (Hertel and Leopold 1963), and appears to move against a concentration gradient (Goldsmith and Wilkins 1964). We have used a modification of the technique developed by Goldsmith and Wilkins in the study of lateral auxin movement (de la Fuente and Leopold 1967) and have made the rather surprising finding that just as gravity enhances lateral movement in the direction of the gravitational pull, so also does it apparently depress the lateral movement in the opposite direction—movement away from the gravitational pull. This situation is illustrated in table 1, by experiments with sunflower hypocotyls and corn coleoptiles. Radioactive auxin was applied

TABLE 1. Lateral Movement of Auxin in Sunflower Stems and Corn Coleoptiles in Response to Gravity

Position of Tissue during Transport	Sunflower Stems (cpm/100 mg)			Corn Coleoptiles (cpm/100 mg)		
	Donor Side	Receptor Side	Lateral Movement	Donor Side	Receptor Side	Lateral Movement
Receptor side up (lateral movement against gravity)	132	35	21%	5211	1313	20%
Stem or coleoptile erect (lateral movement transverse to gravity)	133	61	31%	3366	2639	44%
Receptor side down (lateral movement with gravity)	137	106	43%	1040	2420	62%

Note: In each case, 25-mm sections were notched to a depth of 5 mm at the apical end and auxin-^{14}C in agar applied unilaterally to the uncut side; after 2 hr transport in upright or horizontal positions, the sections were slit the remaining 20 mm and the radioactivity measured in the half-cylinder to which the auxin had been applied (donor side) and in the half-cylinder to which the auxin had not been applied (receptor side). Lateral movement is expressed as the proportion of auxin in 20 mm of tissue which had moved into the receptor side. Data for sunflower when 10^{-6} M indoleacetic acid had been added as the source to the donor side; data for corn with 10^{-5} M as source.

unilaterally to one half of the stem or coleoptile section, and the lateral movement of the radioactivity was measured as the proportion which appears in the untreated half of the section. From the table it is seen that when lateral movement is measured in erect stems, relative values are substantially lower than when lateral movement is measured in geotropically stimulated sections oriented so that the lateral movement occurs in the direction of gravity. Conversely, lateral movement is reduced if the opposite orientation is employed. The existence of a vector in the lateral movement of auxin which is altered by gravity suggests the involvement of an active transport; and further, the lateral movement in the case of geotropically stimulated corn coleoptiles is capable of moving auxin against an apparent concentration gradient—with 62% of the auxin moving to the receptor side of the coleoptile.

Further evidence of the participation of an active transport vector is seen when the inhibitor of auxin transport, 2,3,5-triiodobenzoic acid, is applied to the sections before auxin is supplied. Data for such an experiment are shown in table 2. In this case, it is again true that lateral movement away from gravity is relatively small (18%), and in the direction of the gravity force is relatively large (63%); if the transport inhibitor has been applied, lateral movement is substantial, but it is not responsive to gravity (40% and 43% lateral movement for the two positions respectively). These results support the concept that gravity-sensitive lateral movement involves an active transport; they also suggest that some lateral movement may occur in a passive manner (see also de la Fuente and Leopold 1967).

TABLE 2. Effects of the Inhibitor of Auxin Transport, 2,3,5-triiodobenzoic Acid, on Lateral Movement of Auxin in Response to Gravity

Position of Tissue during Transport	Water Controls			TIBA-treated		
	cpm/100 mg on Donor Side	Receptor Side	Lateral Movement	cpm/100 mg on Donor Side	Receptor Side	Lateral Movement
Receptor side up (lateral movement against gravity)	4369	928	18%	728	482	40%
Receptor side down (lateral movement with gravity)	1322	2217	63%	571	431	43%

NOTE: As in table 1, 25-mm sections of the corn coleoptiles were notched 5 mm and radioactive indoleacetic acid (10^{-5} M) applied to the uncut side; transport was allowed for 2 hr while the coleoptiles were held horizontally, with either the auxin-treated side or the nontreated side up. Lateral movement is expressed as in table 1. Sections were pretreated with TIBA (10^{-4} M) or with water by placing on moistened filter papers for 30 min before transport.

In connection with the question of lateral auxin redistribution, there are two possibilites for redistribution which need not involve a lateral auxin transport system. One is that the gravitational stimulus may bring about an inhibition of polar auxin transport on the upper side of the tissue, resulting in a shunting of auxin to the lower, uninhibited side (Naqvi and Gordon 1966). The other is that it may simply suppress polar transport on the lower side of the tissue; Kaldewey (1963) has established that this is in fact the manner of auxin redistribution in geotropically stimulated Fritillaria pedicels.

Little attention has been given to the problem of the geotropically unresponsive tissues, and there are many examples among stems, stolons, and petioles. We have done experiments on the lateral movement of auxin in the stems and stolons of peppermint, since they are structurally very similar, and have found the stems geotropically responsive and the stolons weakly responsive or unresponsive. We have found that in stems there is a geotropically stimulated lateral auxin movement similar in magnitude to that for sunflower, but in stolons the lateral auxin movement is small and unaltered by gravity.

Turning to the possible involvement of other types of regulators than auxin in geotropism, we might consider at least two growth regulators. Ethylene has been found to be involved in the geotropic responses of pea roots (Chadwick and Burg 1967). Cytokinins have been found to be effective in preventing geotropic curvatures of soybean stems even though the lateral redistribution of auxin occurred (Krul 1967). With each of these growth regulators, it is possible that an alteration of the geometry of growth may be involved. If isodiametric growth is preferentially stimulated on one side of a section, curvature may develop even though similar increases in cell volume are being achieved on both sides of the stem or root section. This may be the nature of the ethylene effect in pea roots. If isodiametric growth is stimulated over the entire cross-section, geotropic responses may be prevented. This may be the nature of the cytokinin effect in soybean stems. A surprisingly different type of stem curvature has been described by Hejnowicz (1967) involving no differences in relative growth of the upper and lower sides of a stem, but rather the development of differences in hydration characteristics in the cell walls.

In conclusion, then, we believe that the classical type of explanation of geotropic curvature does occur in many instances, and that lateral auxin redistribution may occur through the gravity alteration of lateral auxin transport. On the other hand, there are several alternatives for geotropic mechanisms, including redistributions of auxin following alterations in the polar transport system, and involving other growth substances such as ethylene, or even by qualitative changes which do not require differential cell enlargement on one side of the stem.

References

Chadwick, A. V., and Burg, S. P. 1967. *Plant Physiol.* 42: 415-20.

de la Fuente, R. K., and Leopold, A. C. 1968. In *Biochemistry and Physiology of Plant Growth Substances*, ed. F. Wightman and G. Setterfield. Ottawa: Runge Press.

Goldsmith, M. H. M., and Wilkins, M. B. 1964. *Plant Physiol.* 39: 151-62.

Hejnowicz, Z. 1967. *Am. J. Botany* 54: 684-89.

Hertel, R., and Leopold, A. C. 1963. *Naturwiss.* 50: 1-3.

Kaldewey, H. 1963. *Planta* 60: 178-204.

Krul, W. R. 1967. Doctoral dissertation, Purdue University, Lafayette, Ind.

Naqvi, S. M., and Gordon, S. A. 1966. *Plant Physiol.* 41: 1113-18.

Discussion

GALSTON: There are two actions of triiodobenzoic acid (TIBA) that I think are beyond dispute. Winter, in data presented at the Ottawa Conference on Growth Substances, has shown that TIBA increases the amount of immobile IAA in coleoptiles. The other is that there are structural changes in cells which have been treated with TIBA for a long period of time. The microtubules change their orientation from essentially parallel, spirally oriented cellulose microfibrils, to a jumbled mass (Bouck and Galston, *Ann. N.Y. Acad. Sci.* 144 [1967]: 34-38).

LEOPOLD: I don't accept Winter's experiments as purporting to show that TIBA causes an immobilization of auxin. We have reasons to believe that several inhibitors, low temperature, or low oxygen tension interfere with transport itself. With respect to the morphological changes following TIBA applications, these treatments which led to the structural changes you mention were done with very high TIBA concentrations. I think the observations of structural change are very suggestive but not necessarily descriptive of the rapid action at very low TIBA levels used in our transport work.

I would like to mention, in contrast, the effect of TIBA on a nonpolar movement. If you supply auxin unilaterally to the basal end of a coleoptile section, so that the auxin movement is acropetal, the lateral redistribution without geotropism is about 50%—about half goes to the receptor side. If you apply TIBA just before suplying the auxin, about 60% goes to the receptor side. The effect of the inhibitor here has not at all depressed this auxin movement. Similarly, lateral orientation of the tissue with respect to gravity does not alter this lateral movement. So, you see, gravity and TIBA effects on lateral auxin movement can each work on one component of auxin, and neither can work on an apparently passive component. This is consistent with the idea that the gravitational force and the TIBA alter an active pumping of auxin in the plant.

WEIS-FOGH: I have been listening to the evidence given for active transport. Apparently it is taken for granted that an active transport has been proved without any concentrations being given. We don't know whether such concentration gradients truly exist. Do we know whether auxin disappears, is being broken down, in the region where you find an increased number of counts?

LEOPOLD: We've done the lateral redistribution experiments and chromatographed the labeled material to be sure it was still indoleacetic acid. Our chromatographic evidence shows that the redistribution was in fact of indoleacetic acid. Both Dr. Wilkins and my group have shown that lateral movement of auxin can occur against an apparent concentration gradient.

WILKINS: I'd like to emphasize, as at the beginning of my paper, that all is not solved by the demonstration of some lateral movement in some stems. There is the classic case of the response of grass nodes to gravity, where in the horizontal position one finds the initiation of growth of cells in the lower half of the node and no growth of those in the upper half.

AUDUS: There is evidence in the literature that these dark node responses are due to de novo synthesis of auxin in the node.

WEIS-FOGH: I want to ask the botanists whether in addition to movement of radioactive materials, like auxin and other growth substances, they have a problem in trying to estimate the movement of water within the system? If you had a large transport of water you could have passive transport of any other substance. I am not going to suggest that there is an active transport of water. It could be a consequence of any ionic movement. Have these studies been performed?

AUDUS: Well, there have been studies performed on the effect of auxin on the rate of equilibration of tritiated water between plant tissues and surrounding medium. I believe that the rate of equilibration was extremely rapid, so that auxin effects on water movement couldn't in fact account for any kind of growth phenomena induced by auxin.

GALSTON: If you'll permit a statement about another tropism, namely thigmotropism, we have investigated the response of tendrils of the pea plant to touch, involving curvature toward the ventral side after ventral stimulation. We have worked with tritiated water and have shown that if you permit an excised tendril to become labeled with tritiated water and then stimulate it, it will expel water from the ventral side which is shrinking during the curvature (Jaffe and Galston, *Plant Physiol.* 43 [1968]: 537-42). It does not transport water laterally. The pea tendril has two vascular bundles, one on the dorsal side and one on the ventral side, and we know that water movement is mainly in the xylem of the vascular bundle. So it's not too surprising that in the organ there is no measurable lateral transport. I think this phenomenon is roughly equivalent to geotropic curvature.

AUDUS: But in growth it is mass movement of water into the cell which is limited by the rate of expansion of its cell walls, that is, the plastic expansibility of cell walls. I don't think the study with the use of tritiated water, which measures molecular diffusion and not mass flow, would tell us anything about the situation in growth.

13
Linkage between Detection and the Mechanisms Establishing Differential Growth Factor Concentrations

L. J. Audus, *University of London*

I am in the unenviable position of the housewife who has to make sandwiches for her husband's lunch but who has most inferior meat and very grave doubts about the bread. In other words, I am to speak learnedly about a theoretical linkage mechanism between gravity perceptors, which are still far from being unambiguously identified, and hormone redistribution, which is not accepted by everyone as the ultimate cause of differential growth and consequent geotropic curvatures. But perhaps some might not agree that the position is so unenviable since, in the almost perfect factual vacuum, I am at liberty to speculate on this linkage without risk of contravening any established laws or suppressing inconvenient facts. Be that as it may, I hope that out of this physiological jamboree may come a fresh look at a very old problem and a few pointers for future experimental work.

A necessarily meandering survey of a wide-ranging literature has brought me to the conviction that in the course of evolution plants must certainly have evolved more than one set of mechanisms for their perception of and response to gravity. In the past, in the flush of a new discovery, it has been easy and convenient to extrapolate from one organ to another, or from one phenomenon to a related one, and to elaborate in the process a unified and universal theory completely unjustified by the extreme restriction of the available facts. To ride a hobbyhorse of mine, I might quote the extrapolation of the Cholodny-Went theory, now reasonably well founded for the transverse movement of indole-3-acetic acid (IAA) in the *Avena* coleoptile, to the situation in the root. This cannot be justified. For example, there is no unassailable evidence that indole-3-acetic acid controls the normal extension growth of roots, and hence geotropic responses there, or that it is indeed produced in the root tip. Evidence for a consistently polar transport of IAA in all roots is still lacking, and what polarity there is seems to be acropetal rather than basipetal (Yeomans and Audus 1964). Bearing such examples in mind, therefore, I shall not try to draw together any threads or yield to any temptation to formulate a unifying theory. Each group of phenomena will be taken on its own merits, and the implication for the perception-response linkage worked out "in isolation," as it were. If any unifying principle emerges from this piecemeal approach, so much the better.

The Situation in the Root

I begin with the root, since it is here that my own investigations have been concentrated. The Cholodny-Went theory postulates that auxin is produced by the root tip; thence it moves into the extension zone and, under the action of gravity, somehow accumulates in greater quantities on the lower side of the organ. Since the root has a sensitivity to auxins many orders greater than that of the aerial organs, concentrations on the lower side are supposed to be supraoptimal for growth, which is consequently inhibited, and a positive geotropic curvature results.

One of the main supports for this supraoptimal auxin theory was the classical observations of Cholodny (1926) that removal of the root tip caused an acceleration of growth, while reheading with the same tip returned the growth rate towards normal. Several subsequent studies have failed to confirm these claims (see Younis 1954), but the theory stubbornly persists. Similar surgery has been used in attempts to identify the root cap, with its amyloplast statoliths, as the zone of graviperception, but as shown by Younis (1954), removal of a portion of the root tip sufficient to prevent gravity perception also impairs growth and hence response. This unavoidable complication of root decapitation arises from the morphology of the organ, in that the meristem-root cap interface is virtually a spherical surface, and that by simple transverse cuts in one plane it is impossible to remove all the root cap without removing the quiescent center of the root and seriously damaging the meristem itself. If these growth centers are left intact, then so also is some of the root cap.

A happy solution to this impasse came with the accidental discovery that in *Zea mays* it is possible to remove the root cap in its entirety without damaging either the quiescent center of the root tip or the meristem, and that the root cap will regenerate in about 36 hr (Juniper et al. 1966). The application of this decapping technique has enabled us to establish the following facts. (1) In the absence of the root cap the root is completely insensitive to gravity (the situation in coleoptile and hypocotyl will be discussed). (2) The presence or absence of the root cap has no effect whatever on the normal growth rate of the root; that is, the growth rate of a decapped root is the same as that of a vertical normal root, whether or not the decapped root is vertical or horizontal. (3) Geotropic sensitivity returns with the regeneration of the root cap and the statolith starch.

From these facts we can deduce the following with reasonable certainty. (1) The root cap is the only site of gravity perception in maize roots, thus giving very strong support to the amyloplast statolith theory of gravity perception. (2) The root cap in normal vertical roots is not the source of any substance or substances which directly affect the normal extension growth of roots. The situation differs dramatically from that in coleoptile tips, where gravity detection and auxin production apparently takes place in the same cells, and the rate of production of the auxin is not altered by the orientation of the coleoptile.

Since, however, there is fair evidence for the accumulation of an auxin-like growth inhibitor on the lower side of stimulated root tips (Hawker 1932, Boysen-Jensen 1933)—that is, in the lowermost cells of the apical meristem—some kind of information must flow from the root cap to induce this accumulation. Furthermore, this information supply must be asymmetric to the root axis and would flow only when the tips were stimulated. We can speculate about the nature of this information.

First, it could be hormonal, namely, a chemical substance produced asymmetrically in the root cap (that is, predominantly on the upper or lower side). It would need to move back asymmetrically into the growth zone of the root, there ultimately to induce the differential growth response of geotropism. The substance would be produced only on stimulation. It could be the inhibitor itself, which after moving asymmetrically through the meristem, would preferentially inhibit the lowermost extending cells. On the other hand, it might be a substance inactive in itself but capable of triggering the production of an inhibitor in the cells of the meristem. This production of inhibitor would also be asymmetric. The idea of a gravity-induced production of an inhibitor is not new, and there is a considerable amount of evidence that it does occur, together with correlated changes in overall growth rate. This seems to be true for the roots of *Vicia faba* and *Pisum sativum* (Audus and Brownbridge 1957, Audus and Lahiri 1961). Gravity-induced changes in enzyme activity in plant organs are not unknown (see Bara 1965, Konings 1965, Westing 1960). Last, the information could be of the nature of an action potential, also produced and propagated asymmetrically, and again triggering asymmetrically inhibitor production in the meristem. A further necessary condition would be the preferential transmission of the "message" in a direction parallel to the main organ axis, so that asymmetric release from the root cap would result in asymmetric arrival in the meristem. Such a transmission could be propagated in and by the plasmodesmata, which are almost exclusively confined to the transverse walls of the root cap cells (Juniper 1963).

The major problem of linkage which now faces us in the root tip is how an asymmetric distribution of amyloplast statoliths in individual cells of the root cap can be translated into an asymmetrical production of a biochemical or electrical "message" by the root cap as a whole. Such an organ asymmetry can arise from cell asymmetry only if there is an intrinsic radial polarity with which the cell asymmetry interacts in these root cap cells. To make this clearer, let us assume that "message" production is the direct result of the pressure of statoliths on a membrane or membranes in the peripheral layers of cytoplasm lining the tangential walls of the root cap cells. The desired asymmetric "message" production would arise automatically if the membranes on the tangential walls nearest the central axis were less sensitive than those on the tangential walls farthest away from the central axis (fig. 1), the "radial polarity" referred to above. Then cells beneath the main axis will produce more "message" than cells above it and our organ asymmetry is achieved.

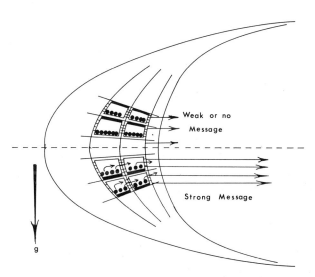

Fig. 1. Method of asymmetric "message" generation by amyloplast statoliths in the root cap. The black cell wall (longitudinal and distal to the main axis) represents the pressure-sensitive region of the cell; g marks the direction of the gravity vector.

Mechanisms whereby statolith contact with a membrane could induce the production of a chemical "messenger" should be considered. The appearance, *de novo*, of a substance in a cell usually presupposes the prior generation of an appropriate enzyme system for the synthesis of that substance, and this in turn might suggest a change in the information reaching the ribosomes from the nucleus, or in the translation of the information already in the polysomes. One could imagine that amyloplast contact with the endoplasmic reticulum could block the flow of messenger RNA to the attached ribosomes, and hence alter the pattern of protein synthesis or even hinder the proper operation of the ribosomes. If the enzymes synthesized on the inner tangential walls differed from those on the outer tangential walls, then we have our source of radial polarity hypothesized above. However, the blockage of messenger RNA flow or of the operation of the ribosomes would be more likely to turn off enzyme production rather than to turn it on. In any case the speed with which such changes in enzyme synthesis could be reflected in the manufacture of its product ("chemical messenger") is probably not fast enough to account for the rapid responses of most geotropically sensitive roots.

An alternative mechanism could be the stimulated release of the "messenger" from a bound form in the root cap cell. Distortion of a membrane by amyloplast contact would be likely to affect its permeability and hence promote such release if the "messenger" were contained inside an organelle. But electron microscopy shows that the only membranes likely to be so affected by sedimenting amyloplasts are the plasmalemma and the endoplasmic reticular membranes lining the tangential walls. Mitochondria and Golgi bodies are somewhat displaced but not otherwise affected by amyloplast movement (Griffiths and Audus 1964). In the absence of a suitable membrane-bounded site for this "messenger" storage, this theory seems unlikely. However, similar membrane distortion might also give rise to an action potential which could constitute the "message" transmitted through the plasmodesmata via plasmalemma or endoplasmic reticulum. Such action potential production by cell distortion is well established in such organs as the sensitive trigger hairs of *Dionaea muscipula* (Jacobson 1965). Action potential generation in geotropic stimulation has not yet been recorded, or even sought so far as I am aware. (It should be noted that action potentials should not be confused with the relatively slow-changing geoelectric potentials which are established across coleoptiles as a result of the asymmetric distribution of auxin.) There are some indications from unpublished work by Mrs. Bruria Shachar in my laboratory that once the threshold stimulus is received by the root cap, the "message" reaches the meristem very quickly; the presentation time (minimum horizontal exposure to give a response) is, within the rather coarse limits possible with current techniques, the same for roots decapped immediately after stimulation as for roots left intact. This situation could be more easily explained by a rapid transmission of an action potential than by a slower transport of a substance; in the latter case one might expect roots decapped immediately after stimulation to have a longer presentation time than that of intact roots.

Studies on Coleoptiles

The *Avena* coleoptile has been studied more than any other organ, and more than one responding system seems to be in operation. It has long been known that the coleoptile tip is both the site of auxin production and the region of maximum geotropic sensitivity. Furthermore, as first shown by Dolk (1936), and more recently confirmed by Brauner and Appel (1960) and by Gillespie and Briggs (1961), gravity seems to induce a transverse polarity in the tip resulting in a movement of the auxin from the upper to the lower side; thus asymmetric distribution of auxin appears to be the main cause of the geotropic reaction. However, this is not the whole story, since blocking of the transverse movement by the insertion of a mica slip still allows very considerable curvature responses (Brauner and Appel 1960). Thus the main axis of the coleoptile, although it appears to produce no auxin, is certainly sensitive to gravity, which will induce physiological differences between upper and lower sides in the absence of the tip (and presumably, therefore, of auxin). Decapitated coleoptiles will show curvature responses to gravity if supplied with auxins (Anker 1954, 1956), and this is associated with the development of a transverse polarity so that auxin (^{14}C-IAA) applied at the tip end will move preferentially to the lower side of the coleoptiles (Goldsmith and Wilkins 1964), where it is translocated basipetally in greater quantities than on the upper side (Gillespie and Thimann 1963). Anker explained the geotropic curvatures he observed in terms of such differential transport, although other interpretations (as differential sensitivity to auxin or differential uptake on the two sides) could account for them equally well. Evidence for sensitivity changes in the hypocotyl of *Helianthus* have been obtained but this will be discussed later.

Mechanisms in the Tip

For the tip a very plausible mechanism has already been proposed (Hertel 1962, Hertel and Leopold 1963), based on the concept of polar auxin transport as a differential secretion phenomenon. There now seems little doubt, from studies of mathematical models set up by Leopold and Hall (1966), that a very small secretion differential between two sides of a cell can, on integration through a column of cells, give almost completely one-way movement through a tissue composed of such cells. Hertel and Leopold suggest that statoliths moving to a lateral wall during geotropic induction could promote auxin secretion from the rigid immobile cytoplasmic layer lining that side of the cell, thus producing a polar auxin transport in the same direction as the movement of the statoliths. This, coupled with an intrinsic longitudinal polarity of cells involving a small excess of basipetal secretion, would result in a general preferential movement of auxin, across the tip and down the lower side of the coleoptile (see fig. 2). Hertel and Leopold regard the statoliths as providing "energy or carriers for auxin secretion". The implication that amyloplast starch might supply "activated sugar" is difficult to accept in view of the extreme reluctance of such statolith starch to hydrolyse (see Rawitscher 1932, Pickard and Thimann 1966). However, the work of Pickard and Thimann (1966) suggests that starch grains may not be the statoliths of

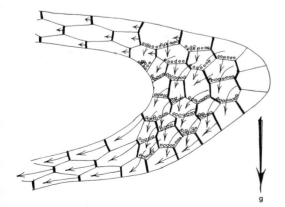

Fig. 2
Method of coupling of statolith sedimentation to differential secretion of auxin in the coleoptile tip. Black transverse walls represent the surfaces of intrinsic basipetal secretion. The arrows represent the general direction of auxin movement, being the resultant of an intrinsic basipetal secretion and a statolith-induced downwards secretion; *g* marks the direction of the gravity vector. Adapted from Hertel and Leopold 1963.

coleoptiles, although it is not easy to imagine what other "particles" might operate as statoliths here. Mitochondria, as providers of ATP, suggest themselves, but observations on root cap cells suggest that they do not sediment appreciably. A differential secretion operated by the weight of the protoplasm itself, a theory of Czapek (1898) recently renewed by Pickard and Thimann (1966), is a similar possibility.

In these speculations attention has been focused on the lowermost boundary of the cell from which an outward secretion is postulated. But might we not consider the uppermost boundary of the cell and postulate the promotion of an inward secretion? There is, after all, some evidence from *Vicia faba* root cap cells that amyloplast sedimentation into the lower part of a stimulated cell results in a significant accumulation of endoplasmic reticulum in the upper half (Griffiths and Audus 1964). Might not this provide enzymes necessary for the release of auxin into the cell from the cell above? Might not secretion be a "pull from below" rather than a "push from above"? So far organelle distribution in the coleoptile tip has not been studied in detail in relation to geotropic induction. It is high time such studies were made.

Mechanisms in the Main Axis

Suggested mechanisms for the linkage of graviperception to the differential movement of auxin in the main axis take us into the realms almost of fantasy. Here more than ever we run up against the frustration of not knowing what the perceptor mechanism is. Certainly there are amyloplast statoliths in the starch sheath surrounding the two vascular bundles but, unlike the situation in the tip, no such visible sedimenting granules have yet been demonstrated in the parenchyma which makes up the rest of the coleoptile and in which auxin transport is supposed to take place. Thus, if we accept starch statoliths as the perceptors, we have the extremely difficult task of explaining how they could govern hormone transport direction in the parenchyma *at a distance*. If we accept the unlikely "protoplast weight" theory, then the problems are much simplified although not eliminated.

There are two phenomena to explain. One is the polar movement of auxin in a direction transverse to the main axis from the upper to the lower side, so clearly demonstrated by Goldsmith and Wilkins (1964). The other is the augmentation of the intensity of polar basipetal transport on the lower side of the coleoptile axis, first demonstrated by Dolk (1936) and confirmed by Gillespie and Thimann (1963). How far the latter is the outcome of the former is not yet clear.

Taking first the induction of a lateral transport polarity, an explanation of linkage of the kind put forward by Hertel and Leopold (1963, see section on tip) would be the most logical if we assume that *all* (or most) cells possess statoliths (that is, either the whole mass of the cell contents or some as yet unidentified but universal "particle"). A promotion of the secretion from the lowermost side of the cell would cause a generalized flow from the upper to the lower side, although, since the coleoptile is a cylindrical organ, movement would not take place uniformly over the whole cross-sectional area of the organ (see fig. 3). If, however, amyloplast statoliths are the perceptors, then, since they are confined to a tubular region around the vascular bundles (see fig. 3), a considerable complexity of postulates would be necessary to explain such a linkage "at a distance." An additional complication is introduced in that the bundle (and hence statolith) distribution is bilateral, and so different mechanisms would have to be postulated for "across the bundles" movement (fig. 3*A*) and "from bundle to bundle" movement (fig. 3*B*). Unfortunately no

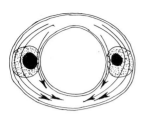 

Fig. 3
Diagrams to illustrate (A) across-the-bundle and (B) from-bundle-to-bundle movement of auxin in relation to geotropic induction in the Avena coleoptile. Stippled areas mark the statenchyma.

A   B

refined analysis of lateral movement, particularly in relation to the bundles, has yet been made; a search of the relevant papers on lateral auxin movement in coleoptile segments does not reveal the orientation of the experimental segments in relation to the bundles. It may well be that further studies will reveal differences in lateral movement in relation to these bundles which would justify theorizing on an "action at a distance" mechanism by these localized amyloplast statoliths; if they do not, the situation, it seems to me, will constitute one of the greatest obstacles to our acceptance of amyloplast statoliths as graviperceptors in the main axis of the coleoptile. But we must reserve judgment until such studies have been made.

When we consider the promotion of basipetal transport in the lowermost half of the coleoptile, explanations of linkage must be equally complex. Supposing first that gravity can be detected by all parenchyma cells and that we can assume, with Hertel and Leopold, that a secretion from the basal cell surface accounts for basipetal polar transport. Our linkage now involves action at a distance, albeit only across a cell, since we must explain how the detection of pressure (i.e., from statoliths or cell contents) by the *outer* tangential walls of the cells *under* the main axis could promote secretion from the *basal* transverse wall of the same cell. Again we could invoke the provision of "energy" or "carriers" provided for this secretion by the outer tangential cytoplasmic layers and induced by the pressure upon them. On the other hand, we could imagine that pressure might be suppressing the production of a secretion inhibitor normally keeping the basipetal polar transport below its maximal possible value. This would be well in line with the "barrier" theory of Kaldewey (1963), which was proposed to explain differences observed in the "speed" of auxin transport on the upper and lower sides of geotropically induced fruit stalks of *Fritillaria meleagris*. Again the essential intrinsic radial polarity would be ensured by the proposition that the inner tangential layers of cytoplasm would have a lower sensitivity to pressure or none at all. These possibilities are illustrated in figure 4.

If, however, amyloplast statoliths are the graviperceptors, it is not entirely illogical to suppose that basipetal transport might be closely associated with the starch statenchyma, or even take place within it. There is some experimental data to support this suggestion. Thus very early studies by Laibach and Kornmann (1933) showed that in the *Avena* curvature test, much greater responses were obtained if the agar block containing auxin were placed over a bundle than if it were placed over parenchyma. On the other hand Anker (1954) showed in both intact coleoptiles and decapitated coleoptiles supplied with indole-3-acetic acid that curvature was much weaker in the plane containing the vascular bundles than in the plane transverse to it. Then again direct studies on transport by longitudinally split portions of *Zea* coleoptiles by Hertel (1962) suggest that the

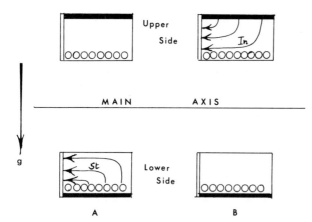

Fig. 4
The method of linkage of the sedimentation of a ubiquitous statolith to the promotion of basipetal transport of auxin on the lower side of the coleoptile. A, promotion by the induced production of a secretion promoter or carrier (St). B, promotion by the suppression of a secretion inhibitor (In). The circles could represent either sedimenting particles or the region of pressure exertion by the cell contents; g marks the direction of the gravity vector.

small region of coleoptiles containing the bundles and statenchyma carries about 30% of the total auxin (IAA-^{14}C) transport (my calculation from Hertel's data). Since the total cross-sectional area of the statenchyma is approximately 15% of the cross-sectional area of the parenchyma as a whole, simple calculation suggests that transport intensity (quantity per unit cross-sectional area) in the statenchyma is approximately 2.5 times that in the remaining parenchyma. Such preferential transport in the statenchyma, if it were modified appropriately by a statolith-linked mechanism, might account sufficiently for the differential longitudinal auxin transport required for curvatures (see fig. 5). The situation obviously calls for further careful studies on the channel of auxin transport, preferably at the cellular level.

If auxin transport did take place in or associated with the statocyte sheath, then to explain our linkage we again would need to postulate an intrinsic cell polarity in the starch statenchyma, not now related to the organ as a whole but to the axis of the bundle. The required pressure sensitivity distribution in the statenchyma cells is shown in figure 5 and will be seen to be confined to the cytoplasmic layers bounding the walls furthest from the bundle in a wide arc on the outside of each bundle. One need not postulate that transport is confined to the statenchyma, only that it takes place there (hopefully in a preferential manner as above), and that statolith pressure on sensitive cytoplasm augments basipetal transport through the induced cells in a way similar to that proposed for parenchyma in figure 4. In this way there might arise a transport differential sufficient to give the normal geotropic curvatures. On the other hand, the promotion of basipetal secretion need not be confined to the detecting statenchyma itself but may extend to neighboring parenchyma cells, for example, by the outward diffusion of a substance or substances responsible for this promotion (see the arrows in fig. 5).

The *Helianthus* Hypocotyl

Yet another aspect of response, clearly demonstrated in the hypocotyl of *Helianthus*, needs to be linked to perception. This is the change in the sensitivity to auxin which arises on induction, whereby part or all of the growing tissue lower than the central axis of the organ becomes more responsive to applied auxin than the tissue above the axis (Brauner 1966). This differential sensitivity is explained by Brauner in terms of the old cofactor theory of Gradman (1930), which postulates the appearance of an auxin cofactor during geotropic induction, but only on the lower

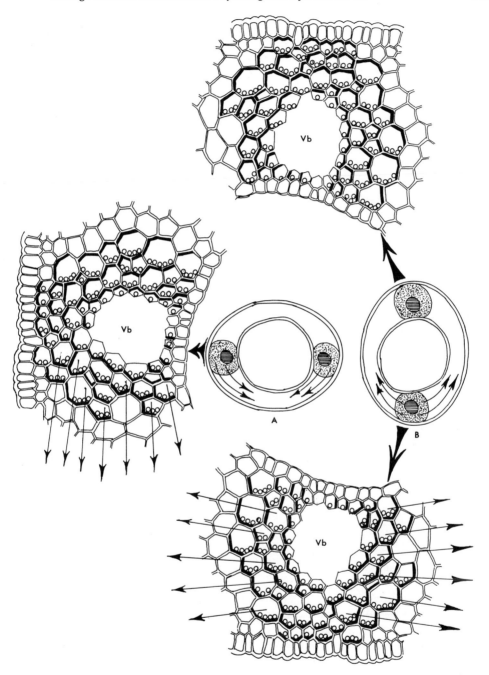

Fig. 5. Method of linkage of graviperception in the bundle statenchyma to the promotion of basipetal transport of auxin in the lower half of the coleoptile. *A*, bundles in the horizontal plane. *B*, bundles in the vertical plane. The black regions on the walls of the statenchyma cells mark the postulated region of sensitivity to statolith pressure. Length of arrows could indicate either the intensity of basipetal auxin transport in the statenchyma or the secretion from the statenchyma of a promoter of basipetal secretion in all neighboring cells. *Vb* = Vascular bundle.

side of the organ. If such a cofactor synthesis is the basis of sensitivity changes, then linkage could follow the same principles as those formulated for root cap perception. Thus, enzymes responsible for cofactor production could be induced or activated in the pressure-sensitive outer layers of cytoplasm of the perceptor cells, whether they are confined to the starch statenchyma that surrounds the stele in the hypocotyl or occur generally in the parenchyma cells (ubiquitous statolith or cytoplasm weight theories). The operation of the former is illustrated in the diagram of figure 6.

Naturally any polar lateral transport or differential basipetal transport underlying geotropic response of *Helianthus* hypocotyl could be linked to perception in exactly the same way as proposed for the coleoptile axis. So far no clear-cut experiments on such transport characteristics have been done for Helianthus, although there are interesting indirect pointers to the relationships in this material. Hartling (1964) has shown that rotation on a horizontal axis appears to cause a movement of auxin toward the periphery of the hypocotyl. This is most rationally explained along lines similar to those proposed for the cofactor theory above. If auxin were transported basipetally in the statenchyma, and were secreted from the outer tangential cell surface only when stimulated by the amyloplast statoliths (i.e., only from the cells when they were under the central axis), then slow rotation would automatically produce an outward flow of auxin in the whole organ (see downward-pointing arrows in fig. 6).

Linkage for Other Proposed Response Mechanisms

Differential transport of auxin and differential production of auxin or a cofactor are not the only explanations put forward for the differences in growth rates in geotropically curving organs. Differential destruction of auxin has also been postulated as underlying any differences in auxin concentration which may be regulating curvature in geotropically induced roots of *Pisum sativum* (Konings 1964). Such differential inactivation by IAA-oxidase could be the reflection of changes in the monophenol (enzyme activator)/polyphenol (enzyme inhibitor) ratios in the cells concerned. Although no direct evidence of such changes is yet available, they could result from parallel changes in enzyme activities in the perceptor cells resulting, as in the previous hypothesis, from pressures operating on their outer tangential boundaries. Again, if perception is confined to the starch statenchyma, then diffusion of oxidase cofactor or inhibitor into neighboring parenchyma would cause the appropriate adjustments in IAA-oxidase activity and hence IAA levels.

Summary

In view of the diversity of structure in organs which detect and respond to gravity, the probability that different linkage mechanisms may operate for different organs is stressed. In the root it may be accepted as reasonably certain that amyloplast statoliths in the root cap are the gravity sensors. Evidence is presented to support the theory that the lateral redistribution of these amyloplasts sedimenting under gravity, coupled with an intrinsic radial polarity of some ultrastructural/biochemical organization of the root cap cells, causes the differential production of a "message" from the upper and lower halves of the root cap. The "message," which may be electrical or hormonal in nature, could induce the production of a greater amount of a growth inhibitor in the lower tissues of the growth zone, thus causing positive geotropic response.

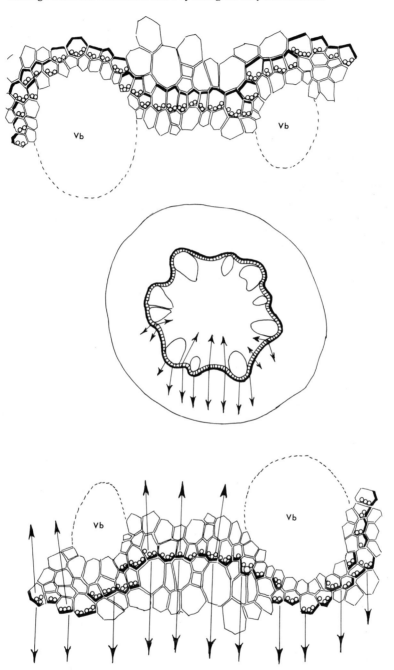

Fig. 6. Method of linkage of graviperception in the starch sheath statenchyma of a *Helianthus* hypocotyl to the production of an auxin cofactor in the lower half of the organ. The thick outer walls of the statenchyma cells represent the regions of pressure sensitivity. The arrows mark the origin and direction of movement of the auxin cofactor. *Vb* = Vascular bundle.

In the coleoptile the role of amyloplasts as gravity sensors is still in doubt. However, the coupling mechanism between gravity sensing and hormone redistribution in the tip seems likely to be the induction of a lateral polarity in the statocyte cells themselves, resulting in a lateral transport of auxin and a differential release into the growth zone. In the main axis of the coleoptile and other axial organs, such as hypocotyls and epicotyls, which are also sensitive to gravity, auxin redistribution seems to result from both a lateral movement and an augmented longitudinal transport on the lower side. Here the linkage mechanisms proposed will depend very much on what the gravity sensor proves to be in those organs. If all cells are sensitive, amyloplasts, which are restricted to the cells of the bundle sheaths, could not be the sensors, and the mechanism of coupling to lateral auxin movement could be similar to that operating in the coleoptile tip. If the amyloplasts are the sensors, then only those cells possessing them in the bundle sheaths could be responsible for lateral transport of auxin, which would then be confined to those cells. A linkage mechanism for promoting the augmented basipetal auxin transport on the lower side, would, whatever the sensor, involve the participation of a cofactor to promote that movement. Again gravity would operate via the sensor by inducing the differential production of that cofactor on the two sides of the organ.

References

Anker, L. 1954. *Proc. K. nederl. Akad. Wetenschap.* C. 57, 304-16.

———. 1956. *Acta Botan. Neerl.* 5: 335-41.

Audus, L. J. 1962. *Soc. Exptl. Biol. Symp.* 16: 197-226.

Audus, L. J., and Brownbridge, M. E. 1957. *J. Exptl. Botany* 8: 105-24.

Audus, L. J., and Lahiri, A. N. 1961. *J. Exptl. Botany* 12: 75-84.

Bara, M. 1965. *Physiol. Plantarum* 18: 1037-43.

Boysen-Jensen, P. 1933. *Planta* 20: 688-98.

Brauner, L. 1966. *Planta* 69: 299-318.

Brauner, L., and Appel, E. 1960. *Planta* 55: 226-34.

Cholodny, N. 1926. *Jahrb. Wiss. Botan.* 65: 447-459.

Czapek, F. 1898. *Jahrb. Wiss. Botan.* 32: 175-308.

Dolk, H. E. 1936. *Rec. Trav. Botan. Neerl.* 33: 509-85.

Gillespie, B., and Briggs, W. R. 1961. *Plant Physiol.* 36: 364-67.

Gillespie, B., and Thimann, K. V. 1963. *Plant Physiol.* 38: 214-25.

Goldsmith, M. H. M., and Wilkins, M. B. 1964. *Plant Physiol.* 39: 151-62.

Gradman, H. 1930. *Jahrb. Wiss. Botan.* 72: 513-610.

Griffiths, H. J., and Audus, L. J. 1964. *New Phytol.* 63: 319-33.

Hartling, C. 1964. *Planta* 63: 43-64.

Hawker, L. E. 1932. *New Phytol.* 31: 321-28.

Hertel, R., 1962. Doctoral dissertation, University of Munich.

Hertel, R., and Leopold, A. C. 1963. *Planta* 59: 535-62.

Jacobson, S. J. 1965. *J. Gen. Physiol.* 49: 117-29.

Juniper, B. E. 1963. *J. R. Micro. Soc.* 82: 123-26.

Juniper, B. E.; Groves, S.; Landau-Schachar, B.; and Audus, L. J. 1966. *Nature* 209: 93-94.

Kaldewey, H. 1963. *Planta* 60: 178-204.

Konings, H. 1964. *Acta Botan. Neerl.* 13: 566-622.

Laibach, F., and Kornmann, P. 1933. *Planta* 21: 396-418.

Leopold, A. C., and Hall, O. F. 1966. *Plant Physiol.* 41: 1476-80.

Pickard, B. G., and Thimann, K. V. 1966. *J. Gen. Physiol.* 49: 1065-86.

Rawitscher, F. 1932. *Der Geotropismus der Pflanzen.* Jena: Gustav Fischer.

Westing, A. H. 1960. *Am. J. Botany* 47: 609-12.

Yeomans, L. M., and Audus, L. J. 1964. *Nature* 204: 559-62.

Younis, A. F. 1954. *J. Exptl. Botany* 15: 357-72.

Discussion

POLLARD: I'm wondering if anybody has done any experiments of this type with plants grown in D_2O with a density of about 1.1? Is geotropism increased if you use D_2O instead of H_2O? If you find that you've got an increased geotropic response in D_2O then you know you're not dealing with a heavy statolith, because you have increased the buoyancy of such things as starch grains. An increased response might indicate a hydrostatic phenomenon, or the function of bodies less dense than the medium. If the response stayed the same you would not know. A decreased response would be compatible with the starch grain hypothesis. So you can simply look to see whether these simple experiments can be done in D_2O and very quickly settle this problem.

AUDUS: I don't think such experiments would prove anything, because we know nothing whatever about dosage response curves of statoliths, that is, the relationship between degree of movement and ultimate response.

HERTEL: One might also suspect such experiments because of the effect of D_2O on growth, and hence on georesponse.

POLLARD: I would like to just say a little bit about that. Such effects are greatly exaggerated. There is very little immediate effect of D_2O on almost any of the processes in the cell. I have worked with D_2O and I have found the differences in enzyme rates very small, so small that they are extremely difficult to measure. The only real difference between D_2O and H_2O is primarily the density and the thermal motion of the molecules.

DAVIS: In many animal cells it is possible to induce long-term electrical changes across membranes by soaking the cells in solutions having unusually high concentrations of certain ions, for example, calcium. I wonder if plant cell membrane potentials can be controlled in a similar way, and if so, whether the geotropic sensitivity is affected. Have experiments of that sort been performed?

LEOPOLD: We have tested numbers of cations on polar transport but never on the georesponse. They have only very small effects on polar transport.

AUDUS: One complicating fact to begin with is that cations have a marked effect on growth response, on the actual growth itself. It is promoted by potassium and very much inhibited by calcium; therefore, you'd be attacking the response mechanism, which would not allow you to ask any questions whatever of the detection mechanism or the transport mechanism.

HERTEL: Our hypothesis mentioned by Dr. Audus can be split in two: a theory and a speculation. The first part was invented to explain how one can go from a polarization of a single cell to the polarization of the whole organ without the necessity for a differential sensitivity of the tangential walls. This first part is essentially the Cholodny-Went theory plus the concept of auxin secretion by the individual cells. The second part, where we talked about statoliths and possible action of starch, was pure speculation. We proposed that some metabolite coming from the starch might be quickly taken up by the membrane and used for auxin transport. Since then I have found no evidence for any covalent linkage to auxin, and some indication that "pressure" of the geosensor is required. Therefore I would rather think of a more sophisticated transduction mechanism at the membrane.

AUDUS: If you are going to accept the fact that you get augmented longitudinal polar transport on the lower side, then your particular hypothesis can't hold.

HERTEL: The postulated increase in the axial transport on the lower side needs more evidence, I think.

14
Bioelectric Phenomena in Graviperception

C. H. Hertz, *Lund Institute of Technology*

Geotropism—the differential growth reaction of plants to gravity—has attracted the interest of plant physiologists for a long time. The studies of plant movements initiated by gravity or centrifugal forces finally led Boysen-Jensen in Copenhagen and Went in Holland to the discovery of the hormone auxin (indolyl-3-acetic acid), which controls the elongation of the cells and thereby plant movements. They advanced the hypothesis, later verified (Pickard and Thimann 1963; Goldsmith and Wilkins 1964; Naqvi, Dedolph, and Gordon 1965), that the influence of gravity on a horizontal plant shoot causes an increase of auxin concentration on its lower side, which increases cell elongation there. These lateral differences in growth produce a curvature of the organ toward its normal orientation with respect to gravity.

However, this interpretation does not explain the reason for the asymmetric distribution of auxin observed, a redistribution that cannot be caused by gravity without intermediary steps. There must exist a geotropic reaction sequence, the first stage of which is sensitive to gravity. There have been two main hypotheses proposed as to the nature of this first step. I will describe them briefly.

At the beginning of this century Haberlandt (1900, 1902) and Nemec (1900, 1902) found that starch particles in certain coleoptile cells could move freely, and their position was influenced by the direction of gravity. For this reason they were called statoliths, and presumed to be the primary gravity receptors of the plant. However, certain facts cast doubt on this theory. First, plants have been found to react geotropically without having such visible statoliths (Pickard and Thimann 1966), and not all geotropic effects could be explained without complicated assumptions (cf. Rawitscher 1932). Second no satisfactory explanation could be given of the connection between the displaced starch statoliths and the asymmetric distribution of auxin.

In 1907 Bose observed that electrical potentials were generated across plant shoots when they were placed in a horizontal position. This phenomenon was investigated in detail (Brauner 1927; Brauner and Amlong 1933) and named the geoelectric effect (GEE). By attaching two nonpolarizable fluid electrodes across a plant shoot he could show that a potential of about 10 mV arose immediately after tilting the plant into a horizontal position. Brauner (1942) also gave an explanation of the effect, which will be discussed below.

The existence of the GEE was later used by Went (1932) to explain the asymmetric distribution of auxin. He assumed that the electric field created inside the plant shoot moved auxin ions from the upper to the lower side of the plant. Thus the primary step in the geotropic reaction chain would be the generation of the GEE, which would in turn lead to auxin redistribution and lateral differences in growth.

To explain the GEE Brauner (1942) made certain experiments with artifical membranes of parchment paper. He placed a plane membrane between two compartments filled with $10^{-3} N\ KCl$, each of which was connected to a nonpolarizable electrode. As long as the membrane plane was vertical no electric potential could be measured between the electrodes. But he could demonstrate a potential of about 15 mV as soon as the system was turned so that the plane of the membrane was horizontal. Brauner explained this effect by the assumption that the ions fell out of solution under the influence of gravity and were separated in the pores of the ion-exchange membrane. Both Brauner (1959) and the author later showed that this was not the case, and that Brauner's earlier results were a consequence of the properties of the electrodes he used. This interpretation has been confirmed by Woodcock (private communication).

Because of these results the existence of the GEE in plants and its role as a primary link in the geotropic reaction chain could be doubted. Therefore Dr. Grahm and the author tried to measure the GEE with a technique very similar to Brauner's. As shown in figure 1, two thin plastic tubes filled with gelatin connect the plant with two calomel electrodes. By tilting the entire arrangement 90° the GEE could be measured. Although a geoelectric effect could be demonstrated, it was found that this type of electrode, for different reasons, did not give reproducible results (short circuits around the plant stem, irritation of the plant caused by the electrodes, etc.). These difficulties can largely be avoided by using a so-called flowing electrode as an electrical contact to the plant (Newman 1963; Wilkins and Woodcock 1965). This electrode consists of two small concentric glass tubes ending at about the same place. By supplying the inner tube with electrolyte and removing this by suction through the outer tube, a fresh drop of electrolyte is maintained at the top of the electrode. Since this electrode does not require a humid atmosphere and touches the plant only very gently, it does not have the drawbacks mentioned above. On the other hand, it allows the electrode-plant system to be tipped only slightly and thus is not suited for the measurement of the GEE. For this reason we developed a new measuring technique whereby the plant is not touched by the measuring electrode at all (Hertz 1960).

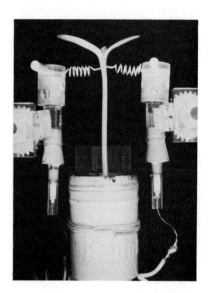

Fig. 1
Measurement of the geoelectric effect by two semifluid electrodes applied to Helianthus shoot. The contact fluid is gelatin contained in spiral tubes of thin plastic.

This measuring device makes use of the well-known vibrating-reed or vibrating condenser principle used for measuring small voltages of very high impedance sources, such as ionization chambers. The principle of the method will be described only briefly in the following. A more detailed description has been given by Grahm (1964).

As shown in figure 2, four coleoptiles of *Zea mays* or Avena are mounted side by side in a small box filled with gelatin. In front of them a flat electrode E is mounted at a distance of 1 to 2 mm. This electrode, which is made to vibrate with a frequency of 250 Hz in the direction indicated, is well insulated and connected to an electrometer tube. Now, if an electric potential exists between the electrode E and the coleoptiles, an AC potential is generated at the grid of the electrometer tube which, after amplification, can be displayed on a cathode ray oscilloscope (CRO).

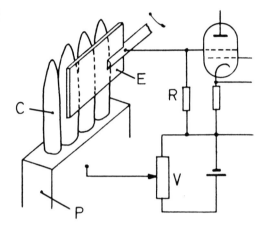

Fig. 2
Measurement of the geoelectric potential by means of the vibrating-reed principle. This method permits measurement of the potential by the vibrating electrode E without touching the coleoptiles C.

In this apparatus the coleoptiles and the vibrating electrode constitute, in fact, a condenser. The "coleoptile plate" of this condenser is connected to a variable voltage by the gelatin and calomel electrode. By varying this voltage the voltage between the coleoptiles and the vibrating electrode can be reduced to zero, which results in the disappearance of the signal on the CRO. If the entire apparatus is tilted 90° and a GEE develops across the coleoptiles, the signal reappears on the CRO. By setting the signal to zero again as above, the magnitude of the GEE can be determined. Since this procedure is quite time-consuming, a servo system has been constructed which automatically adjusts the outer voltage (Grahm and Hertz 1962).

Measurements made with this apparatus, and a more sophisticated version developed by Grahm (1964) showed clearly that the GEE really exists. However, its time dependence and magnitude did not agree with those obtained with less elaborate measuring methods (Brauner 1927; Schrank 1947; Jantsch 1959). As shown in figure 3, the GEE in *Zea mays* did not appear until after a latent period of about 20 min, and then it rose to about 80 mV.

The latent period agreed well with that found by Wilkins and Woodcock (1965) when they applied auxin asymmetrically to decapitated coleoptiles. Further, Jantsch (1959) had earlier found a similar latent period. Grahm and Hertz (1964) and Johnsson (1965a, 1965b) showed that the latent period could be separated into physical and chemical components, with the physical part attributable to the primary gravity reception mechanism.

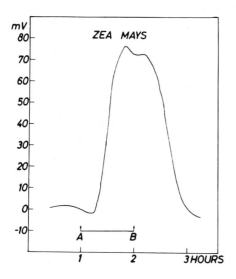

Fig. 3
Geoelectric potential developed in *Zea mays* coleoptiles in darkness (mean value of 3 experiments). The coleoptiles were horizontally exposed during the time interval *A-B*. Note the latent period. From Grahm 1964, by permission.

Grahm and Hertz (1962) showed further that the GEE was sensitive to respiration inhibitors such as NaN_3 and DNP, and that it disappeared in an oxygen-free atmosphere, which was confirmed by Brauner, Dellinghausen, and Böck (1964) and Dedolph, Breen, and Gordon (1965). Further, Grahm and Hertz (1964) found that the appearance of the GEE under different experimental conditions showed a large similarity to the geotropic reaction of the coleoptile, a fact that was pointed out by Wilkins (1966).

Grahm (1964) then demonstrated that the GEE was closely related to the presence of an asymmetric distribution of auxin in the coleoptile. No GEE could be detected in freshly decapitated, auxin-free coleoptiles, but the GEE reappeared as soon as the auxin-producing tip was regenerated. Further, both Grahm (1964) and Wilkins and Woodcock (1965) produced asymmetrical auxin distributions artificially in vertical coleoptiles and showed that the electrical effects observed were similar to the true GEE both as to magnitude and latency time. These measurements confirmed similar work by Ramshorn (1934) and Newman (1963). Because of this Grahm and Wilkins concluded independently that the generation of the GEE was a secondary effect due to the action of auxin, and thus cannot be responsible for the asymmetrical distribution of that hormone under the influence of gravity. This conclusion had been anticipated earlier by Audus (1962).

Naqvi, Dedolph, and Gordon (1965) showed that auxin asymmetries occur in coleoptiles placed in a horizontal position even in a nitrogen atmosphere. This would suggest that the generation of the GEE does require an asymmetrical auxin distribution and the presence of oxygen. However, more work has to be done to clarify this point.

In conclusion, we find that the GEE does not seem to be directly linked to the primary gravity perception of the plant, and thus cannot be responsible for the asymmetrical auxin distribution, as proposed by Went and others. Instead, the GEE seems to be a secondary phenomenon due to the asymmetrical auxin distribution together with other factors. Its significance for geotropism, if any, is not clear. The same situation seems to exist in phototropism, which shows close parallels to geotropism (Johnsson 1965a, 1965b, 1967). A comprehensive review of geotropism, which includes the GEE, has recently been given by Wilkins (1966).

Summary

It has long been known that an electrical potential develops across a plant shoot which is placed horizontally. Since this effect can be shown to be due to the action of gravity, it is called the geoelectric effect (GEE). The paper reviews the older theories on the origin of the GEE as well as its role in the geotropic reaction of the plant.

Since the validity of these older theories was questioned by many physiologists, the GEE has been investigated in detail during the last few years. New measuring techniques have been developed without the drawbacks of earlier methods. With these techniques it was found that theories for the generation of the GEE and its role in the geotropic reaction chain were incorrect, and that the GEE was not a primary result of the action of gravity on the plant. Instead it derives from the asymmetric distribution of plant hormone which takes place as a consequence of the effect of gravity on the plant. The precise reactions leading to the GEE are not yet known.

References

Audus, L. J. 1962. *Symp. Soc. Exptl. Biol.* 16: 197.

Bose, J. C. 1907. *Comparative Electrophysiology*, p. 434. London: Longmans, Green.

Brauner, L. 1927. *Jahrb. Wiss. Botan.* 66: 381.

———. 1942. *Rev. Fac. Sci. Univ. Istanbul, Ser. B.* 7: 46.

———. 1959. *Planta* 53: 449.

Brauner, L., and Amlong, H. U. 1933. *Planta* 20: 279.

Brauner, L.; Dellinghausen, M.; and Böck, A. 1964. *Planta* 62: 195.

Dedolph, R. R.; Breen, J. J.; and Gordon, S. A. 1965. *Science* 148: 1100.

Goldsmith, H. M., and Wilkins, M. B. 1964. *Plant Physiol.* 39: 151.

Grahm, L. 1964. *Physiol. Plantarum* 17: 231.

Grahm, L., and Hertz, C. H. 1962. *Physiol. Plantarum* 15: 96.

———. 1964. *Physiol. Plantarum* 17: 186.

Haberlandt, G. 1900. *Ber. Deut. Botan. Ges.* 18: 261.

———. 1902. *Jahrb. Wiss. Botan.* 38: 447.

Hertz, C. H. 1960. *Nature* 187: 320.

Jantsch, B. 1959. *Z. Botan.* 47: 336.

Johnsson, A. 1965a. *Physiol. Plantarum* 18: 574.

———. 1965b. *Physiol. Plantarum* 18: 945.

———. 1967. *Physiol. Plantarum* 20: 562.

Naqvi, S. M.; Dedolph, R. R.; and Gordon, S. A. 1965. *Plant Physiol.* 40: 966.

Nemec, B. 1900. *Ber. Deut. Botan. Ges.* 18: 241.

———. 1902. *Ber. Deut. Botan. Ges.* 20: 339.

Newman, I. A. 1963. *Australian J. Biol. Sci.* 16: 629.

Pickard, B., and Thimann, K. 1963. *Plant Physiol.* 38: 214.

———. 1966. *J. Gen. Physiol.* 49: 1065.

Ramshorn, K. 1934. *Planta* 22: 737.

Rawitscher, F. 1932. *Der Geotropismus der Pflanzen.* Jena.

Schrank, A. R. 1947. *Bioelectric Fields and Growth*, ed. E. J. Lund. Austin: Univ. Texas Press.

Went, F. W. 1932. *Jahrb. Wiss. Bot.* 76: 528.

Wilkins, M. B. 1966. *Ann. Rev. Plant Physiol.* 17: 379.

Wilkins, M. B., and Woodcock, A. E. R. 1965. *Nature* 208: 990.

Discussion

WESTING: From the data you have presented, Dr. Hertz, I note that auxin-treated, decapitated coleoptiles developed a lesser transverse electrical potential when placed horizontally than did untreated, intact coleoptiles. Does this not suggest that some additional tip-produced electrolyte acts together with auxin to produce the observed geoelectric effect?

HERTZ: We have not investigated this in detail since the differences did not appear to be large.

WESTING: The coleoptile experiments you have described involved very brief horizontal presentations which produced a geoelectric effect that damped out after about 20 min. What would be the geoelectric effect with chronic horizontal placement?

HERTZ: The geoelectric potential stays and decays very slowly.

WILKINS: Using the flowing solution electrodes and a single coleoptile (a row of coleoptiles was used by Hertz), we find a geoelectric potential difference of about 10 mV. The kinetics of our responses were essentially similar to yours.

HERTZ: With flowing liquid electrodes there might develop a short circuit, at least to some extent, across the surface of the plant because of the humid surface. This may explain the differences. Further, the size of the drops touching the plant is appreciable compared with the dimensions of the plant.

KALDEWEY: Did you examine different concentrations of applied auxin, and if so, what was the optimum concentration? And how does this correlate with the growth response curve?

HERTZ: We have not tried different auxin concentrations.

WEIS-FOGH: When you have an electrical asymmetry developing, as you have here, could this possibly reflect an electrical asymmetry across the surface structures? It probably reflects an asymmetry also inside the tissue, and it may mean that the auxin does affect the permeability of the cell membrane. It would be of interest to apply, either externally or internally, auxin or other growth substances to large cells like Nitella to see if they cause changes in ionic permeability.

WILKINS: It is certainly true that indoleacetic acid does alter the permeability of the membranes of plant cells to ions. We have in fact done experiments very similar to the ones you describe in which vertical tissue is decapitated and then supplied asymmetrically, not with auxin, but with some of the diuretics known to affect the permeability of animal cell membranes to ions. These induce a lateral electric effect across the coleoptiles in a similar manner to the application of IAA. Our tentative view is the GEE must be due to differential leakage of ions in the cells of the two halves of the horizontal coleoptile. But of course we do not know which ions are involved. The GEE has nothing whatever to do with growth; it results from an auxin concentration gradient, not growth.

ETHERTON: Some time ago Higinbotham did some work with radioactive rubidium uptake under auxin application and got increased uptake of rubidium. Whether or not that was net uptake or increased leakage, I don't know.

GALSTON: If I remember correctly there is also a potential detectable between the tip and the base of an Avena coleoptile. Since you've shown that the appearance of a transverse potential depends on the presence of auxin, can you say anything about whether that longitudinal potential disappears if you decapitate? You would expect it if it is the same mechanism, wouldn't you?

HERTZ: I cannot really answer your question since we have not made extensive experiments here, except that there is a change with decapitation.

SCHRANK: In the 1940s we found that when you decapitate oat coleoptiles, there is a drop in potential which lasts for 5-10 min. It then returns to the original longitudinal voltage.

GORDON: But this is appreciably before physiological regeneration (resumption of auxin production by the stump) sets in.

GALSTON: So there is a persistance of the longitudinal potential under conditions of auxin deprivation. This would mean that there are different mechanisms involved in the production of the longitudinal and transverse potentials, since the latter is dependent on an auxin gradient.

GORDON: Yet a functional (transverse) auxin gradient, does not necessarily produce an electrical gradient. In our study to which Dr. Hertz referred, we measured the transverse potential across the tip of a horizontally oriented coleoptile that was under nitrogen. No potential developed, even though geotropism occurred. Presumably the tropism stemmed from a transverse gradient of auxin.

WILKINS: In a paper (*Plant Physiol.* 42(1967): 1111-13) in which Mr. Shaw and I reexamined this question, we put corn coleoptiles on their side under anaerobic conditions for 15 hr and there was absolutely no curvature. As soon as we readmitted oxygen the coleoptiles began to bend upwards. Therefore under anaerobic conditions we get neither curvature nor geoelectric effect.

GORDON: I accede that the geotropism we observed probably did take place because of an incomplete anaerobioses. But we still face the experimental observation that *no* electropotential developed across the tip in the same coleoptile. To me this implies that the GEE is not a mandatory cognate of either an auxin gradient or, indeed, of geotropism itself.

15
A Vibrating-Reed Electrometer for the Measurement of Transverse and Longitudinal Potential Differences in Plants

Lennart Grahm, *Lund Institute of Technology*

The direct method for measuring bioelectric potentials in plants is the use of contact electrodes. Various types have been employed. Brauner (1927) and Schrank (1947) used liquid protruding from a small glass tube. Bose (1907) and Jantsch (1959) used metal electrodes, and Clark (1937) used a modified type of liquid contact electrodes containing gelatin in the fluid. The drawbacks of liquid electrodes have been discussed by Wartenberg (1957), McAulay and Scott (1954), and Grahm and Hertz (1962), among others. These authors point out, however, that electrode effects tend to mask the real bioelectric potentials. Electrode effects can also arise from solution of some plant substance in the electrode fluid. Moreover, with liquid electrodes there is the risk of short circuits being established across the surface of the plant.

In later years, modifications of fluid electrodes which avoid some of the drawbacks mentioned have been used by Newman (1963), Parkinson and Banbury (1966), and Wilkins and Woodcock (1965). Some disadvantages still exist, however, particularly the possibility of short-circuiting.

These problems can be solved with the so-called vibrating-reed technique, which enables measurements to be made without touching the plant. This principle makes use of the fact that an isolated capacitor which is charged to a certain potential difference between its plates, and whose capacitance is periodically varied by mechanical vibration, produces an ac signal of the vibration frequency. The amplitude of this ac signal is proportional to the dc voltage between the plates. In such a capacitor (Hertz 1960) one of the plates can be replaced by a row of coleoptiles (see fig. 2 in chap. 14, this volume), the bases of which are held at a certain potential. Any dc change between the base and the surface of the coleoptiles facing the vibrating electrode can be recorded. At first this was done manually by changing the potential at the base of the plants, but soon it was found that it was easier to do it electronically. By using phase detector techniques the compensating voltage could be generated and recorded automatically (Grahm and Hertz 1962).

The apparatus described still has the disadvantage that the measurement is made between the base and the coleoptile surface, which means that the longitudinal voltage is included in the measurement. The longitudinal voltage changes markedly during the first hours after mounting the plants, and also shifts when the plants are placed in a horizontal position.

This disadvantage is avoided in the apparatus shown in figure 1 (see Grahm 1964). In this the plants are surrounded by a U-shaped vibrating electrode. The function of the apparatus can be described in the following way: Assume that the potential at one side of the plants is V_a and the potential at the other side is V_b. The vibrating electrode is at all times close to ground potential. If there is no potential difference between the two sides of the plants (no transverse voltage) but there is a potential difference between the two plates and the vibrating electrode (a longitudinal voltage), we should expect an ac voltage at the isolated electrode as in the apparatus described earlier. But in the present version we have two connected plates, one at each side of the plants. This means that during one cycle of vibration there is a voltage variation which has twice the frequency of vibration, since we cannot tell from the ac voltage whether the upper side of the electrode is close to the upper side of the plants or the lower side is close to the lower side of the plants. Thus a longitudinal voltage gives a signal of double the vibration frequency.

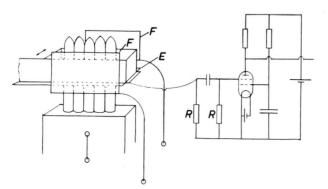

Fig. 1
Vibrating electrode and first amplifier stage in the later, modified version of the electrometer for measurement of the geoelectric effect. E = vibrating electrode; F = feedback plates; R = grid resistance.

Now assume that the longitudinal voltage is zero: i.e., $1/2\ (V_a + V_b) = 0$, but that there exists a transverse voltage, which means that one side of the plants is positive with respect to the vibrating electrode and the other is negative. This would result in an alternating voltage on the isolated vibrating electrode, of a frequency equal to the frequency of vibration; when the positive side of the plants comes close to the electrode there will be a potential on the latter opposite in sign to that produced when the negative side approaches.

Thus, there will be two signals—one of the vibration frequency and the other of double the vibration frequency. The amplitude of the signal with vibration frequency is proportional to the transverse voltage, and the signal with the double frequency has an amplitude proportional to the longitudinal voltage.

As in the earlier apparatus the longitudinal voltage can be compensated for and recorded by phase detector technique. Also, a feedback system was developed for the transverse voltage so that differences in amplification, vibration amplitude, and so on do not affect the results. The feedback system for the transverse voltage consists of two metal plates symmetrically placed with reference to the vibrating electrode and fed with symmetrical voltages derived from a phase detector, working on the vibration frequency (Grahm 1964).

With this apparatus it could be shown that the geoelectric effect is a secondary manifestation of the asymmetric distribution of auxin. This was shown by a comparison of the transverse voltages developed *a)* upon geotropic stimulation; *b)* after vertical, decapitated coleoptiles had been supplied with auxin on half of the stump; and *c)* when, in vertical coleoptiles longitudinal transport of auxin was blocked in half the coleoptile by insertion of a mica sheet. In all these instances similar potentials arise when the measurements are made at the tip. If the measurements are made farther to the base, the latent time to the rise of the potential is almost unchanged for the geoelectric effect, whereas there is an increase in the latent time for the two other types of potentials (fig. 2). The delay corresponds closely to the time for transport of auxin observed by others (e.g., Newman 1959). This may be interpreted as an indication of a transverse transport of auxin along the whole coleoptile *at the same time*; this suggests in turn a perception of the geotropic stimulus along the whole length of the coleoptile.

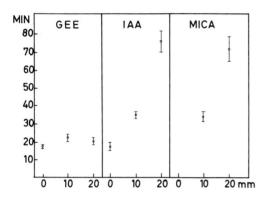

Fig. 2
Left: Latent period for geoelectric effect as a function of the distance from the tip. *Middle*: Latent period for IAA-induced potentials as a function of distance from the section surface. *Right*: Latent period of mica-induced potentials as a function of the distance from the mica plate.

We have described this apparatus, based on the vibrating-reed principle, as useful for the simultaneous measurement of transverse and longitudinal bioelectric potentials without direct contact with the tissues in which geosensing and its physiological sequelae occur. Its use has been extended to the measurement of presentation times (Johnsson 1965b) and to the investigation of the transverse voltages caused by unilateral illumination of the coleoptile (Johnsson 1965a, 1967).

References

Bose, J. C. 1907. *Comparative electrophysiology*. London.

Brauner, L. 1927. *Jahrb. wiss. Bot.* 66: 381-428.

Clark, W. G. 1937. *Plant Physiol.* 12: 409-40.

Grahm, L. 1964. *Physiol. Plantarum* 17: 231-61.

Grahm, L., and Hertz, C. H. 1962. *Physiol. Plantarum* 15: 96-114.

Hertz, C. H. 1960. *Nature* 187: 320-21.

Jantsch, B. 1959. *Zeitschr. Bot.* 47: 336-72.

Johnsson, A. 1965a. *Physiol. Plantarum* 18: 574-76.

———. 1965b. *Physiol. Plantarum* 18: 945-67.

———. 1967. *Physiol. Plantarum* 20: 562-79.

McAulay, A. L., and Scott, B. I. H. 1954. *Nature* 174: 924.

Newman, A. 1959. *Nature* 184: 1728-29.

———. 1963. *Austral. J. Biol. Sci.* 16: 629-46.

Parkinson, K. J., and Banbury, G. H. 1966. *J. Exp. Bot.* 17: 297-308.

Schrank, A. R. 1947. *Bioelectric fields and growth*, ed. E. J. Lund. Austin: University of Texas Press.

Wartenberg, H. 1957. *Ber. dtsch. bot. Ges.* 70: 10-11.

Wilkins, M. B., and Woodcock, A. E. R. 1965. *Nature* 208: 990-92.

16
Geotropic Curvature of Avena Coleoptiles as Affected by Exogenous Auxins

A. R. Schrank, *University of Texas*

Results of recent studies as presented by Hertz (chap. 14) and by Grahm (chap. 15) in this symposium indicate, contrary to earlier observations, that the geoelectric effect in several kinds of coleoptiles starts to develop about 15 min after the beginning of geotropic stimulation. One facet of these studies relates to certain electrical changes caused by externally applied auxin. Both Grahm (1964) and Wilkins and Woodcock (1965) reported that electrical effects caused by indirect asymmetrical auxin application were similar to the geoelectrical effects in magnitude and latency time (see also Ramshorn 1934 and Newman 1963). These observations led to the conclusion that the geoelectric effect is a secondary response arising from the action of auxin. The data I will present reveal some effects of externally applied indole-3-acetic acid (IAA) on elongation and geotropic bending of Avena coleoptiles which relate to the auxin-induced electrical responses.

The results shown in figures 1 and 2 were obtained using 15-mm apical segments of Avena coleoptiles with the apical cells intact. The seedlings used were grown according to the regimen

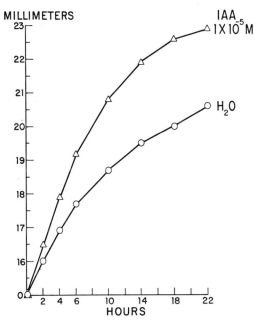

Fig. 1
The effect of 10^{-5} M IAA on the elongation of 15-mm apical Avena segments. The IAA was applied to the intact apices via short pieces of small plastic tubing. Each point represents the average of 20 or more values.

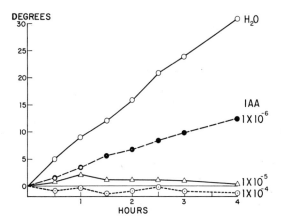

Fig. 2
The effect of the indicated concentrations of IAA on geotropic bending of 15-mm apical Avena segments. The IAA was applied to the intact apices via short pieces of small plastic tubing. Each point represents the average of 20 or more values.

published by Wiegand and Schrank (1959). Other experimental conditions were the same as those described by Schrank (1960).

The 15-mm apical segments were isolated from the seedlings and the primary leaves were removed. After the apices were dipped in acetone for 1 sec and then rinsed in distilled water, a short piece of plastic tubing was fitted to the apex of each coleoptile. Test solutions were applied to the coleoptiles via these tubes. The coleoptiles were held on stainless steel needles so that the basal ends were in contact with aerated distilled water. After a 2-hr rest period, the growth or curvature measurements were started.

The curves in figure 1 show that IAA applied at a concentration of 10^{-5} M mediated a marked increase in elongation. IAA at 10^{-4} M (not shown in the figure) caused essentially the same magnitude of stimulation. Lower concentrations, 10^{-6} and 10^{-7} M, also increased elongation, but not as effectively as the higher concentrations.

Figure 2 shows the effects of several concentrations of IAA on geotropic curvature of similar apical segments. It should be emphasized that concentrations as low as 10^{-6} M strongly inhibit upward bending, and that the data in these two figures have been duplicated by experiments in which the coleoptile segments were completely submerged in solutions of IAA in aerated distilled water.

Several papers in the literature reveal that relatively low concentrations of auxins other than IAA also cause rather extensive inhibition of geotropic bending when applied to the cut basal ends of comparable Avena coleoptile segments. Such compounds as 2, 3, 6-tricholorobenzoic acid (Schrank 1960), n-1-naphthylphthalamic acid (Tsou Ching, Hamilton, and Bandurski 1956), and 4-amino 3, 5, 6-trichloropicolinic acid (Schrank 1968) are effective.

The electrical responses to externally applied IAA observed by the investigators previously cited (Ramshorn 1934, Newman 1963, Grahm 1964, and Wilkins and Woodcock 1965) resulted from IAA concentrations ranging from 10^{-5} M to 0.02%, applied in various media. In some of Newman's experiments lower concentrations were applied, but the magnitude of the electrical changes observed was about the same as the electrical "noise" on the coleoptile.

Since the curves in figure 2 show an extensive inhibition of geotropic curvature by 10^{-6} M IAA, a concentration which apparently is considerably lower than the concentrations used to induce the electrical effects, it appears somewhat premature, on the basis of this type of evidence, to relate these auxin-induced electrical responses to the geoelectric effect.

Acknowledgments

The data presented are from experiments supported by National Science Foundation grant GB-5683.

References

Grahm, L. 1964. *Physiol. Plantarum* 17: 231-61.

Newman, I. A. 1963. *Austral. J. Biol. Sci.* 16: 629-46.

Ramshorn, K. 1934. *Planta* 22: 737-66.

Schrank, A. R. 1960. *Plant Physiol.* 35: 735-41.

———. 1968. *Physiol. Plantarum* 21: 314-22.

Tsou Ching, T.-M.; Hamilton, R. H.; and Bandurski, R. S. 1956. *Physiol. Plantarum* 9: 546-58.

Wiegand, O. F., and Schrank, A. R. 1959. *Bot. Gaz.* 121: 106-10.

Wilkins, M. B., and Woodcock, A. E. R. 1965. *Nature* 208: 990-92.

IV Invertebrate Responses to Gravity

17
Stabilizing Mechanisms in Insect Flight

Donald M. Wilson, *Stanford University*

Insects do not possess specialized gravity receptors analogous to the labyrinth in vertebrates or to the statocysts of various other invertebrates. Instead they use relatively unspecialized proprioceptors to measure the relationship between body parts which may be differentially affected by gravity (see chaps. 19, 20, this volume). Orientation to the gravitational field may also be achieved without any measurement of forces related to the earth's gravitational attraction. For example, the horizon is nearly always related in the same way to the gravitational force; hence visual input can provide cues for the latter. Aside from their use in orientation to the earth's field, proprioceptors or exteroceptors which measure the velocity or accelerations of the whole animal in any angular or translational direction may be of importance in steering insect flight. In this chapter I will discuss some of the known reflex mechanisms by which insects maintain stable flight and show to some extent how these reflexes interact with the preprogrammed flight command.

Reflex Systems

Neck Proprioceptors

The most nearly pure gravity receptor organ in insects consists of the neck proprioceptors and the head in dragonflies (Mittelstaedt 1950). The head is suspended from the thorax on a loose universal joint so that over fairly wide ranges of position of the main part of the body the head acts as a plumb bob. Rotation of the head relative to the body causes asymmetrical bending of sensory hairs in the neck region. In quiet animals asymmetrical input from these sensory hairs results in differential twisting of the wings, such as would correct a rolling motion during gliding. The same reflexes operate effectively during flapping flight. Furthermore, the head position relative to the body may be controlled by neck muscles. A dorsal light reaction is present which utilizes these muscles to rotate the head so that the greatest illumination always falls on the upper part of the eyes. When the head is rotated during this reaction, the neck proprioceptors are asymmetrically stimulated and changed wing movements or postures result in a rotation of the whole animal. In most common situations the dorsal light reaction and the gravity responses mediated by the head in dragonflies will be more or less synergistic. Other insects with loose heads, such as flies, may use similar orientation mechanisms.

Homologous neck proprioceptors in other insects may play a similar role with respect to motor output during flight, but respond to different input modalities. For example, although Thorson (personal communication) found yawing torque in flying locusts not to be very reliably related to horizontally moving visual stimuli, he found activity in the neck muscles and resulting head torque to be precisely regulated by an optomotor reaction. Thus, a moving visual field is followed by the head, the head stimulates neck-hair sensillae, and a reflex relationship between these sensillae and the wing posture can bring the whole animal to follow the visual stimulus. Notice that this organization is similar to that in mammals which connects labyrinthine input to limb motor output via a chain of reflexes involving neck proprioceptors.

The optomotor reaction has been shown in various species to correct both yawing and rolling errors, and probably helps to control pitch as well.

Wind Direction

Another kind of visual orientation in flight has been studied by Goodman (1965). She found that locusts will orient with respect to the roll axis not only so that the dorsum is pointed toward a light source, but also so that a simulated horizon edge (light above and dark below) is kept perpendicular to the dorsoventral axis.

Another source of input helping to regulate flight orientation in insects is a sense of the direction of the wind passing over the head. This sense is again mediated by sensory hairs (Weis-Fogh 1949). Stimulation of these hairs has a nonspecific excitatory effect on the flight control centers of the nervous system, and, in that regard, these sensillae compose part of a positive feedback loop which tends to keep flight going. A puff of air on the head will start flight; flight itself results in a continued wind which maintains flight. The individual hairs and beds of hairs are not equally sensitive to different wind directions (Camhi, unpublished). Wind from right or left causes changes in motor output to the flight muscles which produce wing motions that give a yawing torque toward the wind (Dugard 1967; Waldron 1967). Like the optomotor reaction, this turning reflex helps the animal to maintain a straight flight path, but in this case relative to the air mass. The wing reflexes regulated by wind direction can help to correct transient deviations with respect to gravity orientation. They cannot aid in correcting a very slow drift. Together with some inherent aerodynamic stability of the insect, they might provide a sufficient system for gravitational orientation.

Campaniform Sensillae

There are also proprioceptors, campaniform sensillae in this case, within the wing veins themselves. These are arranged in part, at least, to measure the lifting force. One reflex function of these receptors is involvement in the "constant lift reaction" of flying locusts (Gettrup and Wilson 1964). Weis-Fogh (1956) found that the lift produced by a steadily flying locust was independent of the angle of the body in the pitching plane over a substantial range. He postulated that the wing twisting varied with body angle in such a way that the angle of attack of the wings was relatively invariant. Variation in wing-twist was confirmed by Wilson and Weis-Fogh (1962), but Gettrup and Wilson (1964) found that the regulation of wing-twist in the constant lift reaction was apparent in the fore wings only. A conclusion from this finding was that not only overall lift but also body pitch itself tends to be maintained, since regulation of the lift of only one wing pair gives rise to torque around the pitching axis. Gettrup found that ablation of the campaniform sensillae of the hind wings eliminated the "constant lift reaction," whereas damage to those of the fore wings had different, independent effects. Hence sensillae in the hind wings (which produce most of the aerodynamic power) measure the lift and affect the angle of attack of the fore wings (which are the main control surfaces). The fore wings and hind wings are reflexively integrated. They are probably also aerodynamically integrated (Jensen 1956). The hind wings move through the wake of the fore wings. This wake is shaped by the fore wing posture, and the lift of the hind wings is thus under fore wing control. It is as if the two wings were one with a control surface at the leading edge. According to Gettrup's findings the campaniform sensillae of locust fore wings do not play a role in the lift control phenomenon (which may be the same as pitch control), but they are involved in control of stability in the roll and yaw planes.

Halteres

Flies have no hind wings. Instead they have evolved special organs that resemble tiny gyroscopes. These halteres have sets of sensillae which are undoubtedly homologous to those on the wings of other insects. The halteres are also activated by muscles. During flight the two halteres vibrate synchronously at the wingbeat frequency (Schneider 1953). Rotations of the whole animal give rise to forces in the sensory fields at the base of each haltere which could in principle provide information about the direction and rate of rotation. If the halteres are removed, flies cannot perform stable flight (Fraenkel 1939). Pringle (1948) showed that the halteres are responsible for stabilization in the yawing plane and Faust (1952) gave evidence that they are also involved in control of pitch and roll. The control is via reflexes and not a direct gyroscopic stabilization (Pringle 1948).

If only one haltere is removed, flight is relatively unaffected unless the fly is also blinded (Faust 1952). Pringle (1957) argues that each haltere can provide information about only two axes of rotation, but the combination of input from the two halteres can be used to resolve any rotation into its three components. Apparently, in blinded flies or in flight in the dark, loss of one haltere results in confusion between responses to roll and pitch. Although flies have evolved very special sensory structures for flight stabilization, they still make use of at least one other input, the visual, for this purpose.

The functions of the halteres must have evolved from the more primitive sensory functions of the hind wings, such as the fore wing twist-control reflex in locusts. Notice that in both locust and fly, sensillae of the hind wing (haltere in fly) regulate activity of fore wing muscles. This condition of the more primitive locust already makes haltere evolution permissible. Although this phylogenetic story makes sense, it is certainly not the whole truth. There are other relatively little-known insects in which the *fore wings* have evolved into halteres.

In the above discussion several cases of synergism between different controlling reflexes have been mentioned. In addition to reflex control mechanisms, insect flight is programmed by purely central nervous processes. In the last section of this chapter I wish to show how some reflexes interact with and complement the central mechanisms.

The Central Program

I have already discussed some of the reflexes controlling locust flight. At one time it was supposed that the reflexes constituted the whole control system (Weis-Fogh 1956; Pringle 1957); that is, each wing motion was elicited by a specific pattern of prior sensory input, especially input from the wings themselves. However, one can show that this is not so by surgically eliminating reflex feedback from the wings (Wilson 1961). In fact, even random stimulation of the isolated nerve cord can release a motor output pattern to the flight muscles essentially like that during flight (Wilson and Wyman 1965). The locust thoracic ganglia have built into them an inherent motor score which can play out the flight command whenever it is sufficiently excited, without any special pattern of input. It seems probable that flight is an inherited behavior, and that the central motor score for flight is genetically coded. This hypothesis is very difficult to prove, but for the present I will assume that it is true. If the motor command system can be preprogrammed into the thoracic ganglia, why are there reflexes at all? Clearly there are kinds of information which cannot be anticipated genetically, such as the current air temperature or wind direction. My present feeling about the locust flight-control system is that all the needed information which can be is genetically coded, and that the reflexes supply only that information which must come from current sources.

The best understood reflex in the locust flight system fits this notion. The completely deafferented thoracic nervous system always "flies" too slowly. Four stretch receptors, one at each wing hinge, supply an excitatory feedback which keeps the wingbeat frequency up to suitable values (Wilson and Gettrup 1963). Each locust must fly with a wingbeat frequency which is a function of its size. Since the thoracic skeleton, muscles, and wings form a mechanically resonant system, operation at frequencies other than the natural one would be energetically more costly. But each animal cannot know its own size genetically; size varies with developmental conditions. The stretch reflex provides an input which can supply the missing information about the correct wingbeat frequency for each individual. (For a discussion of this reflex function see Wilson 1967.)

Hoyle (1964) has suggested that genetically controlled behavior might be programmed in the central nervous system in two quite different ways. One way is to structure the nerve networks so that their own interactions give rise to the appropriate output pattern whenever they are sufficiently excited. Such structuring seems to be the case for the flight *motor score* in locusts. The alternative to the motor score suggested by Hoyle is the central storage of a *sensory template* to which feedback following motor activity is compared. If the feedback does not match the template the output is changed. In a sense it is information about the goal of the motor activity which is stored rather than a program for the correct motor activity. Storage of a sensory template seems to occur in one instance of development of bird song (Nottebohm 1967). Control by a sensory template and sensory feedback could also be true for such activities as walking in insects. In walking there is a good deal of adaptation of the limb movements to surface irregularities, limb loss, and other conditions which cannot be genetically anticipated. In fact, there must be heavy reliance upon proprioceptive reflexes in the coordination of arthropod legs during walking. It has long been known that amputation of some legs on an insect, spider, or crab results in a new pattern of movements by the remaining ones, and that the new pattern is adaptive. This phenomenon, called *Plasticität* by Bethe (1930), has been used as the basis for an argument that the whole coordination is no more than a set of integrated reflexes.

The locust central nervous system contains a built-in motor score for flight. Does it therefore lack plasticity in the sense of Bethe? If it does not, then the concepts *motor score* and *sensory template* may not refer to real nervous arrangements, but only to idealizations. Several years ago I made an abortive attempt to demonstrate plasticity in the locust flight control system by recording from certain muscles while the animal flew in front of a wind tunnel, both before and after the ablation of other mucscles or wings. There were no effects of surgery on motor pattern of the order of magnitude of those I expected, and I concluded that the flight control system was very unplastic indeed. Of course, we already knew about the lift control reaction and other minor adjustments which are accomplished reflexively, but these do not seem comparable to the adjustments necessary when a cockroach loses two legs, for example.

A flaw in these experiments was that the locust had *only* the proprioceptive feedbacks to tell it that something was wrong. Because the animal was held still in space both the optomotor and wind-direction feedback loops were opened. Recently I decided that it was foolish to believe that the flight system is not more adaptable than the early recordings had indicated. One finds in nature animals with significant asymmetrical wing damage, yet they do not fly in circles. I began to fly animals untethered in a large room and to look for the effects of unilateral ablation of important control muscles such as the basalar and subalar muscles which regulate the angle of attack. The animals could still fly straight. After removal of a whole fore wing or even a whole hind wing, straight, level flight was possible, although it was much slower than normal. The slowness is not surprising. One hind wing produces about 35% of the total aerodynamic power (Jensen 1956). It *is* surprising that the animals could fly straight, and indeed on many trials they did tailspins. If there had been no adaptation of motor pattern one would have expected a power asymmetry of around 3:1. The adaptation which did occur seems to me to be equal in magnitude to that encountered in the walking systems which supposedly are reflexively run or based on a sensory template comparison. Yet the locust is known to have a built-in motor score.

What sources of input do produce the adaptive changes in motor output which are necessary for stable flight after wing or muscle ablation? The proprioceptors of the wings may help. In the tests on tethered animals the proprioceptive feedback was present, but may have been abnormal, since the animals were not free to roll, yaw, or sideslip (they did have the other three degrees of freedom). Neither vision nor wind-direction sense could give them any information about motor errors. It has been reported already that blinded, but otherwise intact, locusts can fly stably (Gettrup 1966). So can those with the head hairs shaved off or painted over. Animals with the supraesophageal ganglia (brain) cauterized cannot (Weis-Fogh 1956). This last operation may do more than just eliminate sensory inputs which impinge on the head, so the result does not imply that at least one of these inputs is necessary.

In order to sort out the influences of these inputs on normal flight and flight in asymmetrically damaged locusts I flew animals free in a moderately large room which could be illuminated with either white or infrared light. This allowed reversible blinding. The animals gave no visual reactions that I could detect in the infrared illumination. The head hairs were immobilized with a brittle low-melting wax which could be broken away. The wax could be applied and removed several times before noticeable malfunction of the wind responses occurred. Single hind wings were tied down, or a hind wing or fore wing was cut off. Tying or cutting the hind wing seemed to give the same results.

In the animals with all wings intact, blocking either exteroceptive input may result in somewhat weaker flight, but not in unstable flight. Blocking both vision and wind sense may result in instability in animals which were weak flyers anyway, but not in good flyers. The built-in motor score, built-in aerodynamic stability, and proprioceptive reflexes seem to be sufficient for stable flight in structurally intact animals.

In animals with one wing incapacitated, blocking either exteroceptive input resulted in considerable additional deficiency in flight ability, but still on many tests the flight was stable, if weak. Blocking both always resulted in very poor flight, and usually in such apparent discoordination that the behavior could hardly be called flight at all. Yet in two or three out of hundreds of trials on about 20 animals there were suggestions of stable behavior even with both feedbacks blocked. These few cases may be only happenstance, or they may indicate that proprioceptive feedback can mediate sufficient corrections under relatively favorable, rare circumstances.

Of the three types of inputs, optomotor, wind direction, and proprioceptive, I cannot judge which is more important for steering in asymmetrically damaged animals. Clearly all are used and they complement each other. But they, by themselves, are not the basis of organization of the flight control system. Rather, they interact with a central nervous program both by contributing to its overall state of excitation and by modulating details of its output pattern when such modulation is necessitated by organismal or environmental irregularities.

Acknowledgments

The original work reported here was supported by National Institutes of Health grant NB 07631 and Air Force grant AFOSR 1246-67.

References

Bethe, A. 1930. *Arch. Ges. Physiol.* 224: 793-820.

Dugard, J. J. 1967. *J. Insect Physiol.* 13: 1055-63.

Faust, R. 1952. *Zool. Jb., Allg. Zool. Physiol.* 63: 325-66.

Fraenkel, G. 1939. *Proc. Zool. Soc. Lond., Ser. A* 109: 69-78.

Gettrup, E. 1966. *J. Exp. Biol.* 44: 1-16.

Gettrup, E., and Wilson, D. M. 1964. *J. Exp. Biol.* 41: 183-90.

Goodman, L. J. 1965. *J. Exp. Biol.* 42: 385-407.

Hoyle, G. 1964. In *Neural theory and modeling*, ed. R. Reiss, pp. 346-76. Stanford, Calif: Stanford Univ. Press.

Jensen, M. 1956. *Phil. Trans. Roy. Soc. London, Ser. B* 239: 511-52.

Mittelstaedt, H. 1950. *Z. Vergleich. Physiol.* 32: 422-63.

Nottebohm, F. 1967. *Proc. XIV Intern. Ornith. Congr.*, ed. D. W. Snow, pp. 265-80. Oxford: Blackwell Publ.

Pringle, J. W. S. 1948. *Phil. Trans. Roy. Soc., Ser. B* 233: 347.

———. 1957. *Insect flight*. Cambridge: Cambridge Univ. Press.

Schneider, G. 1953. *Z. Vergleich. Physiol.* 35: 416-58.

Waldron, I. 1967. *J. Exp. Biol.* in press.

Weis-Fogh, T. 1949. *Nature* 164: 873-74.

———. 1956. *Phil. Trans. Roy. Soc., Ser. B* 239: 553-84.

Wilson, D. M. 1961. *J. Exp. Biol.* 38: 471-90.

———. 1967. In *Invertebrate nervous systems*, ed. C. A. G. Wiersma, pp. 219-29. Chicago: Univ. of Chicago Press.

Wilson, D. M., and Gettrup, E. 1963. *J. Exp. Biol.* 40: 171-85.

Wilson, D. M., and Weis-Fogh, T. 1962. *J. Exp. Biol.* 39: 643-67.

Wilson, D. M., and Wyman, R. J. 1965. *Biophys. J.* 5: 121-43.

Discussion

GALSTON: Is it fair to say that in your analysis of insect stability the same kind of rules pertain as would apply to an airplane model in a wind tunnel, or do factors come into play here which are not subsumed under the usual rules that aerodynamics engineers talk about? I ask this question because of the "folklore" that says aerodynamics engineers predicted the bumblebee couldn't fly, and yet it flies quite well.

WILSON: The fly has a pilot, so to speak, but the airplane doesn't need a pilot for many of the things it does because it is built to be aerodynamically stable. You can send a glider aloft, and it would traverse long distances relative to what we think a fly could do if it didn't have the feedback involving some kind of piloting mechanism.

WEIS-FOGH: Aerodynamically the bee may still fly on the basis of steady-state aerodynamics. The gist of it is that some insects have fairly small wing areas and their flight can't be understood on the basis of steady-state aerodynamics with fixed wings. But the kinematics is so intricate that it often defeats the first naive impression. The kinematics, that is, the actual movement of the wings through the air, sometimes indicates that normal aerodynamic forces cannot explain the entire performance.

COHEN: What you are saying is that the engineer's analysis of the aerodynamic situation was imperfect.

WEIS-FOGH: It was too simplified. At least one thing can be said here. There may be acceleration phenomena which we do not understand and which in certain medium-sized insects could have an important action, say account for 10 or 20% of the lift actually produced. The more we understand about the system the more comprehensible it is from the point of view of steady-state aerodynamics.

BROWN: You aren't going to say that the bumblebee is really well-designed, are you?

WEIS-FOGH: Yes, of course it is.

18
Flying Insects and Gravity

Torkel Weis-Fogh, *University of Cambridge*

In the preceding chapter, Dr. D. M. Wilson discussed some of the nervous mechanisms responsible for stable flight of insects under normal conditions of gravity. He also emphasized that there is an intricate interplay between inherent stability and operational or reflex stability. However, it is true that we have very little factual knowledge of stability problems of insects and birds in aerodynamic and mechanical terms. Recently, Pringle (1968) has attempted to collect the scattered information and has concluded that some insect groups seem to be inherently stable in flight, but this deduction depends on a string of assumptions concerning the details of the aerodynamic action of the wings. The simple fact remains that detailed information about how wind forces vary and interact during the wing stroke is available only for a single insect, the desert locust *Schistocerca gregaria* (Jensen 1956). According to Jensen's study, the flight of locusts can be analyzed and understood as a sequence of steady-state aerodynamic situations, and less than 10% of the summed lift and thrust, if any, may be caused by nonsteady effects. In a critical review we have pointed out that this is also likely to be true for birds and large insects as well as for medium-sized insects which depend on flapping flight rather than on gliding and soaring (Weis-Fogh and Jensen 1956). In insects like bees and flies nonsteady phenomena may play a relatively greater role, but this has never been proved and their average performance is compatible with ordinary steady-state aerodynamics. When the size decreases, the dimensionless number which characterizes the flow pattern over the wings also decreases; Reynolds number $Re = \rho vd/\eta$ where ρ is mass density, v is flow velocity, d is a representative length, and η is the viscosity. In a flying locust Re is about 2,000, but in very small insects like fruit flies (*Drosophila*), small parasitic wasps, midges, and thrips it is about 100 or less (Horridge 1956; Vogel 1967b). A suitably designed wing will act as an ordinary airfoil down to a value of about 100; that is, the coefficient of *drag* C_D will be many times smaller than the coefficient C_L for the aerodynamic cross-force, i. e. the *lift*. The ratio C_L/C_D is a measure of the fitness of the airfoil and amounts to about 15-20 for a good profile. When Re is reduced to 100 the ratio also decreases, and when it reaches about 20 no vortices can be formed, the drag is the dominant force, and the orientation of the wing is less important. On the other hand, at high Reynolds numbers where C_L/C_D is also high, the flight depends on a very accurate adjustment of the angles of attack (the angle between the relative wind and a chord in the wing profile). These considerations are important when we try to judge the effect of gravity on flying animals, as we shall now do.

Principle of Flapping Flight

Flying animals use oscillating wings to produce both lift and thrust and the main mechanism is apparent from figures 1 and 2. As a rule the downstroke is slower than the upstroke and the main lift and a surplus of thrust are produced during this phase. Relatively little lift and negative thrust are the result of the upstroke (see Jensen 1956; Weis-Fogh and Jensen 1956; Weis-Fogh 1961). In our early experiments with locusts which flew in a wind tunnel there was an indication that some thrust was produced by a backward "rowing" stroke at the beginning of the upstroke, but eventually we found that the effect was due to an inadequate technique of suspending and treating the locust. In fact, the insect tends to place the wings edgewise to the air flow, not as paddles, and to keep the angle of attack as close to zero as possible during the upstroke and about 10° during the downstroke. According to Wohlgemuth (1962), this also seems to be true in freely flying bees, but Nachtigall (1966) has recently claimed that the "rowing" phase is operative in suspended flies (*Phormia regina*). Time will show whether this is so.

In any case, the flight of flapping birds and insects depends on this asymmetry of the two phases of the stroke. Owing to the necessity of compensating for the pull of gravity, the flying machine is highly asymmetrical with respect to the gravitational field. In the majority of insects and probably in all birds, the production of a lifting force which can counteract the pull of gravity depends on a refined adjustment of the angles of attack during the entire wing stroke.

It may be helpful to remember the analogy between the cross-force which an electric wire experiences in a magnetic field and the hydrodynamic cross-force or lift which an airfoil experiences in a wind. In both instances the force acts normal to the direction of flow. For a wing, the direction of flow is determined by the movements of the wing relative to the air, and the main

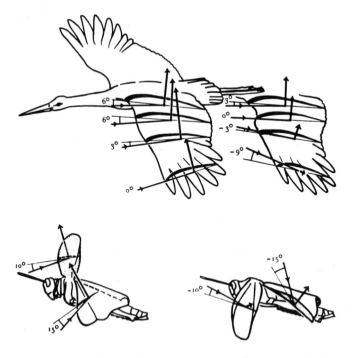

Fig. 1. The wind forces which act on the wings of a stork (*top*) and a locust (*bottom*) when they fly horizontally. On the left side is seen the downstroke and on the right side the upstroke (from Weis-Fogh 1961).

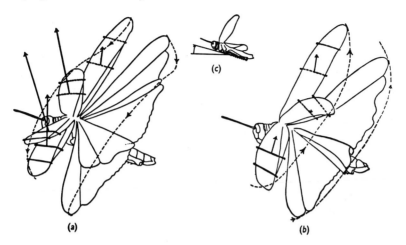

Fig. 2. The wind forces acting on the forewings of a locust in level flight during the downstroke (*left*) and the upstroke (*right*). The broken lines indicate the wing-tip curves mentioned later in the article (from Wilson and Weis-Fogh 1962).

force component is normal to the direction of the relative wind. In flapping flight the relative wind and the cross-force constantly change magnitude and direction, and level horizontal flight is the result of a balance between gravitational pull and the lifting force as averaged over a whole stroke period.

Flight in the "Weightless" State

If, therefore, the gravitational pull is compensated for by a centrifugal force ("weightlessness"), the entire kinematics of the flying animal become upset and it loses all normal possibilities for control. Exactly what happens will depend on the specific construction, but the main events may be imagined as follows:

Consider an insect in level flight in a normal gravitational field. Suppose that we suddenly compensated for this field at the beginning of the downstroke. The insect will then immediately begin to move "upward" in the direction of the lifting force and away from the "horizontal" thrust. This means an abrupt change of the direction of the relative wind so that the angle of attack will change from about +10° to some large negative value; that is, the wings become breaking paddles and not finely adjusted airfoils. If it happens that the animal does not turn around in the pitching plane and loses direction altogether, the chances are that the average tendency over a number of strokes would be for it to move "downward" rather than "upward" because the upstroke is faster than the downstroke. In other words, the animal would be completely confused because few of its normal control reactions would be of any help.

It may be slightly different with hovering species like hummingbirds, syrphid flies, and dragonflies, because they are able to control the wing stroke to an extraordinary extent and they depend mainly on visual cues for their orientation in space. However, it is doubtful whether the control system is sufficiently sophisticated to work when the usual reflex adjustments result in entirely different aerodynamic responses. In principle it should be possible to design a self-optimizing machine where the detector is a gyroscope (as in the halteres of two-winged flies) coupled with the visual system and where the wings are moved so as to bring the animal to the goal irrespective of the absurdity of the aerodynamic situation. It may be possible to conduct an experiment of this kind with a hovering type of bird or insect placed in a spaceship, but it must be large enough to permit free flight. Such an experiment is not likely to give information about the effect of gravity but to tell us whether the nervous and locomotory systems of some animals have developed into an extreme type of goal-seeking, self-optimizing apparatus. In fact, it would be easier and much cheaper to study these problems on the ground.

If we turn to very small insects, Vogel (1966, 1967a, b) has recently shown that *Drosophila* behaves much like a simple actuator disk. Owing to the small Reynolds number, even large changes in the angle of attack do not make "flight" impossible, and insects of this kind will undoubtedly be able to make use of their wings in the "weightless" state; but they will not move in the same direction or with the same speed as under normal conditions.

There is one theoretical and limited exception to these considerations about "weightless flight," since it is conceivable that a bird or an insect which makes use of ordinary airfoil action and known control systems may happen to enter a circular or a spiral path in such a way that the aerodynamic lift is just balanced by the centrifugal force. This means that they move in an orderly way through the air but that they have lost control of where they are going.

The conclusion is that a flying animal which is really adapted to the "weightless state" should be a sort of aerodynamic fish and not a winged creature.

Gravity and Lift Perception

Dr. Wilson has reviewed some of the evidence for a lift-perceiving system in flying locusts. There is no doubt that the aerodynamic lift, or a representative component of the lift, is being perceived by the oscillating wings, since the locust can keep its lift constant by adjusting the wing twist when the pitch is altered by the investigator (Weis-Fogh 1956; Wilson and Weis-Fogh 1962). It is also known that the receptors in question are a group of campaniform sensillae placed on one of the main ribs of the hind wings (Gettrup and Wilson 1964; Gettrup 1965). The main result of the sensory input is that the twist of the forewings becomes adjusted to keep the lift constant, although the entire measured effect cannot be explained only by this mechanism. Gettrup (1966) has recently shown that it is a discrete part of the signal from these sense cells which is essential for the control reaction, namely the firing during the first two-thirds of the downstroke. This makes sense since it corresponds to that part of the stroke during which the main force is produced (Jensen 1956), but it raises a problem. In an oscillating system like that of a locust or a fly, in which the oscillating wings come to a stop at the top and at the bottom, the wing mass is accelerated and decelerated twice during each complete stroke, and I have found that the mass forces involved in most insects are of the same order of magnitude as the aerodynamic lifting forces and are often larger. Since the two force systems have roughly the same direction as the lift, it is obvious that confusion may occur. How is it solved?

As can be seen from a careful inspection of figure 2, the wing tips do not move up and down in one plane but describe a loop. In relation to the animal, the tip moves on the surface of a sphere with the fulcrum in the center. If we define the stroke plane as the great circle which contains the top and the bottom of the stroke, the tips of both pairs of wings move in front of this plane during the downstroke and behind it during the upstroke. In an unpublished study, I have found that the unfolded resting forewings are held elastically in the stroke plane at all angles. In the hindwings, there is no similar elastic equilibrium position and the wings tend to fold back over the body if they are not actively constrained. In both pairs of wings it is therefore possible that the forward movement of the tip during the downstroke is caused by the relatively small component of the aerodynamic lift normal to the stroke plane and that it is this component which is being "measured" rather than the lift as such. Since the mass forces act mainly in the stroke plane, this would solve the problem of confusion. Consequently, Dr. R. Weber and I have analyzed the factors responsible for the wing-tip curve in flying locusts *Schistocerca gregaria* (unpublished). By means of multiple stimulation of the muscles in animals suspended in a wind it was found that the wing-tip curve of the forewings is determined mainly by the action of the muscles responsible for the downstroke. Thus, the basalar muscles pull the tips backward and subalars pull them a bit forward, while the dorsal longitudinal muscles exert a very definite forward pull. The hindwings are pulled forward during the latter part of the upstroke by the action of muscles on the skeletal system, but none of the downstroke muscles are able to pull them forward. Consequently, the forward movement of the wing-tip curve seems to be independent of wind forces in the forewings and to be caused by the wind forces in the hindwings.

To test this conclusion during actual flight, it was essential to monitor the wind forces, particularly the lift, without changing the other factors. This was done by punching small holes in the thin wing membranes, so that the perforated area corresponded to between 3% and 6% of the total wing area. In this way the lifting force is reduced without altering the mass forces significantly, and the main stroke parameters also remain constant. Subsequent flights in the wind tunnel showed that the wing-tip curve of the forewings was not influenced by the perforation, whereas the forward component in the hindwing was greatly reduced during the first two-thirds of the downstroke.

In other words, during the sensitive period for lift perception in the hind wing, there is a significant component of the aerodynamic force which acts normal to the stroke plane and which pulls the wing forward. Preliminary recordings from the nerve innervating the sense organs show that they are particularly sensitive to strains in this direction. The oscillating wings are therefore able to distinguish in a simple way between lift and mass forces, but only if the animal moves its wings as in normal forward flight.

It is interesting to note that the modified hindwings of the two-winged insects, the halteres, are supplied with similar sense organs. The halteres are carefully shielded from the wind and act as detectors of gyroscopic torques; that is, torques normal to the plane of beating (Pringle 1948). In the course of evolution it looks as if the original apparatus for the control of lift has been further developed and refined, finally to become a detector of turning movements and independent of the direct recording of wind forces. It is this type of development of a control function which may transform an animal from an automation into a refined system in which a number of independent input channels are integrated to determine a unified optimum output.

Summary

Except for very small insects, flying animals make use of ordinary airfoil action and are constructed so that the aerodynamic cross-forces balance the pull of gravity. When the pull of gravity is compensated for it is concluded that the animal is accelerated in a direction perpendicular to its normal path, the adjustment of the angles of attack becomes upset, and the wings cannot be used as airfoils. Effective aerial locomotion in the "weightless" state would require a symmetrical construction which resembles a fish more than a winged animal.

Some unpublished experiments conducted in collaboration with Dr. R. Weber have shown that the detection of lift in flying desert locusts seems to depend on a forward movement of the hindwings during the first two-thirds of the downstroke. In contrast to the movement of the forewings, this forward movement is a passive consequence of the aerodynamic lifting force, making it possible for the insect to differentiate between wind and mass forces.

References

Gettrup, E. 1965. *Cold Spring Harbor Symp.* 30: 615-22.

———. 1966. *J. Exp. Biol.* 44: 1-16.

Gettrup, E., and Wilson, D. M. 1964. *J. Exp. Biol.* 41: 183-90.

Horridge, G. A. 1956. *Nature (Lond.)* 178: 1334-35.

Jensen, M. 1956. *Phil. Trans. Roy. Soc. London, Ser. B* 239: 511-52.

Nachtigall, W. 1966. *Z. Vergl. Physiol.* 52: 155-211.

Pringle, J. W. S. 1948. *Phil. Trans. Roy. Soc. London, Ser. B* 233: 347-84.

———. 1968. *Advan. Insect Physiol.* 5: 163-227.

Vogel, S. 1966. *J. Exp. Biol.* 44: 567-78.

———. 1967a. *J. Exp. Biol.* 46: 383-92.

———. 1967b. *J. Exp. Biol.* 46: 431-43.

Weis-Fogh, T. 1956. *Phil. Trans. Roy. Soc. London, Ser. B* 239: 553-84.

———. 1961. In *The cell and the organism*: 283-300, ed. Ramsay and Wigglesworth. Cambridge: Cambridge Univ. Press.

Weis-Fogh, T., and Jensen, M. 1956. *Phil. Trans. Roy. Soc. London, Ser. B* 239: 415-58.

Wilson, D. M., and Weis-Fogh, T. 1962. *J. Exp. Biol.* 39: 643-67.

Wohlgemuth, R. 1962. *Z. Vergl. Physiol.* 45: 581-89.

Discussion

GUALTIEROTTI: Do you have any idea where the sensors are? Are they in the muscles?

WEIS-FOGH: No, they are placed on the underside of the wing; they are campaniform sensillae which are deformed mainly during the middle part of the downstroke. From Gettrup's studies we also know that it is precisely during this part of the stroke that the relevant information is being fed into the central nervous system. He was able to block the sensory response by an anodal block which was put in periodically and which lasted one-sixth of the whole stroke cycle. He could then move the block around in the wing-stroke cycle and see during which periods the lift reaction is maintained and during which periods it disappears. It disappears if the block occurs midway during the downstroke, that is, where the main aerodynamic lift is produced.

BROWN: I have several questions. First, has the experiment been done of adding weight to the body of the flying insect and studying the results? Second, has the barometric pressure been changed to determine whether that changes the flight pattern? And third, if the insect were in a free-fall condition, how rapidly would you be able to detect some inability to fly? Is there something that would occur within a few seconds?

WEIS-FOGH: First of all, all experiments here, apart from the last ones which Dr. Wilson mentioned as free-flight experiments, have been done in a wind tunnel, where the animal is suspended to the aerodynamic balance, and where the resultant vertical force combating gravity cannot be changed by the investigator. The animal itself chooses how much it wants to lift, corresponding to whether it wants to climb or sink; we are not loading it. Second, we have not examined the responses to varying barometric pressure. The last question was, How soon would you, in a satellite, discover an abnormal situation. In the fruit fly I would predict that you would not discover anything; in still smaller insects nothing at all. But in an insect as large as a locust, with its restricted and finely adjusted aerodynamic system, you would probably discover it within the first two wing strokes. That is to say, within a distance corresponding to perhaps 20-40 cm.

BROWN: I asked the last question because I think that in parabolic flight in an airplane you can get not only an approximation of what we call "zero g" but both more than 1 g and something between zero and 1 for short periods, well beyond two wing strokes. So if there is any point in doing that kind of experiment I think the facilities could be made available.

WEIS-FOGH: The thing you could learn from experiments under such conditions is that the majority of flying insects are really adjusted to flight at atmospheric density and in a normal terrestrial gravitational field.

HOWLAND: The multiple inputs into the orientation system suggest, perhaps, a kind of schism between plants and animals, a difference in their behavioral physiology. For example, take a bee, which is quite heavy, really. I think it's true that some bees adjust their flight altitude by the optomotor response. As I recall, bees flying over a perfectly mirror-flat water surface will sometimes fly into it. Here is a bee who shows a response to gravity, but the "gravity receptor" is its eye. This is a curious arrangement, but I think we are going to meet with a very similar situation in fish, which also move around in a three-dimensional environment. I think that a lot of interesting animal physiological experiments have been done by manipulating two or more variables.

COHEN: Provided you can control them adequately, I think you could say this of plants too. One could systematically study the interactions between gravitational force and light.

19
Proprioceptive Gravity Perception in Hymenoptera

Hubert Markl, *Zoological Institute*

For good reasons, many remarkable achievements of orientation have been discovered in social insects such as the bee and ant. The need to forage over considerable distances from a nesting site and to make optimal use of food sources to provide enough food for large colonies makes it indispensable for them to remember where both the nest and a food source are situated. These requirements have challenged their orientational capacities. The part played by the sense of gravity in mastering these tasks, however, has rarely been carefully studied. This has been true even though the unchanging intensity and direction of gravity and its ubiquitous presence afford an excellent system of reference for an animal which does not by chance move in a plane perpendicular to the lines of the gravitational force.

One reason for the small number of studies on gravity orientation in insects is that till very recently graviceptive organs had been known only in a rather aberrant form in a few water bugs (Baunacke 1912). In those groups of insects—like Hymenoptera—where one could expect gravity to play a role in complex orientational tasks, for example, compass orientation, it was not until 1959 that Lindauer and Nedel reported gravity receptors in the honeybee *Apis mellifica*. They described groups of hair sensilla on the neck and on the abdominal joints that were used as gravity receptors. Similar hair plates in other insects had long before attracted the interest of zoologists. Pringle (1938) studied their function as positional proprioceptors in leg joints of cockroaches and pointed out their phasic-tonic response pattern. Mittelstaedt (1947, 1950) described comparable hair plates on the necks of dragonflies and investigated their role in flight control. Although he found that they measured head position relative to the thorax, his evidence indicated that these organs were used only to monitor changes in that position, not the permanent state. In the praying mantis, Mittelstaedt (1957) demonstrated that hair plates in the neck yield part of the sensory input for the complex feedback system directing the stroke. Here again the sensory hairs primarily register the position and changes in the position of head or gaster in relation to the thorax. Since gravity can cause positional changes in these joints when a bee turns in the field of gravity, mechanisms could evolve to keep stimulation patterns constant at these joint receptors and guarantee a constant orientation toward gravity. In other words, a bee's central nervous system has become adapted to "draw conclusions" about the direction of gravity from the information about joint positions given by the hair plates.

That discovery stimulated further investigation into this interesting principle of gravity perception (Markl 1962, 1963). For these studies I selected mainly ants as experimental animals (mostly *Formica polyctena* Foerster), though some more work was done with bees, as will be mentioned. Ants are basically terrestrial, running animals, many of which have to find their way in underground nests or under vegetation too thick to let sun shine through. It could therefore be expected that ants make ample use of gravity orientation. That ants do perceive gravity had been reported several times (Barnes 1929, 1930; Vowles 1954; Jander 1957), though *how* they do it was not elucidated to everybody's satisfaction. I followed Vowles in using a very basic behavioral pattern as a test: an ant's gravity-oriented escape run on a vertical or inclined surface. This reaction is so stereotyped and reliable that ants of any caste, age, physiological condition, or species will perform it even after heavy surgical treatment. It can easily be shown that these escape runs are oriented to gravity. If one tips the running board end over end by 180°, gravity-oriented animals will turn around at once so as to proceed again in the same direction in regard to gravity. This reversal of direction can easily be triggered 30-40 times in succession. The average mean angle of deviation from the former path after tipping is less than 14°. Nondirectional, red illumination must be used to prevent orientation by light.

In order to check the hypothesis that in ants, as in bees, gravity receptors are located at joints, I tested the ability of ants to orient by gravity after successively blocking the movement of different joints by immobilizing them with wax or by severing appendages from the trunk. The outcome was that ants were able to use gravity for orientation as long as they were free to move any of the following joints (fig. 1): neck, joints between thorax and petiole and between petiole and gaster, antennal joints, and joints between thorax and coxae. It is sufficient for gravity orientation if one antenna has free joints and all the other joints mentioned are blocked. This is in agreement with Vowles (1954). However, the antenna is definitely not the only site where gravity receptors are situated in ants.

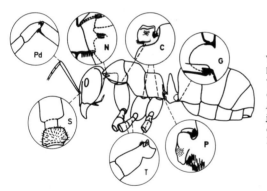

Fig. 1
Worker of *Formica polyctena*. The circles give enlarged views of the different joints bearing hair plates. The lines indicate where the joints are situated. C, coxal joints; G, gaster joint; N, neck; P, petiole joint; Pd, joint between 1st and 2d antennal segments; S, joint between antenna and head; T, joint between coxa and trochanter (2d segment of the leg). (From Markl 1965.)

Some of the joints used by ants for graviception are the same as those used by bees, as described by Lindauer and Nedel (1959). It is interesting that groups of hair sensilla are found on all these joints, which resemble in every respect those described by Pringle (1938) in roaches or by Lindauer and Nedel (1959) in bees.

Thurm devoted a series of papers (1963, 1964a, b) to the fine structure and physiological properties of the single receptor of the most important hair plate in the bee's neck. He found the hairs socketed in a bilaterally symmetrical basal ring, which determines the direction in which the hair can be most easily bent. These preferential directions are distributed in different sensilla of the same hair plate in the neck so as to fit the movement of the head, which loads the hair plate. Response latencies are very short in these sensilla, less than one msec, and the response has a phasic-tonic character. There is a different response level for different degrees of bending, but the response never completely adapts. The physiological properties of these hairs enable them to measure rapidly the precise position of related parts as well as any change of position.

These sensory hairs illustrate that in order to understand the function of a sensory organ, behavioral experiments testing its use by the animal are as important as neurophysiological studies of the single receptor element. Otherwise the graviceptive function of proprioceptive hair sensilla would never have been discovered.

Further evidence supporting the conclusion that the hair plates are involved in graviception is provided by the following experiments. If one singes off the hair plates or cuts the nerves leading to them on otherwise unimpeded joints, as can be done in the gaster joint of ants and in the neck of bees, the ability of the particular joint to participate in graviception is lost (Lindauer and Nedel 1959; Markl 1962, 1966c). On the other hand, in ants of the subfamily Myrmicinae, like *Messor barbarus*, there are three joints between thorax and gaster and only two of them are supplied with hair plates. Only the two with hair plates can be shown to serve in gravity perception. These two lines of evidence show the dominance of hair plates as positional receptors for detecting gravity in Hymenoptera.

Proprioceptors other than hair plates have also been shown to be involved in gravity perception in insects (Vowles 1954; Bässler 1958, 1961, 1965; Wendler 1965; Horn 1970; Jander, Horn, and Hoffmann 1970). Honeybees seem to use proprioceptive input not coming from hair plates, at least when forced to do so (Markl 1966c). Although it seems reasonable to expect insects other than Hymenoptera to use proprioceptive input from hair plates for gravity perception, it cannot be taken for granted, since other specific functions for hair plates have been described several times (Mittelstaedt 1950, 1957; Dingle 1961). It may turn out that the almost exclusive use of hair plates in bees and ants is a late step in an evolutionary series of gradually improving proprioceptive mechanisms for sensing gravity.

Finding such a wealth of joints bearing hair plates immediately led to the question of their relative importance for gravity perception. To compare the performance of uninjured ants with that of ants whose joints had been partially immobilized, tests were used to quantify orientational capacities. I measured the minimal tilting angle of an inclined surface above the horizon at which the ants were just able to orient by gravity. This threshold angle for normal ants lies between $2°$ and $4°$. Also, the distance which a gravity-oriented ant has to run on a vertical surface to cover 10 cm of way was measured. As the number of functional joints decreases, the number and degree of deviations in the ant's path increase. A normal ant runs a linear distance of 10 cm on an average track of 10.4 cm. These tests quantify two boundary aspects of gravity orientation: a threshold sensitivity and the working limit of the steering mechanisms.

To establish the significance of an individual joint one must determine to what extent orientational performance drops as each joint is eliminated. The blocking of any single joint results in only a slight drop in performance, always less than 13% (if a normal ant performs 100%). More instructive is the performance of an ant with only one free joint. Because the antennae turned out to be approximately equally effective in both test situations, they served as a norm for comparing the threshold angles and average path lengths. A series of results showing the relative effectiveness of various receptors is shown in figure 2. The hair plates of the neck are always the most effective and those of the gaster joint the least. Joints at the coxae and gaster perform better close to the horizontal than on vertical surfaces. This is in accordance with their anatomical arrangement. If one compares the average number of hairs per joint with its importance for gravity perception, common trends are obvious. Accuracy of orientation by gravity does not, however, simply depend on the number of sensory hairs available. In worker ants of the same species but of different sizes, orientational capacities are independent of body length, though the number of hair plate receptors is positively correlated with it (fig. 3).

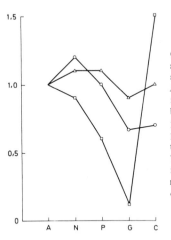

Fig. 2
Comparison of gravity-orientation performance in ants with the number of sensilla along the functional joints. Ants tested had all joints except those shown on the abscissa immobilized with wax (A, antennal joints; N, neck; P, petiole joint; G, gaster joint; C, coxal joints). Triangles represent measures obtained by the determination of the angle to which a plane must be elevated above the horizontal before an orientation to gravity is possible. Circles show the results of measurements to determine the precision with which a gravity-oriented course is maintained. Squares show the number of free sensilla along the joints which were able to function. The values in those ants where only the antennal joints A were free to move were set equal to 1 and were used as reference values for all other results. The performance of gravity orientation is shown to have some correlation with the number of sensory hairs on a functional joint.

Fig. 3
Comparison of gravity orientation performance of normal workers of *Formica polyctena* of different sizes with the number of sensilla of the hair plate on the gaster joint (G, solid circles) and of one of the hair plates of the petiole joint (P_2, open circles). Open triangles represent measures obtained by the determination of an angle to which a plane must be elevated above the horizontal before an orientation to gravity is possible. Solid triangles show results of measurements to determine the precision with which a gravity-oriented course is maintained. Vertical lines give the standard deviation.

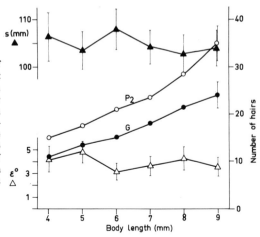

The larger the number of free joints the more effective is the gravity orientation. There is, however, considerable redundancy; an ant with three or four free joints can do at least 90% as well as a normal ant in gravity orientation. Whether the redundancy of graviceptive information is of importance under natural conditions (e.g. when an ant carries heavy loads) has yet to be proved.

There may be other reasons for the multitude of graviceptive joints. Bees and ants are able to use gravity perception to maintain a constant course relative to gravity when they are running on a vertical or an inclined surface (compass orientation). The physiological mechanisms of compass orientation have aroused lively interest in recent years (Jander 1957; Mittelstaedt 1961, 1964). In a most intriguing theory, Mittelstaedt postulated the cooperation of two receptor systems: the response of one is a sinusoidal function and the other is a cosinusoidal function of the animal's angle to gravity. In Hymenoptera the hair plates at some joints are stimulated by a component of gravity which depends on the sine of the animal's angular position, and other hair plates are affected by a cosinusoidal component (fig. 4). By incapacitating either of these groups of receptors it should be possible to upset compass orientation if the receptors are indeed used as the two sources of sensory input postulated by Mittelstaedt. Bees were trained to find their way to a food source by keeping a fixed angle of 60° to gravity while running on a vertical surface (Markl 1966). Then, graviceptive hair plates in these trained bees were rendered nonfunctional. The outcome of these experiments can be interpreted as favoring Mittelstaedt's concept.

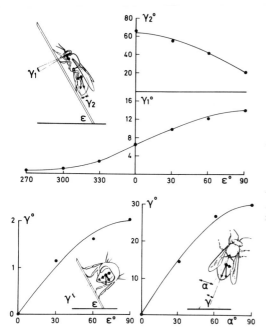

Fig. 4
Deflection of head and abdomen of a freshly killed bee by the components of gravity, indicated by thick arrows in the insets. Deflection in degrees (*ordinates*) in relation to the angle of inclination of a tilted surface (*above and lower left*) or to the angle between the bee's long axis and the vertical direction (*lower right*). The meaning of the symbols a, ϵ, γ is explained in the inset figures. (From Markl 1966c.)

For these hair plates to operate effectively as gravity sense organs one must assume that the relative position of the animal's body parts to each other depends to a certain extent on the position of the body with respect to gravity. To what extent does the actual position of an appendage result from the attitude of the entire animal in space and to what extent does it reflect active control on the part of the animal? Because of its mass and mobility the gaster of an ant is a very suitable "appendage" for getting quantitative answers to these questions. By single-frame analysis of high-speed motion pictures of ants running under different angles to gravity on a vertical surface, one can find out how the angle between the axis of the gaster and thorax is related to the angle between the thorax and gravity. Figure 5 gives results for one quadrant. Figure 6 shows diagrammatically the idealized case. The relation is sinusoidal, which is in accordance with theoretical considerations. The maximum of 15° to 30° bending is considerable. That active control of the gaster position is still involved can be shown by burning off the hair plate on the gaster joint, thus cutting off the sensory part of the feedback control loop. The deflection then almost doubles. This indicates that the deflection caused by gravity is decreased by about 50% through active control. However, the control mechanisms can be overstrained. When the weight of the gaster has been doubled artificially, active control is no longer effective. The gaster hangs as close to the vertical line as the mechanical properties of the joint will allow. Even when the deflections caused by gravity are much smaller in most appendages than in the gaster, the general considerations are the same. This further supports the concept that proprioceptors play a major role in gravity perception.

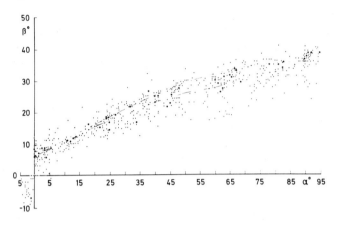

Fig. 5. Deflection of the gaster (β) from its normal position when a, the angle between the thorax and the vertical, is changed. A continuous line connects the average values of β for each 10° of a. Broken line, a fitted sine curve (30.2 sin a + 5.9). (From Markl 1962.)

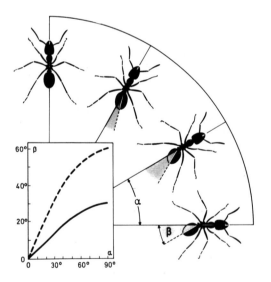

Fig. 6
Deflection of the gaster (β) from its normal position when a, the angle between the thorax and the vertical, is changed. In inset, straight line gives results obtained with a normal ant, broken line those obtained with an ant whose hair plates at the gaster joint had been singed off, thus opening the feedback loop. (From Markl 1965.)

Gravity perception by means of proprioceptors poses singularly interesting questions concerning the central processing of this information. The central integrating mechanisms have to be much more elaborate than for gravity perception with statocysts. Hair plates measure only the position and movement of joints. Whether these changes are due to the action of gravity or to other mechanical forces acting on the appendages, such as muscular contraction, has to be decided centrally to avoid orientational errors. It is conceivable that an insect is able to achieve these distinctions by comparing the input from the hair plates at different joints. There is experimental evidence that only equidirectional change of position at several joints is answered by an ant with a change in its orientation to gravity.

It seems hardly reasonable to think that this complicated sensory system is of no better use than for handling emergency runs in the dark. Owing to von Frisch and Lindauer's (1961) work on the communication of honeybees, we know how they use their ability to orient by gravity in the tail-wagging dance on a vertical comb to indicate the direction of a food source. (On the phylogeny of gravity orientation in bees, see Jander and Jander 1970.) In ants no similar use of gravity orientation has been found. There remains, however, the interesting question whether an ant or any other insect can learn to maintain a specific angle to gravity in order to find its way. To answer this question I trained ants and bees to find a food source from their nest by using only gravity as a compass (Markl 1964, 1966b, c). The ants run from their nest into the middle of a vertical arena with a circular turntable as a central part. Bees fly from their hive to the entrance of a darkroom, where they have to run in a dark alley to the center of the same turntable (fig. 7). They find food at the periphery of the turntable by maintaining a constant angle to gravity on the vertical surface. These Hymenoptera can find their way by learning a compass direction with regard to gravity. After they have run the learned approximate distance, they begin to circle around and thus easily hit the target. There was no indication whatsoever of confounding symmetrical course angles.

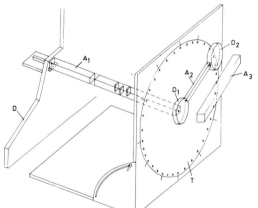

Fig. 7
Experimental setup for training bees on a compass course to gravity. The bees enter the darkroom door (D) at the left, run through A_1 and arrive in the middle of the turntable, T. There they have to make a directional choice in order to enter A_2. After running through A_2 they find food in the center of D_2. D_1 and D_2 are flat lucite dishes, closed from above. A_3 is a lucite alley used to train the bees. It is later replaced by D_1, A_2, and D_2, which permit observation of the bees' directional choices. (From Markl 1966b.)

This seemed to be a simple, straightforward result. However, by applying appropriate statistical tests it can be shown that something is wrong with these gravity-oriented ants and bees. Sometimes they run in a straight line between entrance and feeding place. However, with certain training directions they persistently make rather large mistakes and fail to correct them even when trained over long periods. These errors are not randomly distributed but show some kind of systematic deviation (figs. 8, 9). Surprisingly enough, the errors are different in ants and bees. In both groups the errors made on their way to the food differ from those made when they are running back to the exit of the arena. In bees which are running to the food source, the pattern of orientational errors was similar to a pattern of errors which had been observed in the directional indication of their tail-wagging dances (von Frisch and Lindauer 1961).

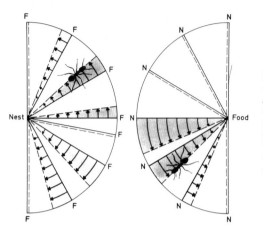

Fig. 8
Persistent orientational errors of ants trained to run in certain gravity-compass directions (indicated by F and N at the periphery of the half-circles). In the left half-circle the average chosen directions of ants running to the food are given by broken lines; in the right half-circle data for ants returning to the nest are given. The concentric arrows emphasize the direction in which the averaged choices deviate from the training directions. (From Markl 1965.)

Fig. 9
Deviation of average chosen direction (*on ordinate*) from the training direction a (*on abscissa*). Upward deviations above, downward deviations below zero on the ordinate. *Triangles*, runs to the food; *circles*, return-runs from the food to the nest. *Solid signs*, honeybees; *open signs*, ants. The dancing errors of honeybees' tail-wagging dances are given by crosses, according to von Frisch and Lindauer 1961. (From Markl 1966b.)

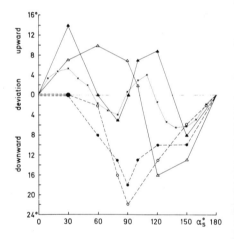

A stepwise improvement in graviception can be traced from the rather poor gravity-compass orientation of homing ants and bees, through the somewhat better performance of ants on search for food, to the precise orientation in bees on their way to a feeding place or when dancing. It was concluded that the basic orientational tendencies to run upward when leaving the nest and downward when homing (Jacobs-Jessen 1959) were interfering with compass orientation, these effects being less influential in bees than in ants. Current investigations on factors influencing the dancing errors (Lindauer and Martin, personal communication) may challenge this view and make it necessary to reconsider this interpretation.

Summary

Proprioceptive hair plates on several joints have been shown to act as gravity receptors in ants and bees. The relative importance of different joints for graviception was experimentally determined and found to be correlated with the development of hair plates at the respective joints. Owing to gravity, the positions of an ant's body parts are altered relative to each other when the animal's position in space is changed. Although these deflections are large enough to be measured (e.g., in joints between thorax and abdomen) they are still kept under feedback control by the ant. Ants and bees can be trained to run a constant-angle course with regard to gravity and thus to find their way in the dark on a vertical or inclined surface.

References

Bässler, U. 1958. *Z. vergleich. Physiol.* 41: 300-330.

———. 1961. *Z. Naturforsch.* 16b: 264-67.

———. 1965. *Kybernetik* 2: 168-93.

Barnes, T. C. 1929. *J. Gen. Psychol.* 2: 517-22.

———. 1930. *J. Gen. Psychol.* 3: 318-24, 540-47.

Baunacke, W. 1912. *Zool. Jahrb. Abt. Anat. Ontog. Tiere.* 34: 179-342.

Dingle, H. 1961. *Biol. Bull.* 121: 117-28.

Horn, E. 1970. *Z. vergleich Physiol.* 66: 343-54.

Jacobs-Jessen, U. 1959. *Z. vergleich Physiol.* 41: 597-641.

Jander, R. 1957. *Z. vergleich Physiol.* 40: 162-238.

Jander, R.; Horn, E.; and Hoffmann, M. 1970. *Z. vergleich Physiol.* 66: 326-42.

Jander, R., and Jander, U. 1970. *Z. vergleich Physiol.* 66: 355-68.

Lindauer, M., and Nedel, J. O. 1959. *Z. vergleich. Physiol.* 42: 334-64.

Markl, H. 1962. *Z. vergleich Physiol.* 45: 475-569.

———. 1963. *Nature (Lond.)* 198: 173-75.

———. 1964. *Z. vergleich Physiol.* 48: 552-86.

———. 1965. *Naturwissenschaften.* 52: 460.

———. 1966a. *Zool. Jahrb. Abt. Anat. Ontog. Tiere.* 83: 107-84.

———. 1966b. *Z. vergleich. Physiol.* 53: 328-52.

———. 1966c. *Z. vergleich. Physiol.* 53: 353-71.

Mittelstaedt, H. 1947. *Naturwissenschaften.* 34: 281-82.

———. 1950. *Z. vergleich. Physiol.* 32: 422-63.

———. 1957. In *Recent advances in invertebrate physiology*, pp. 51-72. Univ. of Oregon Publ.

———. 1961. In *Aufnahme und Verarbeitung von Nachrichten durch Organismen.* Stuttgart:Hirzel.

———. 1964. In *Neural theory and modeling*, ed., R. F. Reiss, Stanford: Stanford Univ. Press, pp. 259-72.

Pringle, J. W. S. 1938. *J. Exp. Biol.* 15: 467-73.

Thurm, U. 1963. *Z. vergleich. Physiol.* 46: 351-82.

———. 1964a. *Z. vergleich. Physiol.* 48: 131-56.

———. 1964b. *Science* 145: 1063-65.

Vowles, D. M. 1954. *J. Exp. Biol.* 31: 341-75.

Von Frisch, K., and Lindauer, M. 1961. *Naturwissenschaften.* 48: 585-94.

Wendler, G. 1965. *Z. vergleich. Physiol.* 51: 60-66.

20
Gravity Orientation in Insects: The Role of Different Mechanoreceptors

Gernot Wendler, *Max Planck Institute, Seewiesen*

Dr. Markl (chap. 19; see also Markl 1966) has discussed whether Hymenoptera use receptors other than hair plates in their gravity orientation. I would like to consider this question in more detail. I shall then discuss the role of two components of the force the insect's body exerts upon the legs and describe experiments to determine which of the two components a stick insect uses for gravity orientation.

As Dr. Markl pointed out, hair plates in the neck, between thorax and gaster, in antennal joints, and between thorax and coxae play a role in the gravity orientation of ants and bees. At first glance the idea is very appealing that insects take parts of their body, like the head or the gaster, or even the whole body as a sort of statolith, using the amount and direction of displacement in the respective joints as indication for the direction of gravity. From the cybernetic point of view, however, such joints are perhaps the worst spots one can find for displacement-measuring gravity receptors, because active movements at the joints always interfere heavily with the gravity information. Think of the honeybee performing a waggle dance while it is measuring its position relative to gravity! In the legs the active movements at the joints are many times greater than the passive displacement by the body weight. Yet information from each of the joints or, in the legs, several joints, is adequate for orienting the ant as Dr. Markl showed. The orientation performance is not as good as it normally is, but it never disappears.

To get rid of the disturbance caused by active movements, the animal could filter out all signals at the frequency of typical joint movements, leaving only the slower changes, which are usually associated with the position or turning of the whole animal. Or the input could be corrected by the efferent command for the movements itself (component w in fig. 1), because the active joint movement is correlated with this command. Unfortunately, the actual position of a joint is determined not only by the efferent command and the load-correlated displacement, but also by irregularities in the environment. It often may occur that the legs of an ant come to rest in a position which does not correspond to the command while the insect is climbing about in the grass. Because of such difficulties the system could only work in a sloppy way.

But why shouldn't insects measure the weight of the parts of the body and the body itself more directly—by strain in the cuticle or tension in the muscles? In fact, the muscles and skeleton have to support those weights all the time, whatever the accidental positions of the appendages. As Dr. Markl mentioned already, insects are well equipped with such sense organs. The muscle organ found in the coxa-trochanteral joint of the honeybee can presumably measure muscle tension (Markl 1965). There are also the campaniform sensillae which measure strain in the cuticle (Pringle 1938*a*, *b*, 1940).

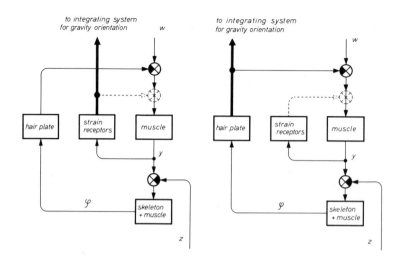

Fig. 1. Two alternative interpretations of the effect of destroying hair plates in insect joints. *Left*, hair plate output is used as gravity signal. *Right*, the strain receptor output is used as gravity signal. In both cases elimination of the hair plates reduces the input to the integrating system. The circles indicate superimposition of signals. Signals entering a black segment are subtracted. W, efferent command; y, muscle tension; z, external forces at the joint; ϕ, joint angle.

Drawing a block diagram of the situation including the inferred function of those receptors, one immediately finds two alternative interpretations of Markl's experiments (fig. 1). They have the following facts in common: ϕ, the angle of the joint, is measured by the hair plate. Let us recall that hair plates so far investigated are parts of feedback loops (Mittelstaedt 1950, 1957; Wendler 1961, 1964; Markl 1962, 1967; Bässler 1965). One example is shown in Markl's experiments with the gaster of ants. The state of excitation of the hair plates affects the output of motor neurons via the central nervous system. In figure 1 the activity of the motor neurons causes muscle tension, y. All external forces at the joint (z) are superimposed on this tension. Any difference between y and z will cause joint movement, which is then transformed by the geometry of the skeleton into a change of the angle Φ, and the loop is closed. Since any change in the hair plate input results in a counteracting movement, we have a negative feedback. Campaniform sensillae show a phasic-tonic response to a step function input (Pringle 1938*b*). In the cockroach leg they are parts of a positive feedback loop which recruits more tension in the muscle to support the animal's weight. We do not know their precise function in bees, so I have indicated this connection by a broken line. The response of the campaniform sensillae depends upon the counteracting muscle tension which is controlled by the hair plates. After the hair plate is removed the counteracting muscle tension is less than normal and consequently the activity of the campaniform sensillae is decreased (Pringle 1938*a, b*, 1940).

The alternative interpretations differ in the signal which is used for gravity orientation. In figure 1 (*left*) the output of the hair plate is used in gravity orientation. This explanation seems to be favored by Dr. Markl. In the second interpretation (fig. 1, *right*) the strain receptors, not the hair plates, are used as gravity receptors. In this case, as in the previous one, the hair plate input would influence the gravity input, this time via the feedback loop. Eliminating the hair plate would reduce the strain receptor input.

In both interpretations, destroying the hair plates affects the gravity input, but the gravity receptors are different. My point is that hair plates certainly do play a role in gravity orientation, but they are not necessarily gravity receptors themselves.

In the following part of this discussion I would like to present a set of experiments designed to study how leg inputs are integrated in stick insects (*Carausius morosus*). The problem is the same in Hymenoptera and many other insects; the solution may also be similar.

The center of gravity of a stick insect (fig. 3) is just behind the coxae of the hind legs. If one pushes the animal laterally at the center of gravity, one does not see a parallel displacement of the body; rather, there is a small rotation about an axis near the coxae of the middle legs. Thus the axis of rotation does not pass through the center of gravity. Consequently the legs must support the body against rotation (yawing) as well as against linear displacement. The question is, from which of the two possible inputs does the animal gain information for gravity orientation?

The animal is placed in total darkness in the middle of a perfectly vertical and perfectly plane disk measuring 1.5 m in diameter. We know that the gravity orientation of stick insects depends upon proprioceptive input not only from the legs but also from the antennae and perhaps from the abdomen (Wendler 1965). To eliminate the influence of gravity receptors other than those in the legs, the antennae are amputated and the body is embedded in a stiff balsa-wood splint (fig. 2). The animal carries an 800 mg harness above the point of rotation. A counterweight is fastened to the harness by a thread that passes over a pulley. The pulley in turn is mounted on a carriage which can be moved on ball bearings on a horizontal track in such a way that the thread between pulley and insect is always exactly vertical. A precision miniature ball bearing allows the animal to turn about its own vertical axis. With this arrangement the animal's movements on the vertical plane are not restricted at all. A miniature light bulb weighing 13 mg is placed on top of the harness; the thread consists of two very thin copper wires, and part of the counterweight is a miniature battery.

The animals are placed in the middle of the disk and start with varying positions of their long axes. While they walk, the trace of the light bulb is directly recorded on film through the open shutter of a still camera. The animals walk successively under four different conditions. The conditions differ in the weight distribution between animal and counterweight, but the total amount of inertia to be overcome by the legs remains constant.

The animals carry a lead collar weighing as much as the animal itself. The collar is situated as far in front of the axis of rotation as the center of gravity is behind it. This arrangement eliminates the torque. The animal's weight is doubled, yet this is precisely balanced by the counterweight (fig. 3*a*). Thus, both torque and body weight are eliminated in this experiment, and gravity receptors in the legs do not receive any information about the direction of gravity. In the second experiment (fig. 3*b*) the torque is eliminated as before, but the linear component of the weight is present since the counterweight is reduced to one-half of that in the first experiment. In the third experiment (fig. 3*c*) the linear weight component is absent, but the torque is present. Conditions are "normal" in the fourth experiment (fig. 3*d*): linear component and torque are both present.

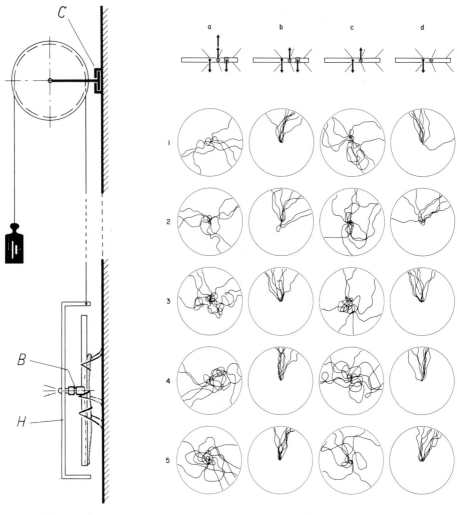

Fig. 2
Experimental setup to control the force on the legs of a stick insect walking on a vertical plane. The weight of the harness (*H*) is always compensated for by a counterweight. *B*, ball bearing which allows free yawing. *C*, carriage for moving the pulley horizontally.

Fig. 3
a-d, the four experimental conditions. In the schematic drawings each black arrow represents one animal weight. Their origins are: *black circle*, center of gravity; *concentric circles*, axis of rotation; *striped rectangle*, lead collar. The front of the animal is to the right. The circles in the lower part of the figure contain superimposed traces of 5 animals (1-5) under the four conditions *a-d*. Up is toward the top of the page.

The five animals tested in all four situations gave consistent results (fig. 3). There is very little, if any, gravity orientation in condition (*a*) where the legs are without gravity information. The animals walk upward provided the linear weight component is available, as in (*b*). There is little or no gravity orientation with only the torque information available as in (*c*). In (*d*), the normal condition, the animals walk upward as they would do without any harness (Wendler 1965, 1966). This means that the animals were oriented if and only if the linear component was available. Thus, under the constraints of the system, the linear component is the necessary and sufficient condition of orientation. From these results I conclude that stick insects use the linear component of the body pressure on the legs rather than the rotatory component (the torque) for gravity orientation.

Summary

In discussing the preceding chapter two alternative interpretations are presented to explain the observation that elimination of proprioceptive hair plates in several joints of bees and ants reduces their capability to orient themselves with respect to gravity. Since hair plates and strain receptors are connected by a feedback loop, both receptor types could be used as gravity receptors. Results of my own experiments with stick insects show that the adequate stimulus for gravity receptors in the legs is the linear component of the body pressure on the legs.

References

Bässler, U. 1965. *Kybernetik* 2: 168-93.

Markl, H. 1962. *Z. Vergleich. Physiol.* 45: 475-569.

———. 1965. *Naturwissenschaften* 52: 460.

———. 1966. *Z. Vergleich. Physiol.* 53: 353-71.

Mittelstaedt, H. 1950. *Z. Vergleich. Physiol.* 32: 422-63.

———. 1957. In *Recent advances in invertebrate physiology*, pp. 51-71. Univ. of Oregon Publ.

Pringle, J. W. S. 1938a. *J. Exp. Biol.* 15: 101-13.

———. 1938b. *J. Exp. Biol.* 15: 114-31.

———. 1940. *J. Exp. Biol.* 17: 8-17.

Wendler, G. 1961. *Naturwissenschaften* 48: 676-77.

———. 1964. *Z. Vergleich. Physiol.* 48: 198-250.

———. 1965. *Z. Vergleich. Physiol.* 51: 60-66.

———. 1966. *Symp. Soc. Exp. Biol.* 20: 229-49.

Discussion (of the Two Preceding Papers)

WESTING: Dr. Wendler, could you explain why you cut off the antennae of your stick insects? I take it you were trying to separate the role of leg receptors from that of antennae in geoperception.

WENDLER: Right. That's also why I embedded the body in the balsa-wood splints, to prevent any bending of the body and stimulating stretch receptors in the abdomen. It's very interesting to see how the antennae work together with the leg receptors. If you have a normal animal walking on the vertical plane, and cut off the antennae, you don't see any difference from what you have observed before. Nevertheless the antennae are involved in the gravity-orientation system. You see their role only if you reverse the information for the legs alone by having the holder now at the center of gravity and adding two body weights as counterweights, thus providing the antennae with the normal input and the legs with the reversed input. Let me extend this to a generalized remark: If you cut off or take away one receptor which you think to be relevant for a certain behavior of an animal, and you do not find any effect on the observed behavior, you can't say that the receptor isn't involved in that function. One example: If you try to perceive the intensity of a light source with both eyes first, and afterwards cover one eye, you perceive the same intensity. Would you conclude that the eye is not used to perceive the light intensity? Now repeat the same experiment with your second eye: certainly the result will be the same. Obviously it is incorrect to conclude that none of the eyes is used for perception of the light intensity simply from experiments removing inputs from single sense organs. Unfortunately there is a corresponding rule: If you take away a receptor in a system where several receptors are interwoven by mechanical devices, and you observe an effect on the response, you cannot conclude that the receptor is really the critical one that directly determines the response. I tried to show this difficulty in my comments about Markl's paper. We should concede, however, that Markl was not as definite in meaning that the hair plates are really the gravity receptors in the strict sense I have presented here.

MARKL: I agree that two different interpretations are possible. For the following reason it seems more probable that the hair plates are used, as shown in your first diagram: In ants there are several joints where no campaniform sensillae or other strain receptors are known. Certainly information on cuticular strain from leg receptors might contribute to gravity perception, but we do not have experimental evidence that this is so in the Hymenoptera. Where this kind of evidence is found, conclusions different from mine will have to be drawn.

WENDLER: I think it is important that there are no strain receptors in those joints. I didn't know that.

HERTEL: What happens in your situation 1, where three bodies are contributing?

WENDLER: In situation 1, three additional body weights are arranged in such a way that the animal cannot perceive any linear force downward or upward, nor any rotational force. As a result the animal does not seem to be able to find its way upward. Please note that my schematic drawings of the four situations are incorrect with respect to the inertia of the whole system: animal–holder–counterweight. In the drawing (*a*), four body weights are involved, and in drawing (*d*), only one. In the experiments however, the total inertia was kept constant throughout the four conditions. In situation (*d*), for example, this was achieved by adding 1-1/2 body weights in the center of rotation upward and 1-1/2 body weights downward.

HOWLAND: In fish we know what happens when we pit the semicircular canals against the otolithic organs in a centrifuge. You get a nice compromise. Also in fish, when you pit the optomotor response against the response to gravity, you have two feedback loops. If one of them happens to lock in, it will become stronger because it is nonlinear. In this way the optomotor response can completely override the response to rotation.

GUALTIEROTTI: How long does it take to learn to compromise?

HOWLAND: It isn't really a question of learning involved; the outcome of whether you get a compromise or whether you get a locking-in on one or the other is going to be completely determined by the nature of the system with which you are dealing. In the semicircular canals or otoliths it is going to be a compromise because it is wired that way. In the optomotor reaction it will probably go to one of the others.

JENKINS: The chigger mite or harvest mite Trombicula is usually considered to be very negatively geotactic. If you put a pencil vertically in a group of mites, they crawl up toward the top; they crawl up a human to his arm in about two minutes. Mites put in the center of a vertical plate crawled upward. However, they would crawl to a light at the top, and if the light were at the bottom they would go down to the light. With two lights of equal intensity they would crawl along the vector. In infrared or dim red light they would move at random. If the mites are conditioned to dim light, they become positively phototactic. In very bright light they become negatively phototactic. The apparent negative geotaxis was in reality a phototaxis. There appears to be no directional reaction to gravity.

21
Primitive Examples of Gravity Receptors and Their Evolution

G. A. Horridge, *University of St. Andrews*

Most lower animals respond to gravity. From the simplest metazoans to man the rapid gravity responses are usually mediated by statocysts, which must have evolved over and over again independently in many phyla, and even in different classes, of animals. The paradox I set out to remove is how statocysts can have evolved. For natural selection to work on an organ there must be a structure which is of value to at least some small extent. A rudimentary statocyst seems to be a contradiction because it could not be effective without being a sense organ which is influenced by tilt.

The uselessness of incipient organs was used in the nineteenth century as an argument against Darwinian selection, especially with reference to the electric organs of fish (surveyed, for example, by Romanes 1892). At that time it was thought that electric organs functioned only as defense mechanisms, so that there would be no *initial* selection pressure to cause muscle fibers to progressively increase their external electric current and aggregate into organs. That paradox has now disappeared because even the weak electric fields of normal muscle can be detected by other fish in special cases (Dijkgraaf 1966), and weak electric organs are important for communication in many species of fish. Therefore large electric organs of value in defense could have evolved slowly from small ones having other functions. Similarly, tracing the occurrence of statocysts in lower animals reveals that they are closely related to the sensitivity of detection of underwater vibration. Furthermore, the primitive and widespread receptor of the displacement of water in vibration, the nonmotile cilium, can readily and naturally be made more efficient as a vibration receptor by converting it into a loaded structure, which is then a primitive statolith.

Many sensory cells bear cilia or modified cilia in the region where transduction may be suspected, but it has not been demonstrated in any example that the shaft of the cilium or the ciliary fibrils are themselves involved in the process of chemical, mechanical, or phototransduction. One possible attitude at present is that the rudiments of the cilialike structures persist because primitively the sensory cell evolved from a ciliated cell. The basal body is an organelle which appears to organize differentiation and which perhaps is conveniently modified in evolution toward the growth of sensory structures. Evolution may have taken this course many times, because even normal motile cilia behave as mechanoreceptors.

A second intention of this chapter is to explore some favorable situations in invertebrates which might increase the general applicability of the finding that underwater displacement receptors are commonly cilia. One ultimate aim is to be able to predict the direction of the mechanical sensitivity in sense organs which cannot be directly analyzed on account of their small size or awkward situation. The general principles, however, necessarily emerge from a description of a number of sense organs, many of which have not been examined previously with the electron microscope.

Nonmotile cilia of sense organs have in fact been known for many years as light microscope objects of dubious structure and function. They were called *Sinneshaaaen, Hörhaaaen, Borsten,* or *Taststifte* by various accurate classical authors (summarized by Schneider 1902). They were quite clearly distinguished from motile cilia, flagella, or *Flimmerzellen*, but were not necessarily recognized as cilia. These sensory hairs have been largely ignored by modern writers and it is now an interesting exercise to run through the old literature in search of sensory "hairs" or sensory "bristles" in animals which have ciliated epithelia but no chitin or keratin. So far all the examples I have examined have turned out to be nonmotile stiff cilia on sensory dendrites, except that arrowworms (Chaetognatha) have nonsensory epithelial bristles as well as nonmotile sensory cilia (Horridge and Boulton 1967).

Motile Cilia

Cilia as Mechanoreceptors

Motile cilia, flagella, and sperm tails respond to bending by movement. Mechanical transmission along the shaft is the basis of at least one theory which explains the curvature of the shaft at different phases of the movement. The theory is that each piece along the shaft responds to the movement of the preceding piece (Machin 1958).

Transmission *between* cilia in a metachronal wave also seems to be normally by mechanical means in many protozoa and in the gills of clams. The demonstration of this depends on experiments in which the velocity of the wave of transmission is shown to depend on the viscosity of the medium (Sleigh 1966). In mussel gills which have been extracted in glycerine, mechanically transmitted metachronal waves occur upon reactivation with ATP (Child and Tamm 1963).

Balancer Cilia of Ctenophores

Ctenophores are gelatinous transparent carnivores of the marine plankton. Their statocyst consists of a calcareous concretion which is held upon the tips of four groups of so-called balancer cilia at the apical end of the animal (fig. 1). Each group of cilia stands at the end of two ciliated tracts of cells, each of which runs along the surface of the animal to a comb row. The eight comb rows bear ranks of large cilia which are the principal organs of locomotion. Steering is accomplished by an asymmetry in the frequency of beat, brought about in the following way.

The balancer cilia are themselves motile, and they beat at a low frequency which depends upon the amount by which they are bent under the load of the statocyst. This is readily observed in a Pleurobrachia which is mounted symmetrically with the four balancer cilia pointing upward on the stage of a microscope. Pushing the statocyst with a needle mounted on a micromanipulator causes decreased frequency of beating of the comb rows toward which the statocyst is pushed, and increased frequency on the opposite side (fig. 2). If the animal were free, with the statocyst under the influence of gravity alone, the asymmetry of frequency of beating would cause a tilted animal to swim downward.

Primitive Examples of Gravity Receptors and Their Evolution

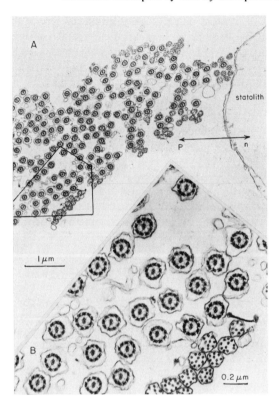

Fig. 1
Transverse section through one group of balancer cilia of the ctenophore showing directions of loading which cause increase in frequency in positively (*p*) and in negatively (*n*) geotropic animals respectively. The area outlined in *a* is enlarged in *b*. Photo by courtesy of Dr. S. Tamm.

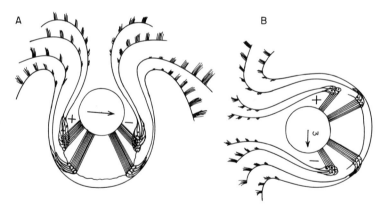

Fig. 2. The action of the statolith and four groups of balancer cilia in ctenophores: *a*, with the apical organ upward, deflection of the statolith by a micromanipulator causes increased frequency of waves of beating of the comb plates on one side and decreased frequency on the other side in a negatively geotropic preparation; *b*, in the normal animal, swimming along on its side, the weight of the statolith causes frequency changes as in *a*. In negatively geotropic animals the lower groups of balancer cilia decrease their frequency as pacemakers; in positively geotropic animals they increase their frequency.

Ctenophores which are undisturbed usually swim upward, but disturbed specimens reverse this geotaxis and swim away from the surface. Since the mechanism is so restricted, the only possible way they can steer downward is by reversing the effect of loading upon the balancer cilia themselves. It is important to note that the reversal involves a change in the receptor cell, which therefore differs from any other directional mechanoreceptor known.

The fact that the sensitivity of the cilia can be reversed in direction seems to show that, in this instance at least, the effect on the frequency does not stem from the asymmetry of the 9 + 2 pattern as seen in cross section. At any one time, however, a physiological directionality certainly exists and cannot at present be explained.

The direction of the beat and of the mechanical sensitivity is at right angles to the line between the central pair of fibrils, which are at right angles to the main axis of symmetry of the animal. The beat is maintained all the way along the tract of cells and ciliated groove, at the head of which stand the balancer cilia. The locomotory wave is a metachronal wave along this tract with the power stroke in the opposite direction from the wave.

Nonmotile Cilia as Mechanoreceptors

Many phyla, perhaps all, have rudiments of the 9 + 2 pattern in some mechanoreceptor structures. Those in the vertebrate acousticolateralis system are reviewed elsewhere (Wersäll, Flock, and Lundquist 1965), those in crustacea (Whitear 1962) and insects (Slifer 1961) are less well known.

The Arrowworms (Chaetognatha; e.g., Spadella)

The arrowworms are slender, transparent carnivores about 1 cm long which grab their moving prey, such as copepods and young fish, from the marine plankton. To do this they use numerous small fans of nonmotile sensory cilia which detect the small displacements of the water set up by a vibrating source (Horridge and Boulton 1967). In the dark they will strike accurately with their jaws at any object vibrating at about 10 cycles per sec, provided that it is a small localized source within range.

The surface of the animal is equipped with numerous groups of stiff cilia, each of which emerges as a fan from a group of bipolar sensory neurons (fig. 3). The dendrite of each neuron bears a single cilium about $1/4\,\mu$ thick and $50\,\mu$ long, which is clearly nonmotile as seen in life through a phase-contrast microscope. Each fan occupies a specific position and is orientated in a definite direction with reference to the axes of the animal's body. If the organs on one side are covered by alginate jelly, the animal fails to strike at prey on that side. These fans of stiff cilia are believed to respond to the water displacements because there is no other appropriate sense organ and they seem specifically adapted to the purpose. They are in some ways analogous to the lateral line organs of fish.

In fine structure the stiff cilia appear normal, but they are orientated more or less in one direction. As seen in transverse section through the base of the fan, the central pair of fibrils in each cilium is orientated so that a line which joins the centers of the central pair of fibrils is at right angles to the direction of maximum sensitivity as inferred from the flat shape of the fan (fig. 4). Because the shaft of the cilium is coupled to the water, the displacements of the water must be felt as a

Primitive Examples of Gravity Receptors and Their Evolution

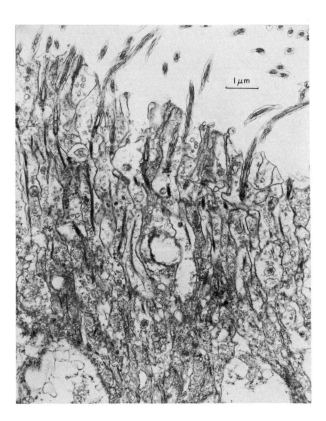

Fig. 3. Longitudinal section through the base of a sensory fan of Spadella (Chaetognatha). Each nonmotile cilium lies at the terminal of the apical dendrite of a bipolar sensory cell. Long roots (period 55 mµ) run the whole length of the apical dendrite.

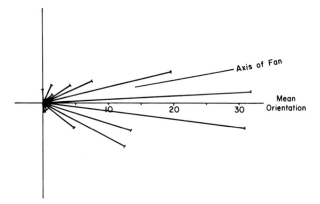

Fig. 4. Frequency distribution of the orientation of the line joining the central pair of fibrils in relation to the axis of the fan, for a sample of 130 cilia of Spadella.

shearing force on the basal region of the cilium. However, the shaft would act equally well whichever way it was orientated, so that the orientation of the shaft is probably no more than an expression of the orientation of the basal body. The next example suggests that the basal body is the region which actually detects the mechanical movement.

Ctenophore Underwater Vibration Receptor

An abundant ctenophore of the Mediterranean, Leucothea, is covered by numerous fingerlike protrusions which are short in the relaxed state (fig. 5). At the slightest vibration of the water in their neighborhood the fingers shoot out to about three times their relaxed length. They work by contraction of circular muscle of the fingers acting on a hydrostatic skeleton of solid mesogleal jelly.

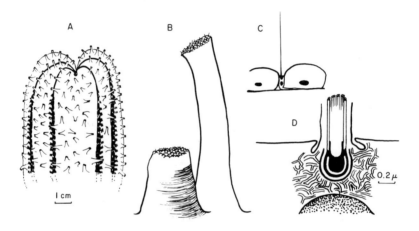

Fig. 5. Vibration receptors on the fingers of the ctenophore Leucothea: *a*, part of Leucothea, showing the fingers distributed on the body surface; *b*, a finger in the relaxed, short condition, and then extended by contraction of the circular muscle acting on the flexible jelly core; *c*, epithelium of the fingertip with large gland cells and nonmotile cilium; *d*, diagram of a basal body as in fig. 6.

The sense organs are sensory neurons which are scattered among large secretory cells over the tip of the finger. Each sensory cell bears a single nonmotile cilium which, as before, must be coupled with displacements of the water. Vibration caused by a drop of water falling 1 cm upon the surface of the aquarium 50 cm away can be effective in setting off the finger response. In an isolated finger it is not possible to touch one of these stiff cilia with a fine probe upon a micromanipulator without setting off the response. This direct test is the best indication of their function, but the stiff cilia are on the only sensory cells that are appropriate as underwater vibration detectors.

In fine structure the stiff cilium is usually not notably different from a normal motile cilium, although some show small peculiarities. The basal body, however, has a remarkable structure not known elsewhere in the animal kingdom. Concentric shells surround a solid core which is attached to the base of the cilium shaft (fig. 6). Around the outer of the spherical shells is a mass of fine tubules which are probably continuous with the outer membrane of the cell. The whole structure appears to be specialized to convert displacement of the water into shear between the concentric shells and thence to depolarization of the outer membrane of the neuron by current flow in the system of tubules.

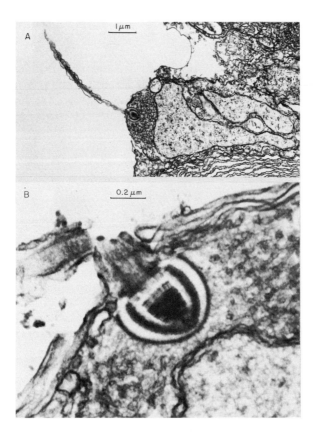

Fig. 6. Vibration receptor cilium of Leucothea, in longitudinal section: *a*, the cilium rooted in a mass of tubules at the surface of an epithelial sensory cell; *b*, enlarged view of the modified basal body.

The Vertebrate Acousticolateralis Gravity and Vibration Receptor

Lowenstein (chap. 24, this volume) has described the alignment of the cilia in the receptor cells of the vertebrate ear. All the acousticolateralis organs, including the lateral line sense organs in fish and amphibia, belong to one family of cells. Typically the sensory cells each bear a single kinocilium and a group of stereocilia. The kinocilium resembles a normal cilium but is probably nonmotile; the stereocilia are thin villi which have rootlets but no central fibrils such as those within cilia. It has been shown by direct demonstration that the direction of maximum sensitivity of the sensory cells of the lateral line of the teleost fish Lota lies at right angles to the line between the central pair of fibrils in the stereocilium (Flock 1965). A shear of the surface in the direction from stereocilia to kinocilium causes depolarization, whereas shear in the opposite direction causes hyperpolarization, as measured by a microelectrode. The structural pattern, including the orientation of the central fibrils, is remarkably constant throughout all vertebrate classes and, until the contrary is proved, the most reasonable assumption is that they function in the same way as the invertebrate examples. In the vertebrates, however, the ciliated primary sensory neurons do not have axons of their own.

The Statocyst of the Octopus

In the wall of the octopus statocyst the rows of ciliated sensory cells are loaded by the statolith which lies at the center. Tilting the statocyst in one direction causes excitation of certain of the sensory cells while tilting in the opposite direction causes inhibition. The direction of the shearing force is at right angles to the line between the central pair of fibrils (Barber 1966).

In contrast to those of vertebrates and the advanced hydromedusae, the receptor cells of the octopus statocyst bear many kinocilia as well as stereocilia.

Coelenterate Nonmotile Cilia and Jellyfish Statocysts

In the superficial ectoderm of sea-anemones, the Hertwig brothers (1879) described ectodermal sensory cells, each with a nonmotile process. From the other results in this survey, it is reasonable to infer that these are underwater displacement receptors. Passano and Pantin (1955) describe the great sensitivity of sea-anemone tentacles to minute underwater vibrations, and Josephson (1961) showed a similar sensitivity in the hydroid Syncoryne. Sensitivity to underwater vibration is also characteristic of hydromedusae, and groups of sensory cells forming small tufts of nonmotile cilia lie around the margin of some hydromedusae, for example, Aglaura and Rhopalonema (Hertwig and Hertwig 1878). As will be seen, these are structures from which statocysts could be evolved.

Jellyfish. Sensory cells of scyphozoan jellyfish are recognizable under the electron microscope because the single cilium springs from the center of a small crater. Unfortunately, this criterion does not hold for other cnidarian groups, where isolated nonmotile sensory cilia have not yet been found under the electron microscope. In hydromedusae, the sensory cilia can be found where they are concentrated in the statocysts and are described below.

The statocysts of jellyfish usually lie in multiples of 8 around the margin of the bell, each organ in its fold of sensory epithelium. The sensory nerve fibers participate in a ganglion of nervous tissue that is also the seat of spontaneity of the beat of the bell. Artificially moving the statolith so that it presses on the sensory epithelium on one side or the other causes the rate of the beat initiated by the ganglion to be accelerated or inhibited. The rate of beating is probably not as significant as the asymmetrical effect of the tilting of the statocyst on the tonus of the bell muscle, so effecting a turning movement by the animal (Fränkel 1925; Bozler 1926). As in the geotactic response of ctenophores, undisturbed jellyfish swim upward, but when disturbed the same animals swim downward. Therefore the asymmetrical steering action of the bell edge, which relaxes more slowly on the side toward which the animal turns is on the upper side in upward swimming animals but on the lower side in downward swimming ones (fig. 7). There must be a switching system in the marginal ganglion, and the sign of the geotactic response depends on whether it is the outer or the inner eipthelium which evokes the local tonic contraction. This choice of alternative behavior patterns, so characteristic of central nervous systems, has not been clearly expressed in textbook accounts of jellyfish statocysts.

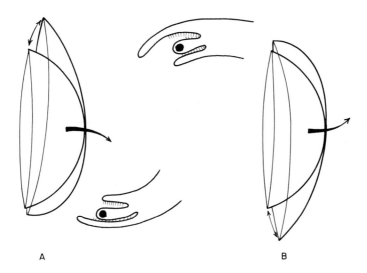

Fig. 7. The asymmetrical gravity response of jellyfish: a, in downward-swimming animals the lower edge remains tonically contracted and so swings through a smaller stroke; b, the converse occurs in upward-swimming animals. As shown by the enlargement of the marginal sense organs, the reactions are caused by the statolith making contact with one of two sensory epithelial folds. In a species such as Aurelia or Cyanea some individuals tend to swim up, some swim down, and some swim along horizontally.

The only new result upon the organization of jellyfish marginal ganglia I can add is that the processes of the sensory cells upon which the statolith bears are nonmotile but otherwise typical cilia, each of which emerges from the center of a small crater (fig. 8). The axons of these cells run into the nearby marginal ganglion, which is a mass of neuropile with few intrinsic neurons. Where the sensory cilia are immediately above or below the statolith, it is evident, from surveying large areas under the electron microscope, that many of the shafts have the line between the central pair of fibrils at right angles to the line of the expected deflection by the statolith.

Some jellyfish, for example, Rhizostoma (Coleman, private communication) are extraordinarily sensitive to underwater vibration and will sink at the approach of a ship or when they come into disturbed water. The receptor is not known but it is clear that the same nonmotile cilia near the ganglia, backed by the inertia of the statolith, are the most likely candidates.

Hydromedusan Statocysts. Most of the hydromedusan statocysts are not yet known in detail, but I have recently concentrated upon the structure of several trachymedusae in which they are well developed and easily orientated under the electron microscope. One can draw up a series of increasing complexity, from Cunina, which has an external organ like a pendulum, through Rhopalonema, with a simple external capsule, to Geryonia, which has a complex internal capsule.

In all these forms the statocyst is organized in the opposite way to that in most other animals, where the stotocyst lies in a cavity which is lined by sensory epithelium. In trachymedusae the ectodermal sensory epithelium surrounds the statolith which is secreted by an endodermal layer below.

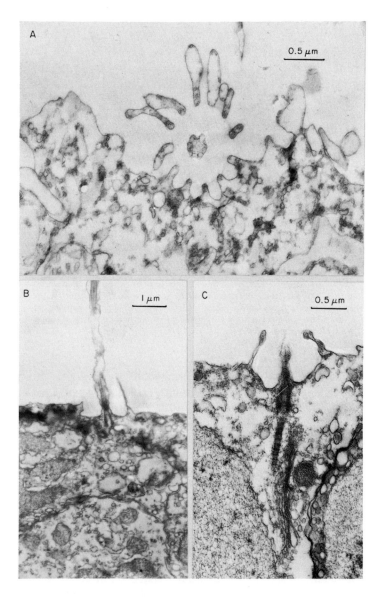

Fig. 8. The bases of sensory cilia of the marginal ganglia of jellyfish (Scyphozoa): a, Nausithoë in TS through the base; b, Aurelia in LS; c, Nausithoë in LS. The ring of processes around the base seems to be characteristic of sensory cilia in this group of animals.

Cunina is a common transparent medusa of the Mediterranean, a few centimeters in diameter. The marginal sense organs, numbering 30-50, spring from the region of the outer nerve ring and project externally around the margin into the seawater (Hertwig and Hertwig 1878). They are surrounded by long nonmotile cilia which are each the single process of a sensory cell. The statocyst of endodermal origin is covered by an ectodermal sensory epithelium which bears more nonmotile cilia.

The general anatomy was worked out by the Hertwigs. My contribution has been only to confirm the details of the body layers, to show that the sensory "hairs" are in fact nonmotile cilia, and to plot their orientation with reference to the direction of loading by the statolith when the animal is tilted. Cunina is particularly favorable in this respect because it is obvious how the deflexion of the organ will load the cilia (fig. 9). In general, the line between the central pair of fibrils is at right angles to the direction of loading. The cilia further from the sense organ are orientated with the line at right angles to the central pair of fibrils running radially along the bell (fig. 9c). It is essential to section the cilia near their bases to avoid misapprehension caused by twist of the shafts.

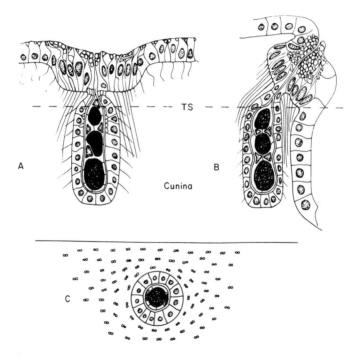

Fig. 9. Marginal statolith of the hydromedusan Cunina (narcomedusae): a, in view looking along a radius of the bell; b, in side view; c, the orientation of the central pair of fibrils in the cilia is a compromise between being tangential to the bell and tangential to the statolith.

The outer surface of the statolith is covered with a sensory epithelium which also bears nonmotile cilia. No function is known for these, but unquestionably this is the most effective site for underwater displacement receptors because of the relatively firm base provided by the inertia of the statolith.

The next stage of development, as found in Rhopalonema, is the enclosed sensory club. The statolith projects as before, but the surrounding ectoderm is raised over it as a roof so that the statolith is enclosed in a cavity filled with seawater. The sensory epithelium still lies around a central statolith, which is secreted by endoderm, as in Cunina. The sensory cells each have an axon at their central end and a long stiff cilium which projects across the intervening space to touch the inside of the protecting hood. Under the electron microscope each stiff cilium is revealed as surrounded by a ring of 8-10 smaller, thinner projections without fibrils, called stereocilia (fig. 10). The central cilium has a typical 9 + 2 pattern in transverse section. The root is small, with no lateral foot on the basal body. The stereocilia, like those of Geryonia, have rootlets.

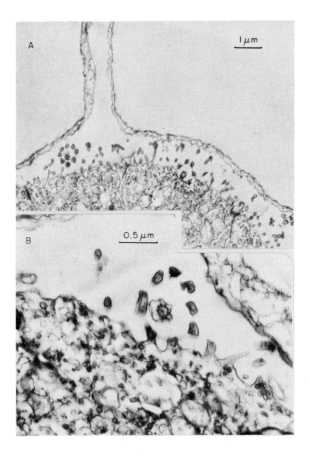

Fig. 10. Statocyst of the hydromedusan Rhopalonema: *a*, low power; *b*, high-power view of one kinocilium surrounded by a ring of stereocilia.

The obvious inference from the anatomy is that the long stiff cilia pick up any relative movement between the statolith and the surrounding hood. If this is so, the kinocilium is the only candidate for the transducer mechanism.

Geryonia (formerly called *Carmarina hastata*), a large transparent trachymedusan of the Mediterranean, has the most complex statocyst known in medusae. The capsule which contains the statolith is wholly within the mesoglea of the margin of the bell. It contains the same elements as in Rhopalonema, but two obvious nerve bundles connect it to the outer ring nerve, and there are numerous stereocilia on each sensory cell (fig. 11). The central statolith is surrounded by endoderm, a thin layer of mesoglea, and an ectodermal epithelium of columnar cells bearing axons. These cells each carry a single kinocilium and numerous stereocilia. The stereocilia project into a clear region which appears to be watery. The kinocilia reach beyond this and are embedded in a peculiarly dense and granular jelly which lies against the outer wall of the capsule (fig. 12). The sensory cells themselves are crowded with mitochondria. Since it is impossible to predict the freedom of movement of the statolith within its capsule of jelly, the motion cannot be correlated with an obvious alignment of the axes of symmetry of the kinocilia. As in other medusan statocysts, the latter appear normal in transverse section, but they have negligible roots.

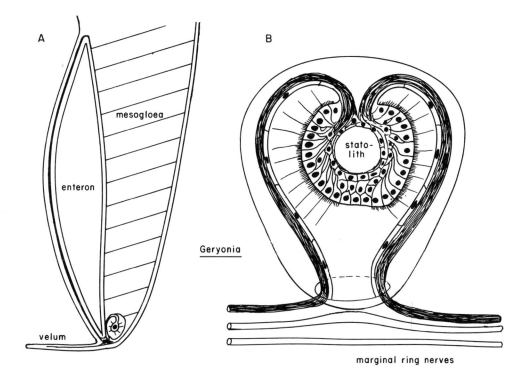

Fig. 11. The statocyst of the hydromedusan Geryonia: *a*, the edge of the bell showing the statocyst in relation to the velum and mesoglea; *b*, the two nerves from the statolith join the outer ring nerve by running along the epithelium.

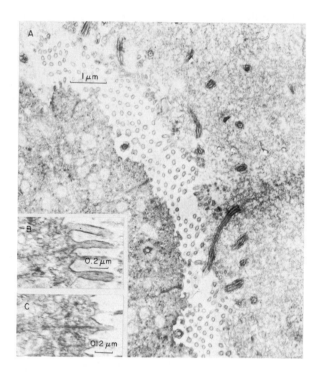

Fig. 12. Kinocilia (*k*) and stereocilia (*s*) along the outer margins of the cells which enclose and secrete the statolith in Geryonia. This is an enlargement of the edge of the statolith shown in fig. 11*b*. The clear area in the region of the stereocilia is probably seawater, and the granular material in the central space is distributed in a much larger volume of water in the living animal before shrinkage has occurred. The insets show fibrils forming roots of the stereocilia.

Discussion

The first problem concerns the location of the transducer. As a vibration receptor a stiff cilium standing alone is efficiently coupled to the medium. Only direct experiment will decide whether the shaft or the basal body, or both, is the sensitive region. The specialization of the basal body in the underwater vibration receptors of ctenophores, as in Leucothea, suggests that the basal body is the primitive mechanoreceptor, and that it has become especially elaborated in this ctenophore example. Further evidence comes from the mammalian organ of Corti, where the kinocilium is reduced to a basal body only.

Second, the distribution of stereocilia is not comprehensible in terms of a transducer action. Even when present, stereocilia are not anatomically arranged in a way which suggests their implication in transduction. They are, however, always associated with secretion of a jelly or contents of a capsule; elsewhere similar microvilli are typical of absorptive or secretory epithelium. The arbitrary distinction between stereocilia and microvilli is that only the former have thin filaments and rootlets.

Roots of the sensory cilium cannot be of universal significance in transduction. Roots are well developed in jellyfish, in ctenophore apical organ cilia, and in the chaetognath and octopus examples. They are poorly developed in hydromedusae, especially in elaborate statocysts. Roots may be straight or may be bent at right angles to the shaft, for no obvious reason. A basal foot at right angles to the shaft is not universal.

The orientation of the line between the central pair of fibrils, at right angles to the direction of maximum sensitivity, can be given no functional meaning. However, it is peculiarly widely distributed, being true for all the examples analyzed. It is probably safe, therefore, to infer from it the direction of sensitivity in organs where the mechanics cannot be studied directly. A point more difficult to explain is why the shaft should be orientated if it acts as a lever on the basal body, while the latter is symmetrical.

An apparently featureless membrane of a vertebrate or crustacean stretch receptor acts well enough as a mechanoreceptor, and even protein fibers are, in insect flight muscle, sensitive to stretch. However, a function of the kinocilium or its basal body as a very sensitive detector is suggested by its persistence in a large number of varied examples. How it works can as yet only be guessed. On account of its large surface area, a cilium is effectively a hole in the cell membrane unless the base of the shaft is plugged against the flow of ionic current. My own hypothesis is that the basal body acts as a control valve: a fall in resistance at this point would allow the cell to depolarize as if a hole $0.2\,\mu$ diameter were punched in the membrane.

This survey of examples of coelenterates and some other lower animals, taken together with consideration of the vertebrate acousticolateralis system, strongly suggests that statocysts evolved as gravity receptors from stiff cilia which were originally underwater vibration receptors. A sensory epithelium bearing stiff cilia is a more effective vibration receptor if it is backed by grains of calcareous secretion. From this a papilla develops and leads to a pendant organ as in Cunina. Such a structure is an effective gravity receptor but it trembles with every movement of the water and in other genera it is protected by a hood or a capsule But statocysts will still respond to vibration and it must be remembered that the whole margin of the medusa bell swings violently during normal swimming. The perception of the direction of gravity, as distinct from vibration, is presumably achieved by filters of long time constant in the integrative part of the nervous system. In the highest animals, speed of response is combined with elimination of vibration effects by burying the organ deeply, by numerous parallel channels in the nervous system, and by other anatomical adaptations.

Summary

1. Motile cilia are mechanically sensitive, and many lower animals have specialized nonmotile cilia upon mechanoreceptor cells.

2. Primitive nonmotile sensory cilia act as underwater vibration receptors.

3. By addition of calcareous particles near their cells, these receptors become more efficient as underwater vibration detectors. It is then a short step to the evolution of statocysts, which otherwise are useless when partially developed.

4. In different medusae one can trace the evolution of stereocilia in company with kinocilia.

5. In all cases where the question can be examined, the direction of mechanical sensitivity of the sensory cilium is at right angles to the line joining the central pair of fibrils. This is useful for prediction in other examples, but no explanation is forthcoming.

6. No particular structure such as rootlets, ciliary shaft, or stereocilia are essential for mechanical transduction in all cases. There are indications that the basal body is the transducer agent, and that the asymmetry is derived from it.

References

Barber, V. C. 1966. *J. Anat.* 100: 685-86.

Bozler, E. 1926. *Z. vergleich. Physiol.* 4: 797-817.

Child, F. M., and Tamm, S. 1963. *Biol. Bull. Woods Hole* 125: 373.

Dijkgraaf, S., and Kalmijn, A. J. 1966. *Z. vergleich. Physiol.* 53: 187-94.

Flock, A. 1965. *Cold Spring Harbor Symp. Quant. Biol.* 30: 133-45.

Fränkel, G. 1925. *Z. vergleich. Physiol.* 2: 658-90.

Hertwig, O., and Hertwig, R. 1878. *Das Nervensystem und die Sinnesorgane der Medusen.* Leipsig: Vogel.

–––. 1879. *Die Actinien, anatomisch und histologische mit besonderer Berücksichtigung des Nervenmuskelsystems untersucht.* Jena: Gustav Fischer.

Horridge, G. A., and Boulton, P. S. 1967. *Proc. Roy. Soc. Land Ser. B* 168: 413-19.

Josephson, R. K. 1961. *J. Exp. Biol.* 38: 17-27.

Machin, K. E. 1958. *J. Exp. Biol.* 35: 796-806.

Passano, L. M., and Pantin, C. F. A. 1955. *Proc. Roy. Soc. Lond. Ser. B* 143: 226-38.

Romanes, G. J. 1892. *Darwin and After Darwin* vol. 1. London: Longmans, Green and Co.

Schneider, K. C. 1902. *Lehrbuch der vergleichenden Histologie.* Jena: Gustav Fischer.

Sleigh, M. A. 1966. *Amer. Rev. Resp. Dis.* 93: 16-31.

Slifer, E. H. 1961. *Int. Rev. Cytol.* 11: 125-59.

Wersäll, J.; Flock, A.; and Lundquist, P.-G. 1965. *Cold Spring Harbor Symp. Quant. Biol.* 30: 115-32.

Whitear, M. 1962. *Philos. Trans. Royal Soc. Lond. Ser. B* 245: 291-324.

Discussion

LOWENSTEIN: Dr. Horridge, you mentioned vibration and statoreception. What about the old hypothesis that there are stimulatory organs?

HORRIDGE: Two aspects are pertinent here. One is that sense organs often have a background discharge and in this sense they are stimulatory organs. The statocyst organs of medusae are associated with nervous tissue that has its own intrinsic rhythm and is responsible for the beat. The beat disappears when all the statocysts are cut off. There is another aspect to their function, in that every movement shakes the receptor and in some way leads to the next movement, so that it may keep itself going by an iterative process. I think the answer is that we just don't know how much this contributes because there is usually no way of doing the necessary experiments.

WEIS-FOGH: Concerning the echinoderm organ with cells that have a higher specific gravity than ordinary cells. Is that due to a vacuole inside the cell which may, for instance, contain a higher concentration of some inclusion? I bring this up for the benefit of the botanists, in that you may have instead of the usual statoliths, specialized vacuoles with a high concentration of an ion, such as sulfate, which is known to control the density of diatoms.

LAVERACK: It certainly is one cell in the interior statocyst. It has an enormous vacuole relative to the size of the cell, with an attenuated cytoplasm along the wall. The nucleus sits up on one side; the rest of the cell seems to be a vacuole with subsidiary small ones around it. As to what's in the middle, I couldn't tell you. Nobody has ever analyzed it, and it would be a manipulative problem to so do. I wonder if they might, in fact, float rather than sink, and have ammonium ions in them as do some of the cranchiid squids, which float with mantle uppermost owing to high concentrations of ammonium ions.

GORDON: If my impression is correct, Dr. Horridge, the ctenophora sense but do not localize vibration, but the Chaetognatha both sense and localize vibration. Does this imply that some of these not too dissimilar sensor groups in the latter are able to triangulate?

HORRIDGE: The ctenophore has no central nervous system at all and each of these fingers acts as an independent organ. There are many fingers. But the arrowworm is much more goal oriented. He sits quite still in the water, and when a copepod swims by he grabs it and takes it in his jaws. A glass thread (or a wire in the dark) vibrated nearby at 50 μ amplitude and 10 cycles per sec will be grabbed in the same way. There is definitely a directional orientation.

GORDON: If one extirpates the fans on one side, what happens?

HORRIDGE: We tried that. We covered them with an alginate glue that is nontoxic. If you cover one side, the animal localizes very well with the other side. Furthermore, if you shake the tank so that all the water vibrates, the animal does not localize at all. It has to have a directional stimulus so that a triangulation is possible.

COHEN: The question has been brought up previously whether plants have any structures analogous to the basal bodies. The answer from the botanists has been, I believe, no. But in

Dr. Sievers's work there were structures he identified as "multivesicular bodies." We see these occasionally in the nervous system of animals and I have not been able to determine their function from the literature. The basal bodies appear somewhat similar in structure to that of the multivesicular bodies. Could these structures not be the common links of mechanoeffector function in the two groups?

LAVERACK: In the sensory cell of Arenicola I have seen a multivesicular body which doesn't look like the basal body of the cilium. Something quite discrete, something quite separate.

COHEN: Do you know what they do?

HORRIDGE: Many persons have seen multivesicular bodies, and nobody knows what they do. They occur particularly in visual cells.

COHEN: Are they anything like the motile cilia that we have seen here in plants?

GALSTON: In some of the ferns, Joseph Gall has taken a look at the components of the mitotic apparatus; there are 9 + 2 oriented fiberlike bodies that may be associated with chromosome movement. These are part of an atypical spindle apparatus.

COHEN: But they are only there for part of the life cycle and could not presumably function as statoreceptors.

GALSTON: That is correct.

LYON: A plant sensor that has at least a superficial resemblance to an animal sensory apparatus is the trigger hairs of the Venus flytrap. These transmit some type of impulse through tissues and produce a mechanical reaction by a rapidly propagated change in hydrostatic pressure.

AUDUS: We just started electron microscope studies of this structure. The cytoplasm in the hinge cell, which is the cell that is deformed when the trigger hair is bent, is very peculiar indeed. It is full of very small vesicles, which may arise from the endoplasmic reticulum.

COHEN: Could they be tubules? What is their diameter relative to the tubules of the cilia you have just seen here?

AUDUS: No, they are not tubules, they are vesicles. They have a much bigger diameter than the microtubules associated with the cell wall. They are not uniform but heterogeneous in size. We haven't actually measured them. There is a suggestion that when you distort the hair these vesicles tend to fuse into a much smaller number of large vacuoles, but this has to be verified. Perhaps they are involved in some way in the transfer of an action potential. There is another feature of this particular structure which we haven't investigated in great detail, and that is the presence of very large pores in the walls. The walls are extremely thick, and their intercommunicating pores are very much larger than plasmodesmata. But again we haven't, as yet, examined enough sections to be able to follow the orientation and the path of these interconnecting structures. I would like to think that they are, in fact, concerned with action potential conduction.

GALSTON: In an equally preliminary sense we have begun electron micrographic studies on tendrils of pea plants, which I think can be considered gravitational orienters in the sense that they pull plants up and away from the earth. When you stimulate one of these by stroking with a glass rod, it responds very quickly. We have ascertained that there is a contractile process occurring in the ventral side of the tendril. Looking at the cells on that side of the tendril one does see fibrillarlike tubules in the cells that we know do most of the contracting. In view of the fact that we've also ascertained that there is considerable ATPase activity in those cells, we are tempted to believe that we have something like a contractile fibril there.

22
Gravity Receptors and Gravity Orientation in Crustacea

Hermann Schöne, *Max Planck Institute, Seewiesen*

In an old and famous experiment, Kreidl (1893) investigated the function of the cyst organs of Crustacea. He experimented with shrimps which, like most Crustacea, have to refill their cysts after molting because the cyst is part of the exoskeleton that is discarded during the molting process. Kreidl kept the shrimps in a tank with iron filings on the bottom. After molting, the shrimps which had iron particles in their cysts responded to directional changes of a magnetic field so that their ventral sides pointed toward the magnet. Kreidl concluded that the cyst organs serve as equilibrium organs. Therefore he called them *statocysts* instead of *otocysts,* as they had been called before when they were believed to have acoustic function.

Gravity receptors deliver information about position in space. This information is used to control the motor activities of the animal in relation to the direction of gravity. I will discuss how this control is accomplished in Crustacea. The material is ordered in four sections:

1. Basic anatomy of the statocyst and the mechanism of stimulation.

2. Statocyst input and compensatory eye movements.

3. Integration of statocyst and proprioceptor information.

4. Orientation of locomotion of the intact animal relative to the gravitational field.

Statocyst Construction and the Mechanism of Stimulation

Morphology and Anatomy of the Gravity Receptor

Gravity receptors are composed of sensitive elements and of masses which stimulate these elements by means of their specific weight. In Crustacea the masses are heavy substances called statoliths. They are attached to the sensory hairs which protrude from the walls of the cyst cavity. In the decapod Crustacea the cyst is located in the basal segment of the two first antennae, the antennules.

The gross morphology and anatomy of the cyst is known from the investigations of Hensen (1863), Prentiss (1901), and Kinzig (1918). Hensen and Kinzig also describe the fine structure of hairs and the innervation. The basic statocyst morphology seems to be similar for all decapod Crustacea.

The structure of the crayfish statocyst is shown in figure 1. The statolith particles which enter by the opening are cemented together and joined to the feathery hairs by means of a secretion. The basal part of the hair is formed like a barrel, which stands on top of a canal in the cuticular wall (fig. 2). More than three-fourths of the circumference of the barrel wall is composed of a thin corrugated membrane; the remaining sector is stiffened by sclerotized layers. This hard triangular part is called "Zahn," meaning tooth. The barrel carries the hair shaft. That part of the hair shaft which is opposite to the tooth protrudes inwards like a small tongue; it is called the *lingula*. It extends into a fine cuticular thread, the chorda, which runs down through the lumen of barrel and canal. Further down it seems to join the endings of the sensory cells. The distance between the hair base and the nervous elements is rather long: at least $100\,\mu$ have to be bridged by the chorda, which seems to serve only as a mechanical transmission structure. This is the picture drawn from the results of Hensen and Kinzig. It contains some blank areas. Whereas Hensen states definitely that the sensory process is connected to the end of the chorda, Kinzig remains doubtful about the kind of connection.

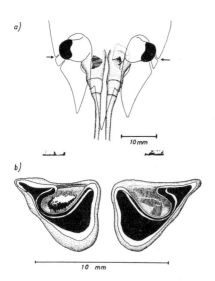

Fig. 1
Statocyst of a crayfish. (*a*) Dorsal view of anterior part of the animal. Bristles covering statocyst opening have been removed on the left side. The outline of the cyst sac is indicated by a dotted line. (*b*) Anterior view of preparation cut through at the level of the arrows in *a* above. The statolith mass can be seen attached to the hairs in the cyst at the left, and the hairs alone are visible in the other.

Fig. 2
Structure of sensory element of crayfish statocyst. (*a*) View of hair from tooth side. (*b*) Cross-sections through regions indicated by arrows in (*a*). (*c*) Side view of hair with indication of canal and chorda. (*d*) Diagram of structure and possible mechanism of action (*a-c* redrawn from Kinzig 1918).

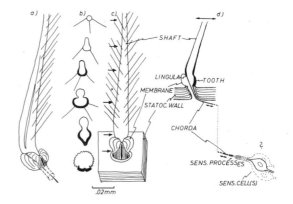

Recently we started an electron microscope investigation of the crayfish statocyst (Schöne and Steinbrecht (1968). Our findings are in basic agreement with those of Hensen and Kinzig and further enlarge the picture. The chorda inserts at the end of the hair shaft. There is no distinct border between the cuticle of the contact area and the material of the chorda. The chorda runs down the canal, bending around the edge of its inner opening, and continues nearly parallel with the cyst wall (fig. 2d). After some 70-170 μ the chorda meets the nerve endings. We discovered three nerve endings connected to *one* chorda. Kinzig had postulated only one sensory cell for each hair. Figure 3 summarizes our findings in this region in a schematic manner. Three sensory processes with blunt tips stick into an intercellular space filled by the granular substance of the chorda end. Proximally, that is, downwards, this material becomes less and less conspicuous. The three processes contain densely packed microtubules. In a distance of about 50 μ the microtubules vanish and the processes constrict in diameter. A typical ciliary structure of nine double filaments arranged in circular order appears in the constricted region of all three processes. A central pair of filaments is lacking. Proximal to the constricted ciliary sector the sensory processes enlarge again. In this preciliary sector they interdigitate and show intimate contact with each other by desmosomes. The processes then diverge from each other and follow a curved path away from the epidermis. The distance between chorda end and curve covers about 70 μ. Each then ends in its sense cell. Thus each process is the dendrite of one sensory cell.

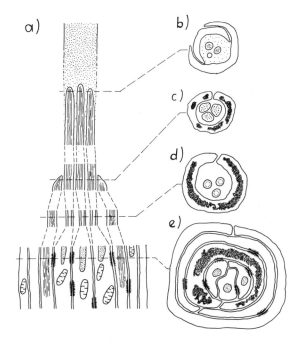

Fig. 3
Fine structure of area of contact between chorda and sensory processes. (*a*) is a longitudinal section. (*b*) Through (*e*) are cross-sections at the levels indicated by the broken lines in (*a*). (*b*) Three sensory processes embedded in the granular material of the chorda, sheaths cells not shown in (*a*). (*c*) Sensory processes containing microtubules surrounded by sheath cells with bundles of microtubules. (*d*) Constricted "ciliary sector" surrounded by sheath cell, bundles of microtubules coalesced. (*e*) Preciliary sector with interdigitation of sensory processes, desmosomes, mitochondria, and central bodies in processes, microtubules in surrounding sheath cell.

Each process contains one electron-dense central body and many mitochondria, which are not found in the upper postciliary sector of the dendrites. Desmosome connections can also be seen between the nerve processes and the surrounding sheath cells as well as between the sheath cells themselves. The sheath cells envelope the dendrites in several more or less concentric layers. In the ciliary region the innermost layer contains a dense body probably composed of masses of microtubules. The body bridges the ciliary area, sometimes forming an almost completely closed tube. The overall structure of the whole apparatus resembles the scolopidia of the chordotonal organ in the legs of Carcinus as described by Whitear (1962), except for the nature of the contact with the apparatus applying the mechanical stimulus.

The Mechanism of Stimulation

We know only the first and last link of the chain which transforms mechanical energy into nervous activity. Initially the mechanical energy caused by a relative change in the direction of the gravitational force is applied to the hairs by means of the statolith mass. The morphological features shown in figures 1 and 2 suggest the following process. Positional changes of the animal cause a shift of the statolith mass along the floor of the cyst. The displacement of the statolith moves the hair about the hinge formed by the connection of tooth and shaft. If, for instance, the hair is tilted towards the hinge side, the opposite end of the shaft with the chorda insertion is lifted. The movement causes the chorda to move around the canal edge and this in turn stretches the attached sensory processes. This seems to be the mechanical action initiating the nervous excitation. The lever construction of the hair indicates a directional sensitivity. Full excitation should be elicited only if the stimulating force acts in that plane of the hair which is defined by the long axis of the hair and a line connecting the tooth and the insertion of the chorda. This assumption coincides with electrophysiological investigations in lobsters (Cohen, personal communication).

If this picture is correct, only a force exerted on the hairs from the side would cause a stimulation. This conclusion is supported by the experiments shown in figure 4 (Schöne 1951). A fine jet of water was directed into a statocyst in which the statoliths had been removed. When the hairs were bent by the water jet, typical righting reactions were observed. The animals showed the same kind of equilibrium reaction that is seen in a normal crayfish which is rolled about its long axis (upper pictures). All appendages participate in this righting behavior: antennae, eyestalks, walking legs, swimmerets, and telson. The movements of the legs and of the eyestalks are most obvious, the legs of the lower side paddle, while those of the higher side are folded against the trunk. Both eyestalks are moved towards the higher side. During the artificial stimulation the animals reacted as if tilted towards the side of the stimulus direction; that is, if the hairs were bent laterally, the legs of that side paddled. If the stimulus direction was changed and the hairs were moved towards the mid-line the reactions were reversed. It should be noted that both kinds of equilibrium reactions, righting to the left and righting to the right, could be elicited from *one* statocyst (fig. 4*b*).

These experiments as well as the anatomical considerations indicate that a force acting parallel to the floor and bending the hairs is the adequate mechanical stimulus. This stimulation is exerted by the shear component of the force caused by the weight of the statolith mass. The geometrical relations can easily be derived from figure 5: shear changes with the magnitude of the gravitational force and with the sine of the tilting angle.

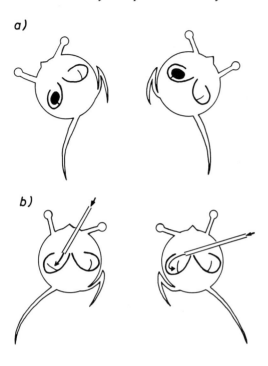

Fig. 4
Effects of statocyst stimulation on righting response of walking legs and eyestalks of crayfishes. (a) Animal with only one statolith tilted to right and left side showing paddling of legs of lower side, folding of legs of upper side and deviation of eyestalks toward upper side. (b) Bending of hair by waterjet from median to side (left picture) and from side to median (right picture) resulting in same righting behavior as in the respective tilting position of (a). (After Schöne 1951.)

Fig. 5
Geometrical analysis of stimulating forces in the statocyst. The shear components change proportionally with the relative weight of the statolith F and with the sine of the tilting angle a in the following way: $S = F \cdot \sin a$. The weight of the statolith is the product of its mass and the acceleration of gravity which is normally 1 g. The relative weight is the weight of the statolith minus the weight of the displaced fluid.

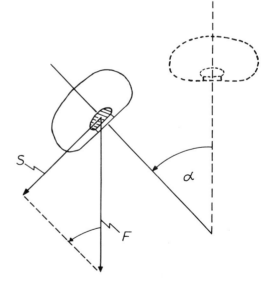

Very little is known about the next step in the transducing chain, the transformation of the mechanical energy into nervous activity. This transformation may occur in the region of the ciliary structures. We know much more about the end of the chain, the spike activity of the sense cells.

Figure 6 shows the results of electrophysiological investigations in the lobster Homarus (Cohen 1955). The lobster is tilted around the transverse axis. The impulse frequency of the single fiber preparation increases when the rostrum is lifted up from the horizontal, reaching a maximum at about 90° of tilt, then decreases again. When frequency is plotted on a linear scale, as in figure 6 the result is a sinusoidal curve. The "bell-shaped" curves of Cohen result if a logarithmic scale is used. This indicates that the discharge frequency increases with the sine of the tilting angle. Because the mechanical force of shear also varies with the sine of the tilting angle, we can conclude that the nervous output is linearly related to the mechanical input. This implies that the transduction from the first to the last step is performed on a linear scale.

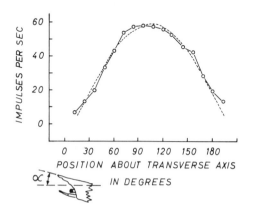

Fig. 6
Recordings from a single position receptor of *Homarus americanus*. The impulse frequency is plotted against the rotation of the animal about its transverse axis (pitch). Each point is the average of measurements taken when the animal was rotated clockwise and counterclockwise. Impulse frequency on linear scale. (Modified after Cohen 1955.)

Statocyst Input and Compensatory Eye Movements

I have already mentioned the compensatory eye movements as part of the equilibrium reactions. I will now concentrate on the quantitative relationship between eyestalk movement and statocyst input in shrimps (Schöne 1954). An animal was placed in an apparatus and turned about its longitudinal axis. In each position the eyestalk position was measured relative to the dorso-ventral plane of symmetry. The bisecting line of the angle enclosed by the eyestalks formed the angle ϵ with the symmetry plane. In figure 7b this angle is plotted against the position of the animal. The data are from animals with only one functioning statocyst. The curve follows a sine function with extreme values at positions of about 60° and 240°. As the drawings on the right-hand side show, the statocyst floor carrying the hairs is at a 30° angle relative to the transverse axis of the shrimp. After a tilt of 60° the floor of the cyst has a vertical position. This implies that the eyestalk deviation is greatest when the shear exerted on the hairs is strongest. These results lead to the conclusion that the eyestalk deviation changes linearly with the force which stimulates the sensory hairs of the gravity receptor.

The effects of the right and left statocyst on the eyestalk position are added algebraically, as can be seen from figure 7c. The addition of curves from animals with the left statocyst intact to curves from animals with the right statocyst intact results in curves which resemble those obtained from animals with both statocysts functioning.

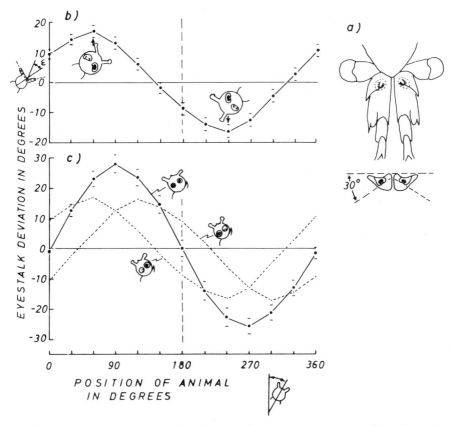

Fig. 7. *Palaemoneters varians*. (a) Statocyst topography demonstrated in an exuvium. Above: view of frontpart of animal. Below: optical section through statocyst region. (b), (c) eyestalk deviation in relation to position of animal. (b) Only left statocyst working, right statoliths removed; (c) solid curve—both statocysts working; dotted curves—either left or right statocysts working, data identical with that of (a). Right-statocyst curve calculated from left-statocyst curve.

Average of 21 animals, brackets indicate three times standard error. Same shrimps have been used in experiments of (a) as in experiments of (b), solid curve. Note that adding up the two dotted curves results in the solid curve of (b), indicating that effects of right and left statocysts are added algebraically. (From Schöne 1954.)

Integration of Statocyst and Proprioceptor Function

Very often the sense organ monitoring the relation of an animal to the gravitational field is lodged in an appendage which can be moved with respect to the body. If a mammal tilts its head or a spiny lobster lifts its antennules, the gravity receptors are stimulated but the animals do not show any reaction to the stimulus.

We investigated this problem in lobster *Panulirus argus* using again the compensatory movements of the eyestalks (Schöne and Schöne 1967). The animal was tilted rostrum up or down about its transverse axis. The eyestalks compensated for the movement by rotating in the opposite direction around the stalk axis. Pointers attached to the stalks indicated the exact position. The angle which the pointers formed with the transverse plane of the animal was plotted against the degree of tilt. Three kinds of positional changes were investigated:

A. The whole animal was moved from 45° up to 45° down;

B. Only the antennules were moved in the same range, the trunk being kept still;

C. Only the trunk was moved, the antennules being fixed in space.

The statocysts should be stimulated in A and B, but the graph of figure 8a shows that reactions occured in A and C. In both of these situations the trunk was tilted. In order to determine if there might be another gravity receptor in the trunk, we blocked the gravity receptors in the antennules by removing the statoliths. No reactions occurred under condition A when the whole animal, including the trunk, was tilted (fig. 8b). This finding excluded the possibility of a gravity receptor in the trunk.

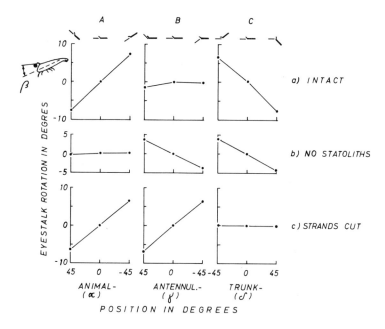

Fig. 8. *Panulirus argus*. Interaction of statocysts and antennular proprioceptors. Eyestalk rotation (β) in relation to position of: (A) whole body, angle α between longitudinal body axis and horizontal, (B) antennules, angle γ between antennules and horizontal, (C) trunk, angle δ between longitudinal trunk axis and (horizontally fixed) antennules. (a) Intact lobsters with normally filled statocysts, average of 15 specimens; (b) lobsters lacking statoliths, average of 5 specimens; (c) lobsters with destroyed proprioceptor strands of antennular basal joint, average of 3 specimens.

In B and C of figure 8b the reactions are almost equal. In both situations the position of the antennules in relation to the trunk is the same. Therefore it was reasonable to look for a proprioceptor which might control this posture. We found a stretch receptor organ at the base of the antennule which suited the prerequisites (fig. 9a). It is an "innervated elastic strand." This kind of proprioceptor is well known in Crustacea from the work of Alexandrowicz and Whitear (1957). The structure seems to be similar to that of the receptors in the distal joints of the Panulirus antennule as described by Wyse and Maynard (1965). It consists of an elastic strand composed of many elastic fibers. Usually one small nerve runs down from the middle of the strand to the antennular nerve. The strand is covered with sensory cells and inserts with rootlets on both ends. At the proximal end these rootlets are attached to the tendon of the main promotor muscle, which lifts the antennula. At the distal end the strand is fastened to the base of the antennule. As can be seen in figure 9b the organ is stretched when the antennule is lifted.

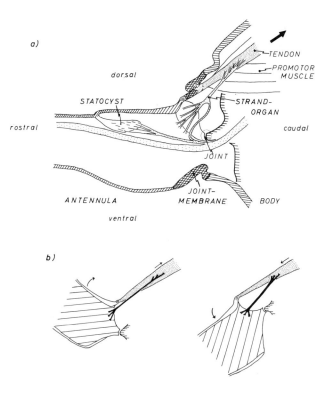

Fig. 9. *Panulirus argus.* Proprioceptor in the basal joint of the antennules. (a) schematic representation of anatomy, (b) model of antennular stretch receptor in the two extreme positions of the antennule, demonstrating function of strand organ.

We tested the function of this organ in animals with cut strands (fig. 8c). The eyestalks did not react under condition C, when only the trunk was moved and the antennules were fixed in space.

But in A and B situations the reactions were equal, indicating that in both experiments the statocysts are fully effective.

These results show that the statocysts and the antennular proprioceptors both affect the equilibrium reactions, but they act antagonistically to each other. If only the antennules are moved, both types of organs are stimulated and their effects cancel each other. But in all cases involving positional changes of the trunk, only one type of organ is stimulated and therefore a reaction is elicited. When the whole body is tilted, the statocysts are responsible for the reaction; if only the trunk is moved, the proprioceptors are used. The two sensory systems interact so that equilibrium reactions are elicited only when the trunk changes its orientation with respect to gravity.

This, however is true only for displacements close to the normal rest position (fig. 10). When the animal is nearly upside down, the additive effects of statocyst and proprioceptor cause an increase of eyestalk reaction instead of cancellation. Within a range close to the rest position, a dorsal displacement of the antennules results in an increase of shear in a proximal direction. When, however, a lobster in the upside-down situation lifts its antennules dorsally, the statocyst stimulation is opposite, that is, shear increases in the distal direction. The proprioceptor stimulation, however, is the same in both situations. The result is a nullification of equilibrium reactions in the former situation and a magnification in the latter. Thus this mechanism of gravity orientation operates adaptively only within a range that is close to the normal position.

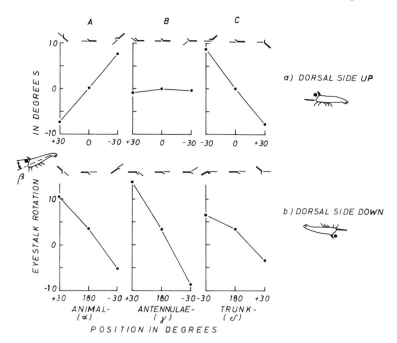

Fig. 10. *Panulirus argus*. Interaction of statocyst and antennular proprioceptor in intact lobsters: (*a*) the dorsal-side-up range of positions. (*b*) The dorsal-side-down range of positions. Same kind of experiments as explained in fig. 8*a*, except that positions were changed at 30° instead of 45° intervals. (*a*) and (*b*) represent the averaged values of the same 3 specimens.

Orientation of Locomotion in Space

An equilibrium reaction is usually defined as a reaction serving to restore the normal position of the animal relative to gravity. But the normal position is not the only position which is controlled by the statocysts. A shrimp swimming downwards also orients with respect to gravity. It corrects deviations from its oblique position in the same manner as it does from the normal position.

Shrimps were investigated in centrifuge experiments in which the magnitude of gravity could be altered (Schöne 1957). Figure 11a shows the apparatus. It could be placed, together with the observer, in the gondola of a big centrifuge (fig. 11b, c). When the shrimp was liberated at the water surface, it started on a downward course. Figure 12a shows the angle between the swimming position and the water surface at various gravity magnitudes. The path was less steep when the gravitational force was increased. In the lower graph the data are plotted in terms of shear. Shear was calculated using the formula: $S = F \cdot \sin \alpha$ (fig. 5). As the curves indicate, all shear values are of the same magnitude; that is, the shrimp behaves so as to maintain a constant statocyst input. The experimental increase of gravity causes a disturbance of this input. The shrimp then reacts with a change of its swimming direction in order to reestablish the input value. This implies that the shrimp controls its downward course by maintaining a particular value of statocyst excitation. This can be called the reference value of statocyst excitation that is valid for that particular course. This kind of gravity orientation can be considered as a feedback control system in which the

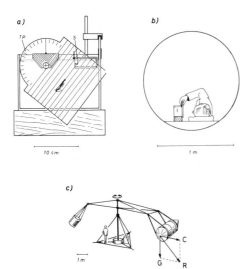

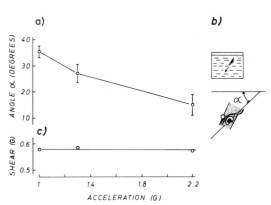

Fig. 11. Design used to investigate swimming direction of shrimps at various gravity magnitudes. (a) Swimming container with device TP for reading of swimming direction, S start. (b) Container and observer in gondola of centrifuge. (c) Centrifuge in motion. G, gravity force; C, centrifugal force; R, resultant force acting always perpendicular to floor of gondola.

Fig. 12. Palaemonetes varians. Downward swimming and statocyst stimulation at various force levels. (a) Swimming direction, angle α as demonstrated in (b). (b) Above downward swimming shrimp; below, angle α between statocyst floor and water surface. (c) Shear component of the statolith at the various force levels, computed using $S(G) = F(G) \cdot \sin \alpha$. Note that shrimps behave so as to maintain statocyst stimulation (shear) at a constant value. (After Schöne 1957.)

statocysts play the role of the recording instruments. The motor activity propels and directs the animal in a particular direction. This results in a corresponding statocyst stimulation which is fed back and thus serves to control the effects of the motor activity.

Many orientation systems work as feedback control mechanisms. It is difficult to describe this kind of mechanism in terms of the taxis terminology of Kühn (1918) and Fraenkel and Gunn (1961). The systems of taxes and kineses refer only to those parts of orientation which involve the correction of deviations that are caused by forced changes of the stimulus direction.

Summary

The relations between statocyst function and motor activities are described at four levels of complexity.

1. Fine structure, morphology, and anatomy of the statocysts and their receptive elements are described. The structural features, combined with experiments using direct stimulation, indicate that the sensory hairs are stimulated when bent by the shear component of statolith weight. Electrophysiological recordings suggest a linear transduction of mechanical input into nerve activity in the receptor system.

2. Equilibrium reactions and compensatory eye movements are dependent on statocyst input. Eyestalk position changes linearly with shear component. The effects of right and left statocysts add algebraically.

3. The interaction of the statocyst and of a proprioceptor was studied in experiments on compensatory eye movements in spiny lobsters. The proprioceptor controls the movement of the antennules which contain the statocysts. When the antennules move relative to the trunk, the proprioceptor output counteracts that of the statocyst to cancel the eyestalk response.

4. Besides the well known "normal equilibrium position," the control of the direction of locomotion also depends upon statocyst function. As shown in experiments on shrimps under various levels of gravitational force, orientation is controlled by a feedback loop in which the statocysts are the recording instruments. Orientation systems of this kind do not fit into the old concept of taxes.

References

Alexandrowicz, J. S., and Whitear, M. 1957. *J. Mar. Biol. Ass. U.K.* 36: 603-28.

Cohen, M. J. 1955. *J. Physiol.* 130: 9-34.

———. 1960b. *Proc. Roy. Soc.* (London) B 152: 30-49.

Fraenkel, G. S., and Gunn, D. L. 1961. *The Orientation of Animals, Kineses, Taxes, and Compass Reactions* (New Ed.), Dover Press, New York: 1-376.

Hensen, V. 1863. *Z. wiss. Zool.* 13: 319-412.

Kinzig, H. 1919. *Verh. Naturhist.-med. Ver.* (Heidelberg), N. F. 14: 1-90.

Kreidl, A. 1893. *S. B. Akad. Wiss.* (Vienna) 102: 149-74.

Kühn, A. 1919. *Die Orientierung der Tiere in Raum.* Gustav Fischer Verlag, (Jena): 1-71

Prentiss, C. W. 1901. *Bull. Mus. comp. Zool.* 36: 167-251.

Schöne, H. 1951. *Naturwissenschaften* 38: 157-58.

———. 1954. *Z. vergl. Physiol.* 36: 241-60.

———. 1957. *Z. vergl. Physiol.* 39: 235-40.

——— and Schöne, H. 1967. *Naturwissenschaften* 54: 289.

Schöne, H., and Steinbrecht, R. A. 1968. *Nature* 220: 184-86.

Whitear, M. 1962. *Phil. Trans. Ser. B* 245: 291-325.

Wyse, G. A., and Maynard, D. M. 1965. *J. Exp. Biol.* 42: 521-35.

23
The Integrative Action of the Nervous System in Crustacean Equilibrium Reactions

William J. Davis, *University of Oregon*

Nearly all animals have evolved *equilibrium reactions* to maintain an upright attitude in the earth's gravitational field. Sense organs, usually including specialized "gravity receptors," register any deviation from the upright position and convert it into a neural "code," which is sent to the central nervous system. There the information is processed and then relayed to the muscles which participate in the equilibrium reactions. These muscles may compensate for the animal's new position by altering a portion of the body or individual appendages without changing the orientation of the entire organism (compensatory responses). This is illustrated by the responses of the eyestalks of a tilted crab (Bethe 1897; Horridge 1966; Burrows and Horridge 1968) or crayfish (Schöne 1954, 1959). Alternatively, the muscles may actively restore the animal to the upright position (righting responses), as in the case of the responses of the limb and trunk muscles when a vertebrate is tilted (Lowenstein 1936).

Some of the sensory and motor systems which participate in the equilibrium reactions of crustaceans have been described by Schöne. I will elaborate on these topics, and then discuss the integrative role played by the central nervous system in converting the gravity sensation into equilibrium reactions.

A Survey of Crustacean Equilibrium Reactions

As indicated by Schöne, the body appendages of crustaceans display a rich variety of equilibrium reactions. When a lobster (*Homarus americanus*) is rotated around the long axis of its body (rolled), the distal portions of both antennulae, both of the antennae, and the large third maxillipeds (mouth parts) are all pointed toward the side tilted downward (fig. 1a). The two eyestalks, on the other hand, are pointed upward, so that the position of the eyes remains relatively constant (fig. 1b). Since these movements do not produce a significant amount of righting torque, they fall into the category of compensatory responses.

The compensatory response of the eyestalks presumably tends to stabilize objects in the visual field, but the specific advantage conferred upon the lobster is not known. The adaptive significance of the remaining compensatory responses, on the other hand, seems more obvious. The antennulae, antennae, and mouth parts are highly specialized sensing devices, covered with numerous chemoreceptors and sensory hairs. By pointing them toward the side tilted downward, the lobster is presumably in a better position to anticipate the dangers or delights of the environment which it may be destined to enter in the next few moments.

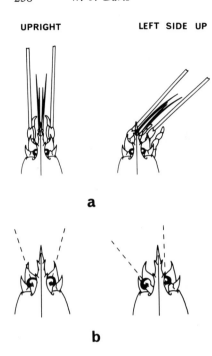

Fig. 1
Compensatory responses of a lobster to tilt about the long axis of the body. The dorsal aspect of the cephalo-thorax is illustrated, with anterior at the top. a) The responses of the antennulae, antennae, and third maxillipeds. b) The response of the eyestalks.

The remaining appendages which participate in equilibrium reactions produce a righting torque when the lobster is rolled. Their movements are therefore classified as righting responses. When the lobster is rolled by tilting the substrate on which it is standing, the claw on the side tilted downward is pressed against the substrate to resist the roll. The walking legs of the two sides react tonically and oppositely, so that the lobster's body remains relatively upright. If the lobster is lifted above the substrate and rolled, the claw on the side tilted upward is lifted and extended forward, while the claw on the side tilted downward is lowered and folded beneath the front of the body. The legs on the side up are also lifted and extended forward, while the legs on the side down perform cyclic, metachronous paddling movements, presumably because of the lack of resistance which is normally provided by the substrate (fig. 2a).

The appendages of the abdomen, the swimmerets (pleopods) and uropods (tail fans), also respond to roll with bilaterally asymmetrical movements. The swimmerets on the side tilted upward perform rhythmic, metachronous rowing movements (figs. 2b and 3) which are directed out toward the side, so that they produce a cyclic righting torque around the long axis of the lobster's body (Davis 1968). The swimmerets on the side tilted downward beat either not at all or straight to the rear. The uropod on the side tilted upward is closed, while the uropod on the side down is opened and extended downward (fig. 2c), so that the rearward water currents produced by the swimmerets on the side down flow against it. These currents presumably generate a force perpendicular to the plane of the uropod, much as the wind produces a force at right angles to a sail; such a force may be resolved into two components, one of which is a righting torque around the long axis of the lobster's body.

Integrative Action of the Nervous System

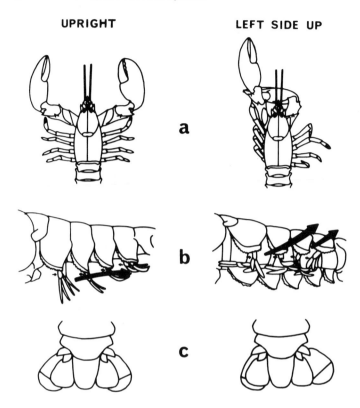

Fig. 2. Righting responses of a lobster to tilt about the long axis of the body. *a*) The responses of the claws and walking legs. *b*) Tracings from single frames of high-speed motion pictures of a lobster's abdomen, showing the response of the swimmerets. Anterior is on the left, ventral on the bottom of the illustration. Heavy arrows show the direction of water currents produced by the swimmerets (from Davis 1968). *c*) A dorsal view of the posterior end of the lobster, showing the response of the uropods. Anterior is toward the top of the drawings.

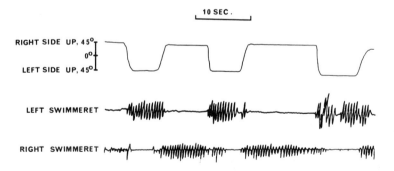

Fig. 3. Response of the left and right swimmeret of one abdominal segment to rotation of the lobster around the long axis of its body. The upper trace shows the angular position of the lobster around its long axis, with 0° corresponding to the upright position. The two lower traces show the movement artifact recorded through fine wire electrodes inserted into the swimmeret muscles. Each complete oscillation on the lower two traces corresponds to one complete cycle of swimmeret beating, and the amplitude of the oscillations is directly proportional to the strength of the swimmeret movement.

Sensory Control of the Righting Responses

One possible source of the sensory information which controls the righting responses is the eyes. When a lobster is blinded, however, the swimmerets and uropods still react strongly when the lobster is rolled. When the "balancing organ," or statocyst receptor, on each side of the lobster is surgically destroyed, however, the righting response of the swimmerets and uropods is abolished (fig. 4c), even though the lobster's remaining sensory capabilities are unimpaired. The righting responses of the swimmerets and uropods are, therefore, controlled exclusively by the statocysts.

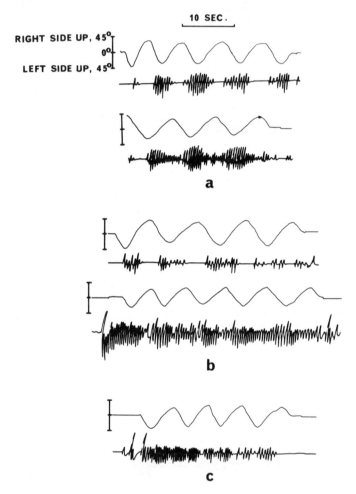

Fig. 4. A series of records taken from one lobster, showing the response of one swimmeret on the left side (lower trace in each record) to rotation of the lobster around the long axis of its body (upper trace in each record). The records were produced as described in the legend to figure 3. The calibration scale for the lobster's position is the same for all records. a) Intact lobster. b) Left statocyst destroyed. c) Both statocysts destroyed. The two records in a show the normal righting response of a swimmeret on the left side during rotation toward the left side up (upper record), and modulation of the frequency and amplitude of spontaneous swimmeret beating by rotation of the lobster (lower record). Destruction of the left statocyst weakens but does not eliminate the swimmeret response (b). Destruction of the left and right statocysts abolishes both the response and the modulation effect (c) (from Davis 1968).

The swimmerets on both sides of an upright lobster are usually inactive. When the statocyst on one side is destroyed, however, the swimmerets on that side tend to beat spontaneously when the lobster is upright. Similarly, the two uropods of an upright lobster are normally held in about the same position, but when one statocyst is destroyed the uropod on the same side tends to close, while the uropod on the opposite side tends to open. These tendencies toward bilateral asymmetry are eliminated by rotating the lobster so that the side with the destroyed statocyst is tilted 20° downward (Davis 1968). In other words, a lobster with one statocyst destroyed "thinks" that it is upright when in fact it is tilted toward the side of the destroyed statocyst.

When a lobster with one statocyst destroyed is rotated back and forth about its long axis, the righting response of the swimmerets and uropods still occurs (figs. 4b, fig. 6). Furthermore, the swimmerets and uropod of a given side respond to roll in the same way when only the right statocyst is present as when only the left statocyst is present (figs. 4b, 5b, 6). Therefore the swimmerets and uropod of one side receive control information from both statocysts. Moreover, when a lobster is rolled in one direction, the sensory information from the right statocyst has the same effect on the swimmerets and uropod of one side as the sensory information from the left statocyst.

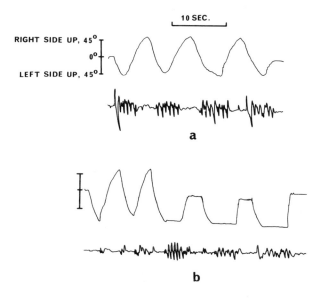

Fig. 5. Records similar to those in figure 4, showing the response of one swimmeret on the left side (lower trace in each record) to rotation of the lobster around its long axis (upper trace in each record). The calibration scale for the lobster's position applies to both a and b. a) Intact lobster. b) Same lobster after destroying only the right statocyst. This operation weakens but does not eliminate the response of the swimmerets on the left side (from Davis 1968).

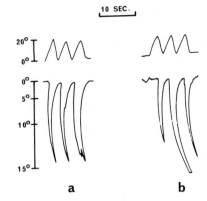

Fig. 6

The closing response of the left uropod (downward deflection of the lower trace in each record) during a 20° rotation of the lobster around its long axis toward the left side up (upward deflection of the upper trace in each record). *a*) Left statocyst destroyed. *b*) A different lobster, following destruction of only the right statocyst. 0° on the upper trace corresponds to the position of bilateral symmetry following unilateral statocyst destruction (see text). The scale on the lower trace shows the magnitude of the closing response, expressed in terms of the change in the angle between the two rami of the uropod. This angle was recorded by means of movement transducers mounted on the rami (Davis 1968).

Since the statocysts are apparently the only source of sensory information for the righting responses, precise information about the afferent inflow from these receptors would permit inferences about the "processing" performed by the central nervous system. The sensory physiology of the statocysts has in fact been studied in detail by Cohen (1955, 1960). When the lobster is upright, the individual statocyst receptor units continually bombard the central nervous system with streams of "spontaneous" impulses. When the lobster is tilted to one side, only the acceleration receptors, or "thread hairs," and the type II position receptors respond. The average discharge frequency of these receptors on the side tilted downward increases while the average discharge frequency on the side up decreases (fig. 7).

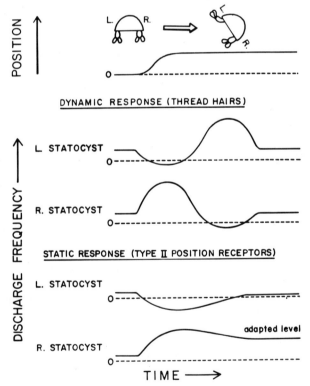

Fig. 7

The probable net response of both statocysts to rotation of a lobster around the long axis of its body, as shown at the top of the picture. This figure was constructed from data on the responses on single statocyst receptor units, as described by Cohen (1955, 1960).

The Integrative Role of the Central Nervous System

When the sensory responses of the statocysts are related to the equilibrium reactions which they control, an interesting feature of the central nervous "processor" is revealed. Cohen showed that when a lobster is rolled in one direction, the afferent responses of the two statocysts are *opposite*. My experiments show, however, that when a lobster is rolled in one direction the right statocyst affects the swimmerets and uropod of one side in the *same* way as the left statocyst. We must conclude that the appendages of one side "recognize" that decreased discharge from the statocyst on the same side has the same meaning as increased discharge from the statocyst of the opposite side. This means that at some site between the sensory input and the motor output the effect of the statocyst signal from one side is reversed.

A simple neural model which can account for the righting responses of the uropods is shown in figure 8. According to this model, each statocyst sends control information to the uropods of both sides. The muscles which open the uropod, however, receive commands only from the statocyst of the same side, while the muscles which close the uropod receive commands only from the statocyst of the opposite side. The position of each uropod is assumed to be maintained by a balance of tension in antagonistic muscles, as has been found in the case of the eyestalk muscles (Burrows and Horridge 1968). Destroying the statocyst on one side would reduce the tonic discharge to the opener muscles on the same side and to the closer muscles on the opposite side. The uropod on the same side as the destroyed statocyst would tend to close, while the uropod on the opposite side would tend to open. This bilateral asymmetry could be abolished by reducing the discharge level of the remaining statocyst, that is, by tilting the lobster toward the side of the destroyed statocyst. Furthermore, according to the model, either statocyst could alone control the righting response of either uropod. All of these properties of the model are in accord with the experimental results.

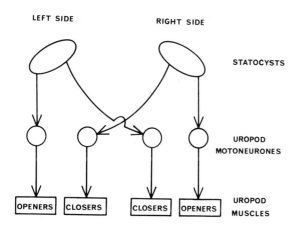

Fig. 8
A hypothetical neural model which can account for the righting responses of the uropods. Arrowheads represent excitatory inputs. The operation of the model is described in the text (from Davis 1968).

According to the model suggested above, the afferent signals from opposite statocysts are routed to antagonistic uropod muscles on one side of the lobster. As a consequence of this suggested "wiring," the required reversal of the effect of one statocyst signal occurs at the level of the motoneurons. Recent work on the uropods of crayfish (Larimer and Kennedy 1969) has shown that an elaborate set of central "command" interneurons controls the attitude of these appendages—thus my model may be *too* simple. In the case of the swimmerets, the existence of endogenous central "oscillators" as well as excitatory and inhibitory command interneurons has long been recognized (crayfish, Hughes and Wiersma 1960; Ikeda and Wiersma 1964; Wiersma and Ikeda 1964; lobsters, W. J. Davis, unpublished data). Repetitive electrical stimulation of an excitatory command interneuron turns on the rhythmic motor pattern underlying swimmeret beating, while stimulation of an inhibitory command interneuron supresses the output. A neural model which incorporates these features and at the same time accounts for the righting responses of the swimmerets is shown in figure 9.

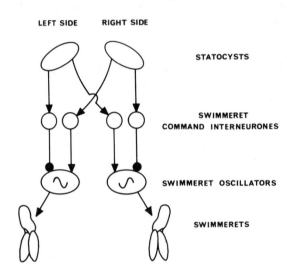

Fig. 9
A hypothetical neural model to account for the righting responses of the swimmerets. All of the components of the model have been experimentally demonstrated and studied. Only the connections between these components are hypothetical. Black circles represent inhibitory inputs, while black arrowheads represent excitatory inputs. The operation of the model is described in the text (from Davis 1968).

In this model the antagonistic swimmeret command interneurons simply take the place of the antagonistic motoneurons in the uropod model. The inhibitory command interneuron of each side is activated only by the statocyst of the same side, while the excitatory command interneuron of each side is activated only by the statocyst of the opposite side. In the intact, upright lobster, both the excitatory and inhibitory swimmeret command interneurons are assumed to be tonically active, so that their effects cancel. Rolling the lobster left side up would send increased inhibitory and decreased excitatory discharge to the oscillators of the right side, and decreased inhibitory and increased excitatory discharge to those of the left side. Consequently, only the swimmerets of the left side would respond with righting movements, as required. Unilateral statocyst destruction would result in an unbalanced turning tendency, in accord with the experimental observations.

Areas for Future Investigation

The above models, based as they are on limited data, cannot deal with several interesting questions. For example, how are the equilibrium reactions to pitch (somersault) mediated? Within each statocyst, individual position receptor hairs are distributed in concentric semicircles on the floor of the statocyst sac (Cohen 1960). Electron microscope studies have revealed that each sensory hair is structurally polarized so that it is probably most responsive to bending toward or away from the center of the circle on whose circumference it lies (Schöne, personal communication). This organizational feature presumably implies that the afferent responses to pitch and roll are mediated by different sets of receptor hairs. Different central circuitry must also be employed during pitch, since both the sensory inflow from the statocysts and the compensating motor responses are bilaterally symmetrical. Under natural conditions any deviation from the upright position undoubtedly contains components of both pitch and roll, and the overall behavioral response presumably represents an appropriate blend. The structural organization of the statocysts seems ideally suited to exert precise control over this blending operation, but whether these sense organs can in fact play such a sophisticated filtering role in the behavior remains to be determined.

A second unanswered question relates to the role of the "thread hairs" or acceleration receptors. Indirect behavioral evidence suggests that they contribute to the equilibrium reactions (Davis 1968), but their relative importance is unknown.

Third, most of my experiments have been performed on the swimmerets and uropods; it is not known whether the proposed models can also account for the righting responses of more anterior appendages. As described earlier, the righting movements of the walking legs can be either tonic or cyclic, and the underlying neural mechanisms may combine elements of both models. Attempts to answer this question may also shed light on how the locomotory movements of the legs are produced.

Fourth, the righting responses discussed here could also be explained by models based on inhibitory cross-connections between excitatory command interneurons, as found between the Mauthner neurons of the goldfish (Furshpan and Furukawa 1962; Furukawa and Furshpan 1967). At present such models are less appealing than the ones proposed here, but only because detailed information on the interactions between crustacean command interneurons is not yet available.

Finally, to what extent can the models proposed here also account for the equilibrium reactions of higher animals? We comparative neurophysiologists usually hope (sometimes secretly) that our investigations of relatively simple invertebrate nervous systems will provide insights into the workings of the more complex systems found among mammals. In the case of equilibrium reactions, this hope seems justified. The many striking functional similarities between the lobster statocysts and their counterparts in vertebrates, the labyrinths, have been noted by Cohen (1955, 1960). Behavioral and ablation experiments also reveal major similarities between lower and higher animals in regard to equilibrium reactions; in both cases, for example, unilateral destruction of the balancing organ causes an unbalanced turning tendency (Burns 1962; Roberts 1967). Of greatest interest, however, are indications that the underlying integrative mechanisms may also bear fundamental similarities. Shimazu and his colleagues have shown that, in cats, a labyrinth of one side inhibits the central neurons associated with the labyrinth of the opposite side (Shimazu, this volume; Shimazu 1967; Shimazu and Precht 1966). These and other experiments (Sala 1963; Cook, Cangiano, and Pompeiano 1968) suggest that neuronal circuitry similar to that postulated here is also operative in the equilibrium reactions of higher animals.

The relative simplicity of the lobster nervous system should permit a more detailed exploration of the neuronal basis of equilibrium reactions than is possible in mammals. For example, recordings from the swimmeret command interneurons have proved possible (W. J. Davis and D. Kennedy, in preparation) and, if performed while manipulating the statocyst receptor units, could provide a critical test of the hypotheses proposed here. Such experiments will undoubtedly contribute to our further understanding of the neuronal mechanisms underlying the responses of an organism to gravity.

Summary

Tilting a lobster on the long axis of its body evokes equilibrium reactions from all of the appendages. These reactions include *compensatory responses*, by which the animal adapts to the tilted position, and *righting responses*, by which the animal actively seeks to restore the upright position. The righting responses of the abdominal swimmerets and uropods, which have been studied in the most detail, are controlled exclusively by the "balancing organs," or statocyst receptors. Either the right or the left statocyst can alone control the righting responses of the appendages of both sides, even though the afferent responses of the two statocysts to roll in one direction are opposite. Neural models based on the bilaterally reciprocal organization of statocyst influences are proposed to account for these findings. Likely areas for future investigation are discussed, and it is argued that the resulting insight may also be applicable to higher animals.

Acknowledgment

This work was supported by U.S.P.H.S. grant 5 R01 NB01624 to M. J. Cohen, U.S.P.H.S. postdoctoral fellowship NB 24, 882, and U.S.P.H.S. grant NS-09050.

References

Bethe, A. 1897. *Arch. mikr. Anat.* 50: 460-546.

Burns, R. D. 1962. *Amer. Zool.* 2: 396.

Burrows, M., and Horridge, G. A. 1968. *J. Exp. Biol.* 49: 251-67.

Cohen, M. J. 1955. *J. Physiol.* 130: 9-34.

———. 1960. *Proc. Roy. Soc. Lon.* B 152: 30-49.

Cook, W. A., Jr.; Cangiano, A.; and Pompeiano, O. 1968. *Pflug. Arch. ges. Physiol.* 299: 334-38.

Davis, W. J. 1968. *Proc. Roy. Soc. Lon.* B 170: 435-56.

Furshpan, E. J., and Furukawa, T. 1962. *J. Neurophysiol.* 25: 732-71.

Furukawa, T., and Furshpan, E. J. 1967. *J. Neurophysiol.* 26: 140-76.

Horridge, G. A. 1966. *J. Exp. Biol.* 44: 275-83.

Hughes, G. M., and Wiersma, C. A. G. 1960. *J. Exp. Biol.* 37: 657-70.

Ikeda, K., and Wiersma, C. A. G. 1964. *Comp. Biochem. Physiol.* 12: 107-15.

Larimer, J. L., and Kennedy, D. 1969. *J. Exp. Biol.* 51: 135-50.

Lowenstein, O. E. 1936. *Biol. Rev.* 11: 113-45.

Roberts, T. D. M. 1967. *Neurophysiology of Postural Mechanisms.* Plenum Press New York.

Sala, O. 1963. *Experientia* 19: 39.

Schöne, H. 1954. *Z. vergl. Physiol.* 36: 241-60.

―――. 1959. *Ergeb. Biol.* 21: 161-209.

Shimazu, H. 1967. In *Neurophysiological Basis of Normal and Abnormal Motor Activities*, ed. M. D. Yahr and D. P. Purpura. Raven Press, New York, pp. 155-76.

Shimazu, H., and Precht, W. 1966. *J. Neurophysiol.* 29: 467-92.

Wiersma, C. A. G., and Ikeda, K. 1964. *Comp. Biochem. Physiol.* 12: 509-25.

Discussion (of the Two Preceeding Papers)

BROWN: It is fair to conclude that there is nothing unique about the gravity-sensing hair cells in the crustacean statocyst?

COHEN: There is nothing about this type of hair cell that sets it apart from other mechano-sensitive hairs in arthropods. There is a directional sensitivity in these receptors that may not be present in all other arthropod hair cells. If a resting hair is moved toward the floor of the cyst the background discharge decreases. If the hair is lifted towards the vertical from the rest position it will increase its discharge frequency up to a point where it becomes overstretched.

BROWN: Is this just in Crustacea or does it occur rather generally in many invertebrates?

SCHÖNE: It should be recalled here that the stimulus-response diagrams of some insect mechanoreceptors (bees) as well as receptor elements of the vestibular endorgans of vertebrates show a directional polarization.

LAVERACK: Hair receptors on the carapace of lobsters also show directional sensitivity.

COHEN: If this is so then there is no known difference, Dr. Brown, between gravity receptors and other hair receptors in arthropods. There is also nothing unique about the acceleration-sensitive hairs other than the fact that they are not in contact with the load. If you mechanically deflect and hold those thread hairs (in the intact animal) that respond only to acceleration they will show a different tonic discharge frequency for each maintained deflection.

BROWN: Then what is unique is that the whole thing is stimulated by a batch of sand and chewing gum.

COHEN: That's right. The special properties are due to mechanical accessories; as far as we know there is no specialization at the cellular level.

GALSTON: Do I understand correctly that the animal gets an impression of its orientation not only from the angle of deviation of the hair from its normal orientation, but from the position of this hair in the entire receptor?

COHEN: Yes. You might ask why an animal like a lobster has approximately a thousand neurons that apparently provide highly redundant information. It could just as well obtain the information with one receptor except for the bell-shaped response curve. And, as Dr. Westing mentioned, how does the animal know whether it is on its back or right side up when it has this type of response curve? What we find is that the range over which individual hairs respond is not identical. If you draw the response curves of several different hairs, there is a considerable spread over the entire stimulus range. There may be some overlap but the curves are definitely not superimposed upon one another. The animal probably tells up from down by depending on the composite input from all of these receptors. A particular unit may fire at the same frequency at two different positions in space because of the bell-shaped response curve. However the total input will differ for the two positions because of the different response ranges of the various receptors.

GUALTIEROTTI: The results do not always appear so nice as they look here. There is a large flutter in the frequency of the response while the behavioral sensitivity to the angle of tilt is so high that it should demand extremely precise information. Therefore, more than one neuron is required to be able to perform the analysis of the input-output transfer through cross correlation of several neurons. This is an additional reason why the animals need a large number of neurons.

COHEN: True. This background discharge is not regular, but appears to be random. There are some interesting questions that can be asked about why random noise has apparently been deliberately built into the background discharge of this system.

JOHNSON: What happens if you remove both statocysts?

DAVIS: You abolish the righting responses of the swimmerets and the uropods. I haven't looked in detail at the sensory control of the other equilibrium reactions, but there is no a priori reason to suppose that they have a different sensory basis. The swimmerets are still capable of beating spontaneously.

JOHNSON: Is there some sort of memory bank?

DAVIS: No. They completely "forget" how to respond when the lobster is tilted. The spontaneous beating following bilateral statocyst destruction undoubtedly reflects the influence of higher nervous "centers" and perhaps other sensory systems.

HORRIDGE: I have some information which is relevant to the eyestalk control. In crabs there are four pairs of muscles (actually there are nine altogether) which work to some extent in two pairs. One pair of muscles are antagonists for tilting forward, one pair are antagonists for tilting sideways, and the other pair are for the horizontal opto-kinetic movements. In addition, when you record during any of these movements you find that the muscles are composed of tonic and phasic fibers. The slow statocyst responses are coordinated by the tonic fibers. It is possible to record while tilting and to plot the frequency of impulses to the tonic fibers in any of these muscles. It is then easy to show that changes in motor output frequency correlated with the statocyst response of the eyes are quite independent of the eye position. This rules out one of the possible feedback loops based upon the eye position itself.

HOSHIZAKI: As I visualize your cross-fed model, it can detect whether the animal is right-side up or upside down.

DAVIS: No, I don't think the model is any more capable of detecting whether the animal is right-side up or upside down than is a single statocyst receptor. According to the model, a righting response is evoked only when the two statocysts respond differentially. It is a "push-pull" system. At 0° and 180° of tilt there is no difference between the inputs from the two statocysts. The model predicts that the lobster cannot distinguish between these two positions. If in fact you hold a lobster upside down, the behavior of the swimmerets is indistinguishable from their behavior in the upright position.

HOSHIZAKI: You don't have a phase shift between left and right?

COHEN: The point is that when the animal is upside down the input from both sides is going to be symmetrical but of a different amplitude than for the right-side up position.

LAVERACK: While all this is going on what the devil are the legs doing? The animal normally stands four square on the floor.

DAVIS: In the adult, lifting the animal off the substrate initiates swimmeret beating. If you then rotate the animal back and forth about its long axis, you evoke righting reactions not only from the swimmerets but also from the other appendages, including the legs. In my experience the righting reactions of the legs are much less certain to occur than the righting reactions of the swimmerets. The swimmeret righting reactions are practically inevitable.

LAVERACK: But when the animal is walking around, it generally depends on its legs rather than its swimmerets.

DAVIS: No, both are used. Continuous beating movements of the swimmerets act as a supplementary device in adult locomotion, especially when the lobster behaves as if he wants to get somewhere badly. Moreover, the statocyst input also reaches the swimmerets when the legs are in contact with the ground. When the lobster is rotated by tilting the substrate on which it stands, the swimmerets still respond with strong righting responses.

HORRIDGE: What is the use to the animal of a mechanism which, when you tilt the animal by 55°, gives only a partial compensation of 10°? If it were really using this mechanism, wouldn't it make almost a 55° compensation of the eye?

SCHÖNE: Compensatory eye movements are quite common within those groups possessing movable eyes. There are, however, no records of full compensation. In humans, for instance, the compensatory effect of the counter-rolling of the eyes is very poor and reaches a maximum of about 6° at a tilt of 60-90°. In cases where good compensation occurs it is restricted to a limited range of tilt positions close to the normal position. Compensation phenomena may be explained in terms of the underlying physiology, but there is no explanation for the biological significance of "failures" in this kind of system to compensate completely.

COHEN: Well, even you, Adrian, cannot compensate fully during nystagmus. You try to keep the same visual field by permitting the eyes to drift in a direction opposite to that of the movement, but you never quite make it and the eyes flip back. The system is most efficient in compensating for small deviations from the primary position, and this may be the most behaviorally significant sector of the response range.

WILSON: As to why experimental animals compensate only partially, we do many constraining things to animals during experiments. We don't let them compensate fully. The lobster would rock back up and the error would disappear completely, but you often don't permit him to do this. I think that whenever you keep part of the feedback loops open, you are not measuring something that is really behavioral.

HORRIDGE: No, that's not true. Because if he moves over to his side and then comes back up again, he needn't have moved his eyestalks at all.

WILSON: Well, he may need to move his eyestalks but he doesn't have to move them exactly the right amount. The animal doesn't walk around much on his side, and all the other things are working to bring him back to the null point.

BROWN: I am not sure that I understand how this gadgetry is supposed to work. It does seem important that the lobster knows how much mass is in the statolith. And if he does, I don't know how he learns this. It presents an opportunity for the experimenter to put different materials in the two sides and I suspect this has been done. What happens?

SCHÖNE: Very often we had Crustacea that showed an asymmetry of the eyestalk response curves for left and right tilt. We found that these responses were usually correlated with differences in mass of the right and left statoliths. In the normal position these animals showed no disequilibrium behavior. Experiments with artificially produced differences of statolith mass led to the conclusion that a calibration process occurs which is expressed in a shift of the symmetry status. It results in a "no-reaction" situation in the normal position. For this calibration (or compensation), a central reference system is required which might be based on tactile, kinaesthetic or proprioceptive cues.

WEIS-FOGH: Isn't it possible that it is precisely the linear proportionality of that particular receptor which makes it possible to have this indeterminancy of the statolith, which can have a big mass and a small mass?

WILSON: The linearity is possible here whereas it is not possible with something like a photic system. In the photic system we have to deal with a range of ten orders of magnitude. It is just not easy to code that much except in a logarithmic or power scale. But the gravity stimulus has a very narrow range.

V Vertebrate Responses to Gravity

24
Functional Anatomy of the Vertebrate Gravity Receptor System

O. Lowenstein, *University of Birmingham*

The earliest record of a vertebrate labyrinth in the cephalaspid fish shows the presence of a relatively large "auditory capsule." In it, two so-called vertical semicircular canals are beautifully exposed. There is no trace of a horizontal canal. The large size of the labyrinth capsule makes it highly probable that otolith-bearing maculae would be found to lie below the semicircular canals. In fact the labyrinth very likely shows the general development characteristic for the cyclostomes of the petromyzon type (Stensiö 1927). As will be shown later, the otolith organs in this type of labyrinth are concerned with the responses to gravitational stimuli and to vibration. A similar dual function has always been attributed to the various types of invertebrate statocyst. Both types of stimuli represent linear acceleration, unidirectional in the one case and oscillatory in the other. Identical in effect to those of gravity are all stimuli connected with linear translations in any plane of space, the effective stimulus being the vectorial resultant between gravity and positively or negatively accelerated linear movement.

There exists, of course, the long-standing problem of whether the semicircular canals are in fact sensitive to angular acceleration only, as has been generally assumed in the past, or whether they can under certain circumstances respond to simple linear acceleration, or at least to rotating vectors of linear acceleration. If the latter alternative were to be shown to be true, the cupula-covered sensory endings on the ampullae of the semicircular canals would be relevant to our theme. However, this question is at present totally unresolved. All kinesthetic and myotatic sensory structures respond to gravitational stimuli, but they too fall outside my terms of reference. I shall confine my remarks to otolith-bearing labyrinthine end organs, and among them to those that have been shown to respond to positional changes.

Anatomy

The Otolith Organs

Fundamentally, all otolith organs are designed on the statocyst principle. There is a sensory epithelium consisting of sensory cells with an apical system of hair processes and of supporting cells. Above this epithelium lies a mass of crystalline matter, usually composed of calcium salts. This is specifically heavier than the surrounding endolymph and this otolithic mass will therefore "seek" the lowest possible level. In doing so it will slide along the sensory epithelium, whenever this slants from the truly horizontal level. Two types of otoliths are known. The first are solid and of a constant shape characteristic of the species and, in fact, of the specific otolith organ within the labyrinth. Thus, for example in the minnow *Phoxinus laevis* the lapillus in the utriculus is nearly spherical, the sagitta in the sacculus is arrow-shaped with winglike ridges, and the astericus in the lagena is flat and star-shaped with a horseshoelike ridge on one side (Wohlfahrt 1932). They show annual growth rings by means of which the age of a bony fish can be fairly accurately assessed.

The second and more common type of otolith is made up of a pastelike mass of crystalline "statoconia" in a ground substance. The overall size and shape of such statolithic masses can be fairly constant, but they appear to show plastic flow. The solid otoliths are anchored to the membranous wall of the labyrinth by systems of connective tissue strands functioning like guy ropes. Such structures are sparingly developed in connection with pastelike otolithic masses. In the latter case the otolithic mass of one otolith organ can be continuous with that of another. In the elasmobranch labyrinth such continuity has been observed between the otolithic masses in sacculus and lagena.

The general shape and layout of otolith organs in the various classes of chordate animals has been well described in the classic work of Retzius (1881), and his findings are remarkably accurate considering the restricted range of techniques at his disposal. I am best acquainted with the otolith organs in the cyclostome and elasmobranch labyrinths, and it may therefore be appropriate to illustrate important aspects of the functional morphology of otolith organs by my own and my collaborators' findings in these two groups of lower vertebrates. As our work on the cyclostomes is as yet largely unpublished, what I say about the ultrastructure of the hair cell will refer chiefly to the situation as described for the elasmobranch *Raja clavata* (Lowenstein, Osborne, and Wersäll 1964) and for the bony fish *Lota vulgaris* (Flock 1964). A diagrammatic view of the elasmobranch labyrinth is shown in figure 1.

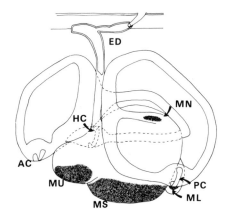

Fig. 1
Diagrammatic view of elasmobranch labyrinth. *AC*, crista of anterior ampulla; *ED*, endolymphatic duct; *HC*, crista of horizontal ampulla; *ML*, macula lagenae; *MN*, macula neglecta; *MS*, macula sacculi; *MU*, macula utriculi; *PC*, crista of posterior ampulla.

The Sensory Cells

The epithelium of the macula is made up of sensory and supporting cells. The supporting cells surround the sensory cells and are apically sealed to them by tight desmosomic junctions. The functional significance of this may well be the establishment of an ion-tight seal separating the endolymphatic labyrinthine space from the subepithelial intercellular spaces in which the dendritic endings of the first-order sensory neurones make synaptic contact with the body of the sensory cells. The chemical composition of the endolymph differs substantially from that of the intercellular fluid, which is similar to that of the perilymph, the cerebrospinal fluid, and the blood. The endolymph has a remarkably high potassium content, which is balanced osmotically in the elasmobranches by a correspondingly lower urea content. In other vertebrates, however, the higher potassium content goes hand in hand with a sodium content so low that the endolymph approaches the ionic constitution of the cell interior. I have pointed out elsewhere the physiological consequences of this state of affairs (Lowenstein 1966).

The ultrastructure of the labyrinthine sensory cells is by now well known and need not be described here in detail (fig. 2). Two aspects, however, are worthy of recall. First, this so-called secondary sensory cell has neuronal status insofar as it possesses within its cell body synaptic structures for the chemical transmission of excitation through its membrane to the dendrites of an afferent sensory neuron. Second, the hair bundle at its apical pole, consisting of an array of sterocilia and a single kinocilium, shows directional polarization. The topographic position of the kinocilium and the direction of the plane through its two central filaments, as well as the position on its basal body of the so-called basal foot, can be shown to determine the direction in which mechanical deformation leads to excitatory and inhibitory changes at the synapses and in the afferent pathway leading from them. Deformation of the hair bundle toward the kinocilium is excitatory, away from it inhibitory.

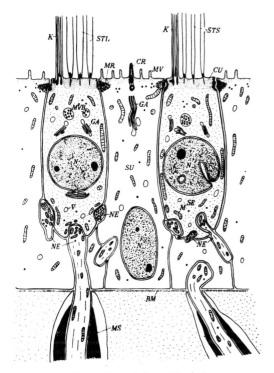

Fig. 2
Schematic drawing to represent the general structure of the sensory epithelia in the labyrinth of the ray. Two sensory cells are shown. One has a hair process composed of large-diameter, and the other, a hair process composed of small-diameter stereocilia. Nerve endings make contact with the base of the sensory cells. The supporting cells bear on their free apical surfaces numerous microvilli and a short ciliary rod. *BM*, basement membrane; *CR*, ciliary rod; *CU*, cuticle; *GA*, Golgi apparatus; *K*, kinocilium; *M*, mitochondrion; *MS*, myelin sheath; *MR*, membrana reticularis; *MV*, microvillus; *MVB*, multivesicular body; *N*, nucleus; *NE*, nerve endings; *SE*, sensory cell; *STL*, large-diameter stereocilia; *STS*, small-diameter stereocilia; *SU*, supporting cell; *V*, vesicle. (From Lowenstein, Osborne, and Wersäll 1964.)

Functional Considerations

Armed with this information we can now attempt to form an idea on how an otolith organ functions. The otological literature is replete with speculation (Magnus 1924; Quix 1935; Mygind 1930). However, it is interesting to note that the first hypothesis on otolith function formulated by Breuer (1891) may now be considered to be more closely in tune with the most recent morphological and physiological findings. Breuer postulated that the otolith glides tangentially over the sensory epithelium when the spatial orientation of the head changes from the normal.

Most of the hypotheses were formulated on the assumption that all three otolith organs (two in mammals) functioned as gravity receptors. A series of investigations on representatives of all vertebrate classes showed that this need not be so. It could be shown that the elimination of the sacculus need not have any effect on equilibration (Lowenstein 1932, 1936). Conversely, it was found that the utriculus can control all known positional reflexes during tilts about any axis.

Hair Cell Topography and Function in the Elasmobranch Labyrinth

In the elasmobranch labyrinth gravity reception is localized chiefly in the utriculus and in the lagena. The sacculus presents an interesting picture of dual function. There is one continuous otolith, but the macula sacculi responds to vibration in the anterior two-thirds of its surface. The posterior third adjacent to the lagena is a gravity receptor. Its response picture closely resembles that of the utriculus rather than that of the lagena, to which it is antagonistically opposed. This is understandable, as the surface of the macula lagenae lies in a diagonally outward facing nearly vertical plane opposed to the sacculus macula, which lies in a longitudinal and nearly dorsoventral plane. For the sake of completeness it may be mentioned that an unloaded dorsal offshoot from the utriculus macula was shown to respond to vibration and not to spatial displacement (Lowenstein and Roberts 1949).

Let us then assume the directional arrangement of the unilaterally polarized hair cells to be of fundamental functional significance. In this case, the population of hair cells in each macula represents a pattern of response directionalities indicating that none of them responds to uniform otolithic shearing forces tangential to its whole surface. The maculae are subdivided into different regions of often diametrically opposed hair-cell orientation, as shown in figure 3. Excitatory or inhibitory responses occur simultaneously or in succession as the otolithic mass flows along the macula surface under the influence of a gravitational or inertial stimulus.

It was shown that the hair cells in the elasmobranch utriculus give excitatory responses to tilts around all axes; that is, to diagonal, longitudinal, and lateral tilts. In accordance with this fact and with our basic hypothesis, we find in the utriculus macula hair cells whose kinocilia point outward interspersed with others pointing inward. However, the longitudinal axis of the utriculus itself runs at an angle to the longitudinal axis of the skull, pointing forward and outward, and the functional axis of the hair cells appears to run radially perpendicular to the periphery of the oval-shaped utriculus macula. It is therefore clear that the arrangement of the hair cells in the elasmobranch utriculus resembles that described for the utriculus macula of Lota by Flock (1964). They differ in that in Lota oppositely directed hair cells are separately assembled into a marginal inward looking and a central outward radiating field, whereas in an elasmobranch they are interspersed. In both instances the hair cell directions fully account for the response picture described for the elasmobranches by Lowenstein and Roberts (1949). The preponderance of responses to side-up tilting over those to side-down tilting reported by these authors is not borne out by the configuration of the hair-cell map, but was in all probability a consequence of selection of recording sites enforced by anatomical circumstances. Characteristic response pictures must therefore be expected from single units in the utriculus on tilts about all possible axes.

Fundamentally, therefore, it is perfectly feasible to postulate that the utriculus alone is capable of controlling all postural responses (Lowenstein 1932, 1936). However, the evidence from the work on elasmobranches shows that in these animals, at least, the lagena has an important role as a graviceptor. During tilts its responses run counter to those of the utriculus, inasmuch as the majority of units are found to have maximum discharge rates in or near the "normal" position of the skull in space. It is likely, in my opinion, that further scrutiny might disclose more units, the maximum activity of which occurs in the upside-down position. This guess is based on the presence in the nearly vertically extended macula lagenae of hair cells pointing dorsally interspersed with others pointing ventrally. It is also likely that the lagena may respond more sensitively to fore and aft than to lateral tilts. But at present this is guesswork.

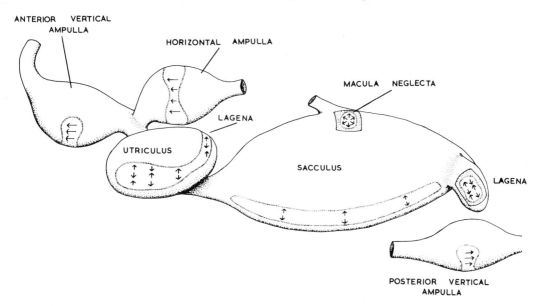

Fig. 3. Diagrammatic representation of the polarity of sensory hair bundles found in the cristae and maculae of the left labyrinth of the ray. Parts of the dorsal wall of the sacculus above the macula neglecta and of the posterior wall of the lagena have been cut away to show their two sensory areas. In this schematic rendering of the sensory hairs the orientation of the hair bundle is symbolized by an arrow, the arrowhead indicating the position of the kinocilium. (After Lowenstein, Osborne, and Wersäll 1964.)

This is the proper place to make quite plain what justification there is to make such extrapolations from topography to function. The basic hypothesis concerning the functional significance of hair-cell polarization owes its formulation to the fortunate circumstance that in the cristae of the semicircular canals all hair cells point in the same direction, and that the differences in the response picture between horizontal and vertical canals are strictly correlated with corresponding differences in the topography of their hair cells (Lowenstein, Osborne, and Wersäll 1964). The extrapolation of this hypothesis rests on nothing more than the assumption that the functional parameters of hair cells in cristae of semicircular canals and maculae of otolith organs are the same.

The macula sacculi lies in a nearly dorsoventral plane on the ventromedial aspect of the recessus sacculi. The hair-cell map is simple. Two populations of hair cells are divided by a longitudinal line. Above it they point upward, below it downward. There is very little overlap along the dividing line. A scrutiny of the gravity responses from the posterior part of the sacculus shows similarities with those obtained from the utriculus. It might be expected that further electrophysiological mapping might show a preponderance of responses to lateral tilting, as it is difficult to point to a topographically suitable substrate for a response to pure fore-and-aft tilts. A further extrapolation from topography would be to expect good maxima nearer the normal and upside-down position during full circle lateral tilts, in contrast to the lagena, in which these maxima may preponderate in the course of fore-and-aft tilting. However, at present this again is pure speculation.

The survey of the situation in the elasmobranch labyrinth may now be followed by a résumé of results of work on the labyrinth of the lamprey *Lampetra fluviatilis* (Lowenstein, Osborne, and Thornhill 1968). The labyrinth of the lamprey contains an otolith-covered sensory epithelium clearly divisible into areas which may or may not be true homologues of the maculae of the gnathostome labyrinth. An anterior horizontally disposed macula, the largest of the three, one median vertical macula, and a posterior horizontally disposed macula contain populations of hair cells each with a characteristic pattern of distribution of hair-cell directions, as seen in figure 4. The interesting feature of these three maculae is the continuity of the sensory epithelia and also of

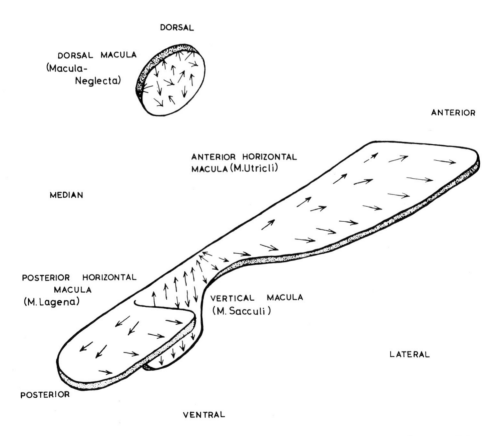

Fig. 4. Diagram to illustrate the general orientation of the sensory cells of the macular areas of the right labyrinth of the lamprey. The arrowheads indicate the position of the kinocilium. (From Lowenstein, Osborne, and Thornhill 1968.)

the overlying otolithic mass. The epithelium of the vertical macula is connected with that of the anterior and posterior horizontal maculae by upward-twisting regions in which the hair-cell pattern is transitional. In the vertical macula, the hair cells are arranged in two populations divided by a longitudinal line above and below which they point upward and downward respectively. There is little overlap. This hair-cell arrangement is identical with that of the gnathostome sacculus, of which it is the most likely homologue. The arrangement of the hair cells in the anterior and the posterior horizontal maculae are strikingly symmetrical. In the outer halves of both horizontal maculae the hair cells point toward the lateral periphery. In the medial half of the anterior horizontal macula they point forward, and in the medial half of the posterior horizontal macula, backward. There are groups of diagonally directed hair cells. In the anterior horizontal macula they are found chiefly along the border between the forward- and outward-pointing populations and point diagonally forward and outward. In the posterior horizontal macula the diagonally directed cells point backward and outward and lie along the border between the backward- and the outward-pointing populations. The hair-cell maps of the anterior and posterior horizontal maculae show therefore a remarkable degree of symmetry in opposition and they are considered to be the homologues of the gnathostome utriculus and lagena maculae respectively.

The macula communis of Myxine is less complex at first sight. It lies on the medial aspect of the labyrinth capsule and lies in a nearly vertical plane. Its anterior and posterior ends curve around, becoming more horizontal in the process. There is no anatomically different middle region. However, the hair distribution corresponds closely to that on the three maculae in the lamprey, if one imagines these to be projected more or less into a single plane (Lowenstein and Thornhill 1970).

What does all this signify in terms of function? The similarity between the horizontal maculae in the lamprey and the anterior and posterior ends of the macula communis in Myxine encourage the hypothesis that they may function in much the same way as they have been demonstrated to do in the elasmobranch *Raja clavata*. The orientation of the hair cells on these two maculae can account for excitatory responses to fore-and-aft, lateral, and diagonal tilting in opposite directions. Whether the vertical macula participates in gravity reception cannot be ascertained, as the localization of the source of impulse responses in electrophysiological experiments on the isolated labyrinth is not sufficiently accurate. There is, however, circumstantial evidence for the localization of vibration sensitivity in the vertical macula as well as in the so-called macula neglecta of the lamprey labyrinth (Lowenstein 1970). In both these sensory epithelia is found an aberrant type of hair cell, the hair bundle of which consists of an exceptionally tall and stiff kinocilium accompanied by a bundle of exceptionally short stereocilia. Clear-cut responses to vibration have been recorded from the eighth nerve of the lamprey. No such specialized hair cells are found in the macula communis of Myxine and there is no evidence that the organ responds to vibration. Furthermore, a macula neglecta is absent in Myxine. It appears therefore that the vertical macula may resemble the sacculus macula of the elasmobranches in having a dual function. The specialized hair cells in the vertical macula are dispersed among the rest and not confined to any specific area of the macula.

Few ultrastructural maps exist with respect to the various maculae in the higher gnathostomes. Early speculations on the relations between the disposition of these maculae in space and the origin of the various postural reflexes of eyes and limbs await reappraisal in the light of such new evidence. Here is an important field for further research. One may, however, be justified in assuming that the utriculus is generally the chief receptor for gravitational stimuli (Lowenstein 1932, 1936). There are exceptions. There is evidence that in the herringlike bony fishes at least a part of the utriculus rather than the sacculus may be the organ concerned with hearing (Wohlfahrt 1932, 1936), and it was shown that in the flatfish *Pleuronectes platessa* and *Platess flesus* the chief gravitational responses derive from the sacculus with the possibility of participation by utriculus or lagena or both (Schöne 1964).

Whereas von Frisch and Stetter (1932) believed that in the minnow *Phoxinus laevis* both sacculus and lagena may function as sound receptors, there is definite evidence that in the bony fish *Gymnocymbus tereutzii* and in the frogs *Rana sylvatica* and *Rana palustris* the lagena participates in the control of postural equilibrium (Schoen 1950; MacNaughtan and McNally 1946). It is interesting to note that removal of the lagena in these cases leads to instability in or near the normal position. The electrophysiological findings (Lowenstein and Roberts 1949) point to a similar functional range for the lagena in the elasmobranches.

In summary it may be said that the otolith organs of the vertebrate labyrinth respond to linear accelerations in general. They are therefore all potential gravity receptors. In addition, they may respond to linear translation, centrifugal stimuli, and rotating linear vectors during constant speed and accelerated rotations and finally to oscillatory linear accelerations in the form of vibrational and acoustic stimulation.

The mode of functioning and the functional range for the various otolith-bearing maculae in the vertebrate labyrinth have in the past been deduced from their spatial arrangement and from the effects on equilibration of separate operative eliminations of one or the other otolith organ. More recently electrophysiological work has made possible the recording of impulse responses directly from the afferent nerves serving individual maculae. This evidence is, however, confined to the lower vertebrates. Finally, accurate electronmicroscopic mapping of hair-cell orientation in the various maculae in cyclostomes, elasmobranches, bony fish, and the Squirrel Monkey (Spoendlin 1965) forms the basis for a new approach to the localization of gravitational responses in the various maculae, resting on the assumption of the functional significance of the directional polarization of the hair processes of the labyrinthine sensory cells.

I hope that this type of work can soon be extended to include the higher vertebrates.

References

Breuer, J. 1891. *Arch. ges. Physiol.* 221: 104.

Flock, Å. 1964. *J. Cell Biol.* 22: 413-31.

Frisch, K. von, and Stetter, H. 1932. *Z. vergl. Physiol.* 17: 686.

Lowenstein, O. 1932. *Z. vergl. Physiol.* 17: 806-54.

–––. 1936. *Biol. Rev.* 11: 113-45.

–––. 1966. *U.S. National Aeronautics and Space Administration* SP-115: 73-90.

———. 1970. *Proc. Roy. Soc., Ser. B* 174: 419-34.

Lowenstein, O.; Osborne, M. P.; and Wersäll, J. 1964. *Proc. Roy. Soc., Ser. B* 160: 1-12.

Lowenstein, O.; Osborne, M. P.; and Thornhill, R. A. 1968. *Proc. Roy. Soc., Ser. B* 170: 113-134.

Lowenstein, O., and Roberts, T. D. M. 1949. *J. Physiol.* 110: 392-415.

Lowenstein, O., and Thornhill, R. A. 1970. *Proc. Roy. Soc., Ser. B.* 176: 21-42.

MacNaughtan, I. P. J., and McNally, W. J. 1946. *J. Laryngol. Otol.* 61: 204-14.

Magnus, R. 1924. *Körperstellung.* Berlin.

Mygind, S. H. 1930. *Arch. Ohren. Nasen Kehlkopfheilk* 124: 238-47.

Quix, F. H. 1925. *J. Laryngol. Otol.* 40: 425-511.

Retzius, G. 1881. *Das Gehörorgan der Wirbelthiere*, vol. 1, Stockholm.

Schoen, L. 1950. *Z. vergl. Physiol.* 32: 121-50.

Schöne, H. 1964. *Biol. Jarh.* 4: 135-56.

Spoendlin, H. 1965. NASA SP-77: 7-22.

Stensiö, E. A. 1927. *The Downtonian and Devonian vertebrates of Spitsbergen.* Part I. *Cephalaspidae.* Skrifter Omsvalbard og Nordishavet. Norske Vidensk. Akad. Oslo.

Wohlfahrt, T. A. 1932. *Z. vergl. Physiol.* 17: 659-85.

———. 1936. *Z. Morph. Oecol. Tiere* 31: 271-410.

Discussion

MITTELSTAEDT: I wonder about the consequences of this back-to-back arrangement of the kinocilia. If it is true that bending toward the kinocilium increases the discharge rate of the respective sensory cell, and if you record from randomly selected fibers while tilting the animal, you should get in about 50% of the cases fibers with increasing spike rate and in the other 50% fibers with decreasing spike rate.

LOWENSTEIN: Originally we reported excitatory responses solely on upward tilting. I have reinvestigated this and it is connected exclusively with access to the macula. I have detected excitatory responses to downward tilting, and when you have a multifiber preparation consisting of, say, 5, 6, or 10 fibers you get, in fact, both increasing and decreasing spike rates.

25
The Gravity Sensing Mechanism of the Inner Ear

Torquato Gualtierotti, *University of Milan*

A major question in the consideration of gravity sensing mechanisms in organisms is whether specialized receptors really exist. Are there specialized structures in the examined organism (whether plant or animal, invertebrate or vertebrate, uni- or pluricellular systems) which respond to differences in magnitude or direction of gravitational force? Moreover, is the evoked activity of such receptors proportional to constant levels of linear acceleration or, possibly, only to the transients from one value of linear acceleration to another?

A second relevant problem is concerned with the way in which information about the relation of the organism to the gravitational pull is coded in the afferent neurons and presented to the first level of analyzers in the nervous system. Extending the problem above this level would intrude on the more general field of second-order information processing within the central nervous system, an area beyond the scope of this meeting. Therefore, my discussion of the above-mentioned questions will be limited to the most specialized gravity sensing mechanisms, the utricular and saccular otolith cells of the inner ear.

Functional Characteristics of the Statoreceptors

There is no doubt that the otolith cells are the mechanoelectric transducers for the detection of gravitational information. The anatomical characteristics of the two kinds of receptors shown so far (fig. 1) have been already described by Professor Lowenstein. Functionally, three different events can be described in each receptor: 1) the mechanical alteration provoked by the gravitational component; 2) the corresponding energetic changes in the receptors; and 3) the consecutive electrochemical events at the synaptic junction that fire the information along the nerve fiber. The first mechanism is situated in the system, including the statoconia surrounded by a protein statolith membrane, the endolymph, and the receptor's hair. This system acts as a density difference accelerometer, as the density of the statoconia and the statolithic membrane that acts as a whole is twice the density of the endolymph—1.9-2.2 and 1.02-1.04 respectively (Trincker 1962). The statoconia shift along the decline with every movement of the head and act on the hair, beginning the energetic and electrical change.

Although some doubts still exist about the relative roles of the stereocilia and the kinocilia in the transformation of the mechanical deformation into electrical change (Lowenstein 1966, 1967), the prevalent opinion is that the shearing movement of the statoconia and the corresponding bending of the hair process provide adequate mechanical stimulus. There are still some different views: Sasaki et al. (1961) maintain that pressure perpendicular to the macula is the acting force. A new hypothesis was presented by Dohlman (1960), namely, that the excitation of the hair cells is not

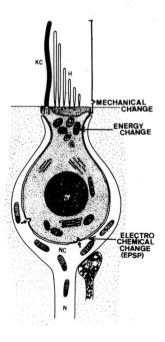

Fig. 1
Functional changes in one flask-shaped gravitoceptor: (1) At the upper end, the statoconia (not shown), the hair, the basal membrane of the stereocilia (H) and the foot of the kinocilium constitute the site of the mechanical alteration. 2) At the mitochondria (M) level the energy change takes place. 3) At the nerve chalix (NC) junction the electrochemical event produced the EPSP. Note a) two synaptic areas (arrowed) and a probable efferent nerve ending (NE2) (modified from Ades and Engstrom 1965).

due to hair bending but, at least in the cristae of the semicircular canals, to a transduction from fluid movement into static charges in the large surface of the hair, thus inducing a change of potential in the cell.

De Vries (1956) has measured directly the shifting of the single otolith of the Ruff as a function of linear acceleration up to 11 g (fig. 2). He found a value of approximately 200 μ for 5 g (maximum displacement) and of about ±70 μ for ±1 g. The displacement is less for smaller otoliths (from amphibia on up to mammals), although the entire mass of the statoconia acts as a single body (Trincker 1962). The work of Vilstrup (1951) indicates that during extreme positional changes with gravity acting as a sole force only tangential displacement occurs, of about ±15 or 30 μ in total.

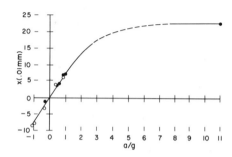

Fig. 2
Displacement of the otolith of the Ruff as a function of the applied linear acceleration: on the abscissa, acceleration in g, on the ordinate, displacement in mm (modified from De Vries 1956).

In any given position of the head the statoconia mass will reach a corresponding given position along the decline and stay there. This might be an argument in favor of a direct response to a constant linear acceleration, at least in the mechanics of the system, as against a response to the transient only (Lowenstein 1966); this would restore the mechanical state of the hair, and consequently the excitatory state of the cells, even if the statoconia maintain their shifted position.

The mechanoelectric transduction process inside the receptor and at the synaptic junction is still largely unknown. Some indirect evidence, however, exists about its nature. In the central part of the vestibular cell, immediately below the cuticular plate, there is a large concentration of mitochondria. The deformation of the cuticle plate due to the bending of the stereocilia, or a polarized compression of the root of the kinocilium, might produce a chemical and possibly an electrical change at the high energy level of the mitochondria beneath. The ultrastructure of the connection between the hair cells and the afferent fiber shows a typical presynaptic structure in the cell membrane. Synaptic bars surrounded by vesicles seem to indicate the release of a chemical transmitter toward the neuron terminals. This will produce a postsynaptic excitatory potential (EPSP) that will start the firing along the initial part of the afferent endings. The synaptic junction of the sensory cell nerve endings will therefore behave like an ordinary neuron synapse.

Unfortunately, there is no experimental evidence of the EPSP as no one has been able to stick a micropipette in a vestibular receptor and measure the membrane polarization. However, Flock at the Bell Laboratory (pers. communic.) was able to measure the membrane polarization of some large cells of the lateral line of the Necturus maculosus. He found a resting potential of 40-60 mV and depolarization and hyperpolarization during induced movement of the cilium in opposite direction. But he could not record any action potential. The receptors of the lateral line system are similar enough to the ones in the vestibule to take these results at least as an indication of an EPSP.

The depolarization or the hyperpolarization of the synaptic membrane are interpreted as due to an alteration of sodium-potassium transfer across the membrane. It is unlikely that the electrical change is generated directly at the apex of the cell. In fact, it has been shown that the outside concentrations of Na^+ and K^+ in the endolymph are very close to the ones inside the receptors: no Na^+ - K^+ mechanism can work in these conditions, as the cell top has to be hermetically sealed (Trincker 1965). There is no convincing explanation of the excitation transfer from the hair zone to the synaptic area. Trincker (1965) has suggested that a large enough potential might be generated by the deformation of a thin layer of mucopolysaccharides surrounding the hair bundle through the hermetic seal of the upper cells. The charge might be transferred capacitatively inside, therefore producing the generator potential. But this is still mere speculation, although the polysaccharides are found in high concentration in the endolymph (Dohlman 1960).

Output of the Statoreceptors

The output of the transducer system, namely the firing along the corresponding nerve fiber, has been well studied and is the best answer to our early question on the gravity sensing mechanism. It is in fact the proper channel of information to reach the first analyzer. While it is irrelevant to consider how many receptor cells correspond to a single channel, it is important that the response analysis be made for the primary neurons. Recordings from the nuclei introduces a complexity which tends to make it more difficult to analyze the firing and to have it interpreted (Gernandt 1949, Duensing and Schaefer 1958, Shimazu and Precht 1965).

The response to linear acceleration recorded from the primary neurons consists always of a change in the characteristics of the pulse train along the nerve fiber (fig. 3). At steady state under constant acceleration, however, the main factor to be considered is accommodation. If 100% accommodation is not achieved after an indefinite period of time, the receptors can be considered as being truly sensitive to gravity. In fact, in this case a different activity will correspond to a different gravity vector. Accommodation is present in all vestibular units studied so far (Ross 1936, Adrian 1943, Ledoux 1949, Lowenstein and Roberts 1950, Coppee and Ledoux 1951, Lowenstein 1956, Rupert et al. 1962, Trincker 1962, Lowenstein 1966), but it varies widely from one unit to the other. Most of the units show only partial accommodation and some to a very small degree. Hence a definite relation exists between the responses of the units and the constant acceleration applied (Adrian 1943), although there are some reports of a 100% accommodation of all units of the maculae utriculi (Cramer 1962).

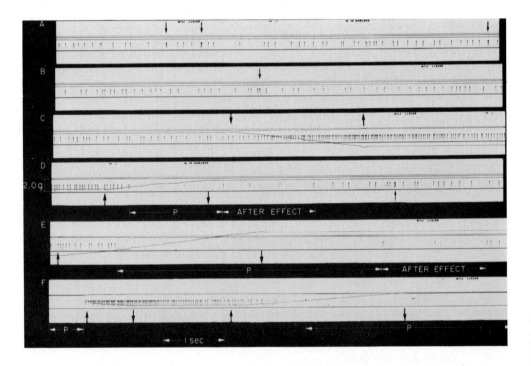

Fig. 3. Lower record: resting discharge (A-B) and evoked response (C) of a single fiber of the eighth nerve in the bullfrog corresponding to a gravitoceptor. The three upper tracings indicate the linear accelerations in the x, y and z axis. In C: the head-down tilt provokes an increase in the average rate of firing. In D, E and F back tilting suddenly stops the discharge (P = pause). This is followed by a slower firing (aftereffect). The pause seems to be related to the speed and duration of the back tilt. Note the irregularity in the rate of firing: very short intervals, practically double firing, are arrowed in A and B. Note also the accommodation effect at a steady tilt in C and D, indicated by a spontaneous decrease of the rate of firing. Calibration of the accelerometer and time in the record amplitude of the spike: 500 uV.

Lowenstein and Roberts (1950), for instance, reported that a unit from the isolated labyrinth of the Raja had a resting discharge of 6 per sec. It went up to 16 per sec during nose-up tilting. After 30 sec it decreased to 11 per sec, and then held constant for the remaining 20 min of observation. Units with very limited accommodation show a truly stationary state during constant stimulation (fig. 4). It is possible to conclude at this point that gravisensors do exist in the vestibule.

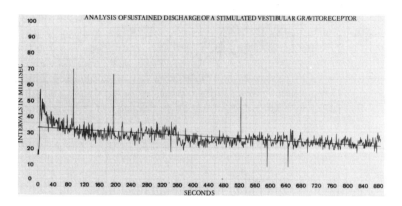

Fig. 4. Analysis of a sustained discharge of a stimulated gravitoceptor in the bullfrog. At 0 time the end of the tilt: the head maintains the acquired position; note slight accommodation at the very beginning of the curve. Stationary state is observed from the 300th second of recording on. On the abscissa, duration of discharge in seconds. On the ordinate, intervals in milliseconds. Each point is obtained averaging 100 msec of discharge.

The statoceptors have been classified in different ways according to their response. In the frog, Ross (1936) distinguished a receptor type, group "i," which responds when the head is tilted out of normal position. A second receptor type, group "ii," responds only when the head is tilted back toward the normal position. Lowenstein (1956) also confirms the existence, in the ray, *Raja*, of "out of position" and "in position" receptors, the latter showing a fast accommodation. Hiebert and Fernandez (1965), recording from the nuclei of the cat, classified their responses in four groups: steady increase of frequency, steady decrease of frequency, no change, varying frequency. Adrian (1943), in the cat, found results comparable to those of Ross (1936), and concluded that no significant difference exists between amphibians and mammals as far as the gravitoceptors are concerned. Rupert et al. (1962), also in the cat, essentially confirmed Lowenstein's data (1956), distinguishing one group of gravitoceptors proper, and one group responding only to transients. In summary, two kinds of gravitoceptors are described. One responds to "out of position" displacements with a graded response in a preferred direction and significant differences in the adapted firing rate at different stationary levels of tilt. The other is the "in position" receptor and responds only near or at the normal position of the head. It is rapidly adapting and is insensitive to the direction of the tilt.

Once the conclusion is reached that the sensory cells of the maculae of the utriculus and sacculus are capable of measuring directly constant linear acceleration, and therefore gravity, the second problem remains: what kind of sensory coding is used to transmit the information to the analyzer? According to classical theory, sensations are transmitted from the sensors up to the first analyzer by means of a change of the rate of firing proportional to the stimulus. Such a mechanism might still be considered valid for units which show fairly constant discharges, both under resting conditions and during excitation. These units have been described in the isolated vestibule of the Raja (fig. 5) (Lowenstein and Roberts 1950, Lowenstein 1956, 1966). Trincker (1952) also described regular firing rates in mammals. However, in the vestibule, the irregularity of both spontaneous firing and single cell response to controlled acceleratory stimuli has been observed by many authors, starting from Ross (1936), in the severed frog head, and by Adrian (1943), in the decerebrated cat. Records from papers by Gernandt (1949, 1950) and by Rupert et al. (1962) show the same variability in the resting discharge and in the response to stimulation as described above (cf. fig. 6). On the other hand, these authors also describe units with a fairly constant rate of firing. Most of these studies, except that of Rupert et al., dealt with decerebrated or deeply narcotized animals. This might be the reason for the large variability observed. Bizzi et al. (1964), however, recorded from the vestibular nuclei of an intact cat free to move, and reported irregular discharges from the lateral vestibular nuclei, although the medial vestibular nuclei showed a remarkable constancy in the rate of firing.

The existence of regularly firing units, with a simple frequency of firing-stimulus relationship, cannot be denied. Evidence exists, however, that irregular resting discharges and irregular evoked firing are very common in the vestibules of all species, just as they are widespread in the sensory systems in general. For instance, irregular firing has been found even in the muscle spindle receptors. Stein and Matthews (1965), working on 200.000 interspike intervals from 13 muscle spindles from the soleus muscle of the anaesthetized cat, found in the primaries a standard deviation of 3.5-6.8% of the mean (coefficient of variation 0.035-0.068). A large variability is found in the activity of the primary neurons on the acoustic side of the inner ear (Weiss 1964, Kiang 1964). The same irregularity is reported for the resting discharge, in darkness, of retinal units (Kuffler et al. 1957) and of somatic afferents (Werner and Mountcastle 1965).

The cause of irregularity may be intrinsic in the cell or extrinsic (Moore et al. 1966). The intrinsic irregularity might be due to a number of processes. Fluctuation of the junction membrane potential provoked by thermal agitation has been suggested by Fatt and Katz (1952). Molecular agitation in the mechanical receptor substance may also be a source of noise (Katz 1950).

Another mechanism which might produce a large variability both in the resting discharge and during excitation is the convergence of several receptor cells onto the same afferent. Branching of the fiber takes place, in fact, both in the unmyelinated ending and in the myelinated part of the fiber. Whereas the first can be considered as a synergic mechanism providing from a number of cells the amount of excitation of the axon necessary for the onset of the spike, myelinated branches should show some occlusion phenomena with possibly the antidromic invasion of the quiescent receptors from the convergent excited ones. The pulse discharge might therefore be subjected to a complex interplay within the cluster of cells connected to the same fiber. Consequently, a large irregularity in the fiber firing might result.

Buller (1965) considers the mechanism for the generation of firing rate to be independent of the source of random fluctuation. The latter only modifies the regular discharge pattern. It is difficult to say whether the vestibular mechanoreceptors are the site of an intrinsic randomness in the firing. Their sensitivity is so high that the existence of a true threshold is doubted (Jongkees 1960).

The Gravity Sensing Mechanism of the Inner Ear

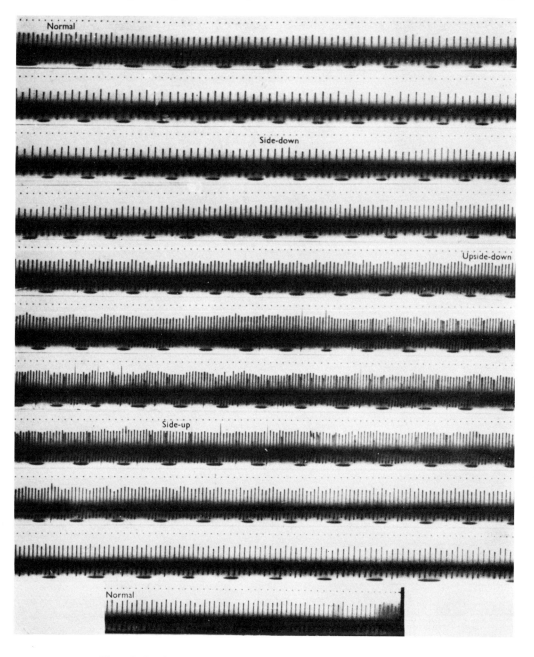

Fig. 5. Isolated labyrinth of the Raja. Utriculus: recording of the response of a 2-fiber preparation to a full circle lateral tilt. Time marker 24/sec. Speed of rotation approximately 10°/sec. Note the regularity of the firing in contrast with the record in figure 4 (from Lowenstein and Land 1950).

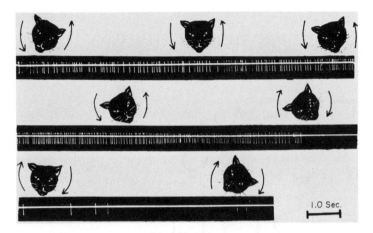

Fig. 6. Chronic cat in normal conditions. Response of an otolith unit to the tilting of the head: the direction of the movements is indicated by the arrows. Note a) irregular firing, b) increase in the average rate of firing during left tilt, c) the pause at the onset of the back tilt. This record is very similar to the one shown in fig. 4 in the frog (from Rupert et al. 1962).

Data in the literature show that the receptors of the vestibular apparatus which respond to linear or angular acceleration are sensitive to accelerations of less than 1 cm per sec^2. Buys (1940) found that an acceleration of $1°/sec^2$ was enough to cause nystagmus in man ($1°/sec^2$ corresponds to a tangential acceleration of about 0.08 cm/sec^2 at the human labyrinth). Ter Braak (1936), observing photographically the ocular deviations in rabbits, found a much lower threshold ($0.1°/sec^2$). Groen and Jongkees (1948), working on thirty healthy human subjects with three different methods, found that the average threshold value to produce a rotational sensation was $0.5°/sec^2$ (corresponding to a tangential acceleration of about 0.04 cm/sec^2 at the labyrinth).

In the intact frog the threshold at $1 g$ level for the single otolith unit during linear acceleration is well below 1 cm/sec^2 for a duration of acceleration of less than 5 msec (fig. 7, lower record). In 1908 Mulder was the first to point out that the product of stimulation time and acceleration level to reach the threshold of rotational sensation, is constant. Groen and Jongkees (1948) found that it took approximately 30 sec to bring about a rotational sensation at an acceleration of $0.5°/sec^2$ ($0.04 \text{ cm/sec}^2 = 1.2$). With a stimulus lasting 100 msec, at labyrinth level an acceleration of 12 cm/sec^2 ($0.1 \text{ sec} \times 12 \text{ cm/sec}^2 = 1.2$) would be needed. The results obtained by monitoring directly the otolith unit in the frog indicate, however, a much lower threshold, namely $0.005 \text{ sec} \times 1 \text{ cm/sec}^2$. This is understandable, as the sensitivity at the receptor site must be much higher than the one involving the complex overall mechanism of sensation.

The result of such a high sensitivity is that extrinsic factors producing fluctuations in firing must be taken into consideration as well as the intrinsic randomness. As the gravitoceptors appear to be highly sensitive to vibration, the particular position of the head must also be considered; physiological movements of the head actually produce a modulated excitation of the labyrinth receptors.

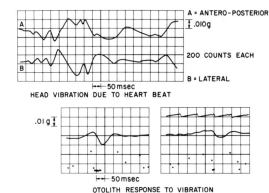

Fig. 7
Upper record: vibration of the head in a human subject sitting in a chair comfortably. Lower record: response of a single otolith unit of the bullfrog to vibration comparable in intensity to the one in the upper record. Upper tracing, acceleration profile. Lower tracing, each dot corresponds to an interval. The value is read from the "0" line to each consecutive dot.

Head Accelerations Acting on the Vestibule

We have measured the movements of the head in man under various conditions. Five healthy subjects were used. Two accelerometers were fixed on a cast of the head of each subject to obtain a tight fit; the accelerometers were placed in different positions in order to measure accelerations in different planes. EKG, vibrocardiogram and carotid pulse were recorded simultaneously. The accelerations of the head were measured while the subjects were instructed either to stand still with their two feet close together and their eyes open or closed, or to sit comfortably resting against the back of the chair, with their eyes open or closed. Only the observations obtained for the sitting position will be discussed, as a number of artifacts due to larger movements are avoided when in this position.

The results showed that beside the maintenance of balance, the most consistent periodic source of movements of the head appeared to be linked to the cardiac cycle (fig. 7, upper record). In effect, during rest the acceleration profile shows two peaks with an interval of 300 msec. The acceleration peak value of single cycles ranged from 15 to 25 cm/sec^2, the two waves being of similar value, with a duration of approximately 100 msec. A number of minor oscillations indicated further periodic acceleratory components beside the one described.

In a second series of tests the effect of exercise and fatigue on head movements was studied. During exercise the heart rate and blood pressure increase proportionally to the physical work performed. Consequently, the head movements related to the cardiac cycle should increase also. It was found that this was indeed the case, even with mild exercise. The subjects were instructed to squat twenty times consecutively. The two peaks of the acceleration were still present but their values were increased by a factor of 2, namely 40-50 cm/sec^2. As reported above, these two waves seem to be related to the cardiac cycle, particularly to the carotid pulse. It appears that the first head movement follows the rise in pressure in the carotid arteries while the second movement follows the decrease in pressure in the carotids and the closure of the aortic valves.

It seems clear from the above observations that the vestibule is subjected to a continuous background of periodic excitation. One possible consequence is a decrease in the inertia of the system of the maculae, since the statoconia mass will be in a state of continuous vibration. Secondly, it will upset, to some extent, receptor accommodation by maintaining excitation of the macula.

Efferent Systems

In the intact animal another potential source of firing fluctuation of the gravitoceptors may be due to central control of the receptor cells. The existence of efferent fibers in the eighth nerve has been postulated since the end of the last century. Not until 1927, however, was the existence of centrifugal fibers in the eighth nerve shown by van Gehuchten. Ernyei (1935), discussing the existence of unmyelinated fibers in the vestibular nerve, advanced the hypothesis that they might be part of an efferent system. In 1955 Petroff stated explicitly that efferent fibers do exist in the trunk of the eighth nerve. Rasmussen (1946) had suggested that part of the olivo-cochlear tract, through the cochleo-saccular anastomosis of Hardy, might reach the sacculus, thus supplying it with efferent fibers. But Petroff (1955) described, in the vestibular neuro-epithelium of cat and monkey, efferent fiber projections reaching the maculae sacculi and utriculi, as well as the three semicircular canal cristae. By serial sections he was able to prove that such fibers are central in origin. This was later confirmed by Rasmussen and Gacek (1958) and Gacek (1960). Rossi and Cortesina (1962), using a histochemical method, have described an efferent cholinergic cochleo-vestibular system composed of five tracts, four of them direct and the fifth crossed.

Strong evidence of an efferent system impinging on the vestibular receptors is found in the electron microscope studies (Wersäll 1956, 1960, Engström 1958, 1965, Spoendlin 1965). Iurato and Taidelli (1964) showed that two types of nerve endings occur in the vestibular receptor cells; one type was heavily granulated and might represent the endings of an efferent system.

The efferent vestibular system was described in detail at a recent symposium on vestibular organs. The centers and the descending pathways in the medulla and the distribution of the bundle of efferent fibers in the eighth nerve and its branches were shown. Smith (1967), through degenerative studies following intramedullar sections of the vestibular fibers, suggested the existence of two kinds of efferent structures in the macula utriculi of the Chinchilla: the button terminals, synaptic junctions on the chalice of the receptor cell and on the terminals of the sensory fibers, and the buttons en passant, connected mainly with the nerve fibers and the button terminals. This author also reported that the origin of the efferent fibers was near the lateral vestibular nucleus; these fibers thereafter descend in the vestibular nerve. Gacek (1967) characterized in detail the distribution of the efferent system in the medulla, in the eighth nerve and inside the vestibule. It has been found that some 400 efferent fibers, or approximately 10% of the total, exist in the vestibular nerve of the cat. Rossi (1967), with degenerative and histochemical methods, described in the cat the position and the characteristics of the two nuclei which originate the efferent fibers in the region of the medulla and pons.

Although the efferent system is now well enough described anatomically, there are few physiological studies as to its significance. Ledoux (1958) observed a reduction of the nerve potentials of one semicircular canal when the contralateral one was stimulated. Sala (1962) showed that the stimulation of the contralateral vestibular nuclei produced spikes in the vestibular nerve; the response disappeared with a medial cut in the IV ventricle floor. He also demonstrated that the spontaneous activity of the vestibular nerve decreased simultaneously with the appearance of the evoked efferent spikes. It was concluded that the suppression was mediated by the efferent system, through which the vestibular receptors might be directly controlled. Similar conclusions were reached by Fluur and Mendel (1963); they reported that the efferent system is mainly inhibitory. Schmidt (1963) recorded efferent impulses from the free end of nerves detached from the ampullae, sacculus, utriculus, and lagena in the leopard frog. These impulses originated from any ampulla and a variety of extralabyrinthine proprioceptors. None were evoked in response to the stimulation of the auditory or otolith organs. Recently Wersäll was able to record in the frog the activity of efferent nerve fibers (pers. communic.), and I have also been able to obtain similar records. The activity is maintained after the cutting of the connection with the labyrinth, and shows a typical modulation of firing. Figure 8 shows an example of such discharges in the eighth nerve of the bullfrog. The analysis of this activity shows a regular modulation in the firing, resulting in a 5-peak histogram (fig. 9).

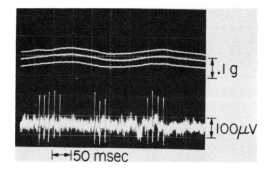

Fig. 8
Activity of an efferent fiber recorded from the eighth nerve of a bullfrog (lower record). The three upper tracings correspond to x, y and z acceleration. Note the modulated firing quite independent of the artificially provoked vibration. Calibration and time in the record.

Fig. 9
Histogram of the activity of the unit in figure 8 (see text).

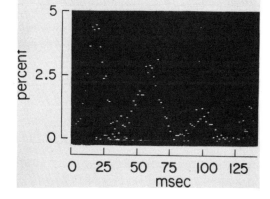

Stimulus/Response Ratio

The responses of otolith receptors have to be considered in terms of dynamic and static factors: the dynamic factors are the exaggeration of the response (if the speed of the applied accelerating stimulus is above a certain range) and the abaptation. The exaggeration of response is, very approximately, a function of the speed of the change, although no information exists up to now as to which function it is. If the animal is rotated very slowly but continuously (at or below 1°/sec) the dynamic factors are minimized and the stimulus response ratio can be studied.

Most mechanoreceptors are known to follow the Weber-Fechner law. This law is expressed by the equation $\Delta I/I = C$, where I is intensity of the stimulus, ΔI the smallest detectable difference in intensity, and C a constant. It signifies that the smallest difference in the stimulus intensity bears a constant relation to the absolute value of the stimulus intensity. Therefore, the increase of sensation follows a step-like pattern, each step being a quantum of the stimulus corresponding to a quantum of sensation. It is expressed by the equation

$$S = a \log I + b \quad \begin{array}{l} S = \text{sensation} \\ I = \text{stimulus} \end{array}$$

This logarithmic stimulus-response ratio has been verified for a number of mechanoceptors, such as the frog muscle spindle (Von Leuvwen 1949), and other receptors, such as the omnatidia in the eye of Limulus (Hartline and Graham 1932). A logarithmic term in the stimulus-response ratio would be consistent with the existence of ionic equilibria across receptor cell membranes.

As far as the otolith cells are concerned, Lowenstein (1967) concludes that the increase in activity is a linear function of the acceleration, that is, of the sine of the vertical rotation angle, over a considerable range. This does not exclude a logarithmic relationship. In fact, the initial branch of a logarithmic curve can be readily approximated as a linear ratio. The log stimulus-response ratio approximates well our own data on the frog (fig. 10a, b). The linear ratio suggested by Lowenstein might represent the onset of a logarithmic curve. This could result if the unit under study were functioning at the edge of the response field. In this case the unusually broad range of response, over the full 360° of rotation, might be related to the unit being minimally sensitive to the stimulus because it is at the edge of the receptive field. A second possibility is that the response itself, being far from maximum, does not reach the knee of the logarithmic curve and therefore appears to be in linear relationship to the stimulus. In our results, from mapping the otolith units that were maximally sensitive at the center of the field, no range was found extending above $0.7 g$ (45° of tilt from the horizontal). The sensitivity decreases and the range increases progressively up to the edge of the field (fig. 11b).

Sensory Coding

The coding of the vestibular messages or, more generally, of the sensory systems which are the site of large fluctuations in the firing rate, is complex and difficult to understand. Negrete et al. (1965) have delineated several mathematical models to describe the impulse code in the visual systems of

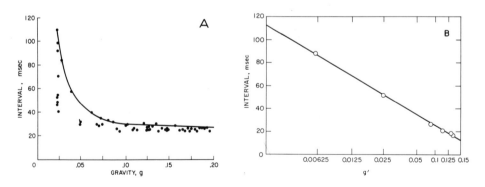

Fig. 10. Diagram showing the stimulus-response ratio of a single otolith unit in the bullfrog, taking in account the envelope of the curve only: in A on a linear and in B on a logarithmic scale. Note a nearly perfect fit with the Weber-Fechner law (see text).

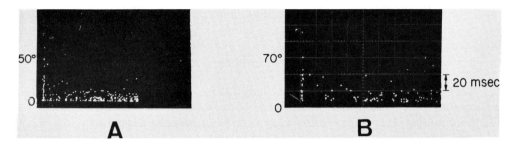

Fig. 11. Diagram of the stimulus-response ratio during the same tilt (12°-1°/sec) of a single otolith unit in the bullfrog at the center (A) and near the edge of its field. The horizontal rotation (marked on left) between A and B is 20°.

men and crayfish. Two of these were: the detailed pattern code, in which every impulse is significant and the average frequency code, in which the mean of the activity of a number of units is considered necessary for meaningful information. Negrete et al. (1965) favor the latter hypothesis. Models of an analysis mechanism at different levels above the peripheral sensory system have been presented. A gate which allows the information to go through on the basis of a convergency pattern of numerous irregularly responding units is postulated by Wall and Melzack (1965) for the skin receptors, and by Desmedt (1965) for the auditory system. A centrifugal control system plus a feedback from the peripheral activity itself are supposed to regulate this gate.

The second leading idea on the information transfer is based on the "edge" theory. Whitfield (1965) emphasizes the importance of such edges in the auditory information processes. According to this hypothesis, the central analyzers obtain their information by their ability to recognize the number and the pattern of active versus passive channels reaching them and each channel originates in a primary sensory cell or group of cells. The precise characteristics of the activity in each channel are of minor importance, provided that they are sufficiently different in the active and passive channels. The fact that sensory units seem to be not only the site of excitation but also of inhibition is very useful for this concept. In fact, the main task of the inhibitory process is to define the edges between excited and unexcited units, surrounding the active ones with a circle of sensors, the firing of which is completely suppressed. Retinal function appears to be consistent with the above model (Barlow and Levick 1965).

Study of the resting discharge and the evoked response of the otolith receptor type that does not show regularity of firing leads to the following conclusions:

1. During spontaneous activity and during excitation the rate of firing varies through such a large range that a mechanism of information based only on frequency modulation cannot work. Long intervals between spikes are found only at low stimulus levels or in the resting discharge. However, short intervals between spikes, characteristic of the excited state, are commonly found in the long bursts often seen in the resting discharge. It would be very difficult for the analyzer to determine if a few spikes occurring at short intervals are related to a particular stimulus level.

2. For the same reason, a mechanism based on time-averaging of the discharge in a single unit cannot work. If only information from a single unit is involved, the analyzer should average a large number of intervals to differentiate between long bursts during rest or fast firing during excitation.

number of intervals to differentiate between long bursts during rest or fast firing during excitation. This process would require some seconds, and would not be compatible with the rapid responses demanded during equilibration responses.

3. The results seem to agree to some extent with the edge concept mentioned above (Whitfield 1965). The change in the parameters of the activity is particularly sharp between units which are stimulated increasingly or at a constant level, and units in which the stimulation is decreasing. The main rate of firing increases in the former while the latter is subjected to a pause (fig. 11). In this way contrast during any given movement of the head is sharpened. For fast movements the pause is also present in quiescent units, with suppression of spontaneous firing. Thus, for fast movements, contrast is even more pronounced owing to the additional pause of quiescent units.

There are some elements, however, which do not completely fit into the edge theory. When analyzing the response to acceleration of an otolith unit, two different conditions must be considered: the transient resulting when the stimulus changes, and a steady state stimulus, such as a gravity component or a constant linear acceleration. The two conditions of stimulation (transient and steady) have to be considered separately.

Transients. During a transient a graded response, both negative and positive, is observed. Both show a profile of response that is related to the rate of the change (fig. 12). The positive response, following an increasing stimulation, consists of an increase in the number of short intervals which is the more marked the greater the rate of change. The negative response consists of a pause, the duration of which is proportional to rate and duration of the decrease of the stimulus. Even the resting discharge of a quiescent unit shows a pause if a sudden tilt is applied. Such events may be used by the central analyzer to determine speed and duration of the movement in respect to the gravity vector. The edge theory does not take into account these graded responses to acceleratory events. The described phenomena do not take any part in its "pattern" analysis, even if a third channel with "negative" response is considered.

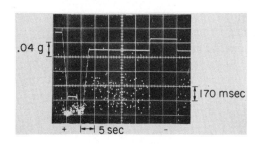

Fig. 12. Positive and negative response to a rapid tilt (20°/sec) and a back tilt, respectively, of a single otolith unit in the bullfrog. The consecutive intervals are shown as the distance between the "0" line and each white dot. The acceleration resulting from the tilt is indicated by the continuous line.

Steady State. After the end of a transient and a period of accommodation, a stationary state in the unit firing is observed which is significantly different at different levels of excitation. It is also different from the resting discharge. The graded response does not agree with the edge theory, as the theory does not take into account a change in the unit activity proportional to the amount of stimulation.

A second mechanism may therefore be hypothesized by which the information is passed on the higher level centers according to the input density (cf. Gualtierotti and Alltucker 1966). Let us imagine the existence of a time gate in the level immediately above the peripheral organs on which a number of otolith units converge. Let us suppose that such a gate opens, allowing the information through, only when two or more spikes reach it within a given time. Naturally the probability of transfer in this system will be a function of the density of the spikes in each channel (fig. 13).

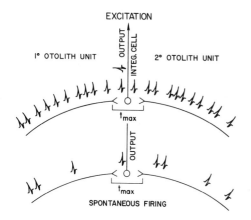

Fig. 13
Schematic of a possible mechanism for central analysis of the information incoming from the otolith unit. Two conditions are shown, one during spontaneous activity with long intervals and high variability (lower sketch), and one during excitation. The long intervals have disappeared and although there is still some variability, all intervals are shorter than the maximum time for the gate to open. t_{max}: maximum time at the gate that allows the spikes coming along the otolith nerve fiber and converging on the analyzer (integrative cell) to trigger the analyzer to activity. This happens when the spikes coming from the two pathways reach the integrative cell within t_{max}. For a full discussion see text (from Gualtierotti and Alltucker 1966).

The "edge" and the "gate" theories may be combined. In fact edges are provided by the excitation plus inhibition effect. Considering even only one excited unit partially surrounded by inhibited ones, a channel will result, the field of which will be sharply defined by a ring of suppressed responses of inhibited channels. This would enhance contrast. Alternatively, positive (excited) fields might be alternated with negative (inhibited) fields; the resulting pattern might serve as the basic element for the central analyzer to work on. Within each field, however, a graded response is observed, as the response of each unit is proportional to the logarithm of the stimulus. This will result, in each field, in an area of maximum response surrounded by an area of progressively fading evoked activity. The edge mechanism might provide stationary information on steady head position. The gate mechanism would add more knowledge on the kinetics of the transients, like the speed and the direction, following the head movements and the history of preceding events.

In conclusion, there is no doubt that true gravitoceptors exist in the inner ear, and that their activity is a function of the gravitational vector in all directions. The mechanism by which the mechanotransducers, having acquired the information of a gravity force pattern in each head position, transmit such information to the first analyzer is still, however, an unresolved problem.

Summary

The existence of true statoceptors in the vestibule is discussed. The main index for receptors responding to gravity is indicated as lack of accommodation of the evoked activity, or at least as the presence of only a partial accommodation over an indefinite period of time of constant linear acceleration. Various observations show that true statoceptors are found in the vestibule, according to the accommodation standard. The basic characteristics of the statoceptors do not seem to vary significantly in mammals in comparison with lower vertebrates.

A second problem discussed is the sensory coding by which the statoceptors send information to the primary analyzers. Some cells seem to fire at a surprisingly constant rate. For these a simple stimulus-frequency relation may be assumed as a satisfactory sensory code. The majority of the receptors, however, show a great variability both in the resting discharge and in the evoked discharge. The problem of the sensory information is therefore more complex. The origin of the randomness of firing is classified as due to two factors, one intrinsic to the cell and one extrinsic. The intrinsic factor might originate in the thermal noise in the nerve ending and/or in fluctuations in the cell membrane. It is difficult to pinpoint such an intrinsic factor with the high sensitivity of the statoceptors to linear acceleration and to vibration. One of the most important extrinsic factors is the vibration of the head following the heart beat (head ballistocardiogram). Accelerations more than ten times the threshold of the vestibular receptors are recorded from the human head. These accelerations are further increased during and after physical exercise. A variable excitatory background is therefore always present under normal conditions.

The high "noise" of the statoreceptors disqualifies a simple stimulus-frequency relation. A number of theories of information processing are summarized. The edge theory proposed for the auditory system might be applied to the vestibular apparatus, based on the positive and negative responses of the gravity receptors (positive response = increase in the rate of firing, negative response = prolonged suppression of the discharge when the stimulus decreases). However, the graded responses obtained during various levels of stimulation, do not agree completely with the edge theory. A gate theory is also described; the gate would open according to the density of the spikes in the converging channels. A sensory code which utilizes both "edges" and "gates" is also considered.

Acknowledgment

The study of human vestibular response to acceleration was performed in cooperation with Dr. F. Bracchi at the NASA Ames Research Center, Moffett Field, California.

References

Ades, H. W., and Engstrom, H. 1965. Proc. 1st symp. on the role of the vestibular organs in space exploration, Pensacola, Fla.: NASA SP-77, p. 25.

Adrian, E. D. 1943. *J. Physiol.* 101: 389-407.

Barlow, H. B., and Levick, W. R. 1965. *J. Physiol.* 178: 477-504.

Bizzi, E.; Pompeiano, O.; and Somogyi, I. 1964. *Science* 145: 414-15.

Buller, A. J. 1965. *J. Physiol.* 179: 402-16.

Buys, E. 1940. *Ann. d'Otolaring.* 3-4: 109-15.

Coppee, G., and Ledoux, A. 1951. *J. Physiol.* 114: 41P.

Cramer, R. L. 1962. *Aerospace Med.* 33: 663-66.

Desmedt, J. E. 1965. *Sci. Proc.* 4: 242-44. Tokyo: IUPS.

De Vries, H. 1956. In *Progress in Biophysics* 6: 207-63.

Dohlman, G. 1960. *Confin. Neurol.* 20: 169-80.

Duensing, F., and Schaefer, K. P. 1958. *Arch. Psychiat. Nervenkr.* 128: 225-52.

Engström, H. 1958. *Acta Otolaryng.* 49: 109-18.

Ernyei, J. 1935. *Arch. F. Ohrenheilk.* 141: 343.

Fatt, P., and Katz, B. 1952. *J. Physiol.* 117: 109-28.

Fluur, E., and Mendel, L. 1963. *Acta Otolaryng.* 56: 521-22.

Gacek, R. R. 1960. In Neural mechanisms of the auditory and vestibular systems. Springfield, Ill.: Ch. C. Thomas, pp. 276-84.

———. 1967. Proc. 3d symp. on the role of the vestibular organs in space exploration, Pensacola, Fla.: NASA SP-152, pp. 203-13.

Gernandt, B. E. 1949. *J. Neurophysiol.* 12: 173-84.

———. 1950. *Acta Physiol. Scand.* 21: 61-72.

Groen, J J., and Jongkees, L. B. W. 1948. *J. Physiol.* 107: 1-7.

Gualtierotti, T., and Alltucker, D. 1966. Proc. 2d symp. on the role of the vestibular organs in space exploration, Ames Res. Ctr., Moffett Field, California: NASA SP-115, pp. 143-49.

Hartline, H. K., and Graham, C. H. J. 1932. *J. Cell. Comp. Physiol.* 1: 277.

Hiebert, T. G., and Fernandez, C. 1965. *Acta Otolaryng.* 60: 180-90.

Katz, B. 1950a. *J. Physiol.* 111: 248-60.

———. 1950b. *J. Physiol.* 111: 261-82.

Kiang, N.Y.-s. 1964. *Acta Otolaryng.* 59: 186-200.

Kuffler, S. W.; FitzHugh, R.; and Barlow, H. B. 1957. *J. Gen. Physiol.* 40: 683-272.

Iurato, S., and Taidelli, G. 1964. Congr. Electron. Microscopy. Prague.

Jongkees, L. B. W. 1960. *J. Laryng.* 74: 511.

Ledoux, A. 1949. *Acta Othorhinolaryng. Belgica* 3: 335-49.

———. 1958. *Acta Otorhinolaryng. Belgica* 12: 111-346.

Lowenstein, O. 1956. *Brit. Med. Bull.* 2: 110-14.

———. 1966. Proc. 2d symp. on the role of the vestibular organs in space exploration. Ames Res. Ctr., Moffett Field, California: NASA SP-115, pp. 73-90.

Lowenstein, O., and Land. 1950. *J. Physiol.* 110: 392-415.

Lowenstein, O., and Roberts, T. D. M. 1950. *J. Physiol.* 110: 392-415.

Moore, G. P.; Perkel, D. H.; and Segundo, J. P. 1966. *Ann. Rev. Physiol.* 28: 493-522.

Mulder. 1908. Thesis. Utrecht (from Groen and Jongkees 1948).

Negrete, J.; Yankelevich, G. N.; and Stark, L. 1965. *Quart. Progr. Rept. Mass. Inst. Technol. Res. Lab. Electron.* 76: 336-343.

Petroff, A. E. 1955. *Anat. Rec.* 121: 352-53.

Rasmussen, G. L. 1946. *J. Comp. Neurol.* 84: 141-219.

Rasmussen, G. L., and Gacek, R. R. 1958. *Anat. Rec.* 130: 361-62.

Ross, D. A. 1936. *J. Physiol.* 86: 117-46.

Rossi G. 1967. Proc. 3d symp. on the role of the vestibular organs in space exploration, Pensacola, Fla.: NASA SP-152, pp. 213-25.

Rossi, G., and Cortesina, G. 1962. *Minerva Otorinolaringol.* 12: 1-63.

Rupert, A.; Moushegian, G.; and Galambos, R. 1962. *Exp. Neurol.* 5: 100-109.

Sala, O. 1962. *Boll. Soc. It. Biol. Sper.* 38: 1048.

Sasaki, H; Yamagata, M.; Wanatabe, T.; Ogino, K.; Ito, M.; and Otahara, S. 1961. *Acta Otolaryng.* Suppl. 179: 42-55.

Schmidt, R. S. 1963. SAM-TDR-63-66, 11 pp.

Shimazu, H., and Precht, W. 1965. *J. Neurophysiol.* 28: 991-1013.

Smith, Catherine. 1967. Proc. 3d symp. on the role of the vestibular organs in space exploration. Pensacola, Fla.: NASA SP-152, pp. 183-203.

Spoendlin, H. H. 1965. Proc. 1st symp. on the role of the vestibular organs in space exploration, Pensacola, Fla.: NASA SP-77, pp. 7-22.

Stein, R. B., and Matthews, P. B. C. 1965. *Nature* 208: 1217-18.

Ter Braak, J. W. G. 1936. *Pflüg. Arch. Ges. Physiol.* 238: 319-26.

Trincker, D. 1962. *Symp. Soc. Exp. Biol.* 16: 289-316.

Van Gehuchten, P. 1927. *Rev. Otoneur.* 5: 777.

Vilstrup, T. 1951. *Ann. Otorhinolaryng.* 60: 974-81.

Von Leuvwen. 1949. *J. Physiol.* 109: 142.

Wall, P. D., and Melzack, R. 1965. *Sci. Proc.* 4: 234-241. Tokyo: IUPS.

Weiss, T. F. 1964. *Tech. Rept. Mass. Int. Technol. Res. Lab. Electron.* No. 418.

Werner, G., and Mountcastle, V. B. 1965. *J. Neurophysiol.* 28: 359-97.

Wersäll, G. 1956. *Acta Otolaryng.* Suppl. 126: 1-85.

———. 1960. In Neural mechanisms of the auditory and vestibular system. Springfield, Ill.: Ch. C. Thomas, p. 48.

Whitfield, I. C. 1965. *Sci. Proc.* 4: 245-47. Tokyo: IUPS.

Discussion

COHEN: Is the change in frequency always downwards in these slowly adapting receptors or does it sometimes oscillate? One can get misled by looking at a single unit when the receiver may be dealing with tens of thousands of units. These fluctuations that you see in one unit may be smoothed out by the receiver.

GUALTIEROTTI: Actually the discharge frequency in these units is oscillating around an approximately constant value. In effect, if we look at the regression line while analyzing a long recording of one hour, the slope tends to zero. Sometimes it goes up very slightly but it is practically horizontal.

COHEN: Normally if a discharge oscillates around some mean frequency for 15 or 20 sec after coming into the steady-state stimulus situation, this unit is classified as a "non-adapting" or "tonic" receptor. You have to deal with this problem statistically. It may be misleading to look at one unit because the properties of the single unit may not encompass the properties of the whole system. The ambiguities and imperfections of the single unit may be resolved by simultaneous integration of information from many units.

GUALTIEROTTI: Yes, but how? If I had time to talk about coding, the conclusion would have been that there is no information possible except through cross correlations of the activity of a number of different units.

COHEN: This is true only when the discharge is random.

GUALTIEROTTI: I agree.

26
Semicircular Canal and Otolithic Organ Function in Free-swimming Fish

Howard C. Howland, *Cornell University*

The angular orientation of a fish is determined by many factors, and its relationship to the force of gravity is certainly one of the most important of them. Horst Mittelstaedt's bicomponent theory describes this relationship in terms of two gravity receptor inputs and their interaction with an internal "commanded" orientation. I want to make a useful critique of his theory and cast it in a format which will be applicable to a wide variety of orientation problems. My experimental animals are fish. First I will describe fish angular orientation in general and then I will briefly review von Holst's work on the dorsal light reaction. After criticizing Mittelstaedt's theory as it applies to fish I shall turn to my own work and place it on the same petard.

The Angular Orientation of Fishes

At any moment in time the angular orientation of a fish is determined by a number of environmental variables, the immediate history of the fish's reaction to these variables, and its own internal state. Light, gravity, the pattern of the optical environment, motions of this pattern, and accelerations due to turbulence are all relevant inputs to control systems such as the dorsal light reaction, the optomotor reaction, the "gravity reaction" (by which I mean the tendency of the fish to assume a characteristic position with respect to the gravity vector), and the semicircular canal feedback loops.

The fish itself, as well as the parts of some of its receptor organs, has inertia; that is, once its angular orientation begins to change, that change continues until counteracting forces work for long enough to decelerate the fish and stop the change. But the fish does not always react in the same fashion to the same set of inputs, and this alone makes it necessary to postulate an internal state, a "command," whose existence is reflected in the behavior of the fish.

Because the angular orientation of fish is the result of the simultaneous operation of so many different subsystems, almost all attempts to study it begin with an exclusion of some of these subsystems, the better to study the others. Thus, in dorsal light experiments the pattern of the optical environment is suppressed as far as possible. The "command value" of the system is either inferred or averaged out, and the fish is measured at its steady-state value where the recent accelerative history of the fish is known because the fish has been at rest. These are some of the techniques which von Holst (1950) used in his dorsal light reaction studies.

It may be remembered that he first became interested in the dorsal light reaction because the angular position of angelfish (*Pterophyllum scalare*) was observed to vary with the "appetite" of the fish—that is, to reflect the "command" value of the system. Subsequent centrifuge studies showed that for constant light intensity and a constant angle of light to body in the transverse plane of the fish, the fish oriented in such a way that the shearing force component acting on its utricular otoliths was kept constant. Moreover, von Holst interpreted this to mean that the shearing force is the true input of the otolithic organs. He writes:

Änderung der mechanischen Feldstärke bei konstanter Helligkeit und konstantem Winkel zwischen Lichtrichtung und Fischauge bewirkt jeweils eine solche Änderung von a [the angle between the dorsoventral axis of the fish and the apparent vertical], dasz die an den Statolithen auftretende Scherungskomponente genau gleich grosz belibt. Daraus folgt unmittelbar, dasz allein der Scherung der rezeptoradaquate Reiz ist. Auf Druck, Zug, sowie hydrostatische Druckanderung spricht der statische Apparat nachweislich nicht an.

In fact it does not follow from von Holst's experiments that shearing is the adequate stimulus for the otolith organs, though this is a very plausible interpretation of them. To show that his theory does not *necessarily* follow from his experiments we need only bring forth an alternative scheme to achieve the same effect. Mittelstaedt's bicomponent theory will serve here, for not only does it model correctly all of von Holst's experiments, but, since it assumes two receptor organs oriented in the horizontal and vertical planes, they could just as well be pressure receptors following a cosine law as shearing receptors following a sine law.

Mittelstaedt's Bicomponent Theory

The bicomponent theory allows an animal to assume a stable position in all possible directions with respect to a reference direction, regardless of the total forces acting on its otoliths. It provides the animal with a sinusoidal restoring force when it deviates from that direction. Because the theory correctly models von Holst's experiments as an extreme case (when the sine component of the command vanishes because the command is zero), we cannot use von Holst's experiments to refute the bicomponent theory. In order to test it we must look for situations where the command value is non-zero. Most fish swim belly down most of the time. When they do stand on their heads or their tails or swim on their sides they ususally do this in an optically patterned environment. Thus, a priori, the probability of finding a fish on which one could demonstrate the action of the bicomponent theory with respect to gravity would appear to be low.

Previously, the standard test for a non-zero command was to place an animal in a centrifuge in a homogeneous visual environment and vary the force acting on the otoliths. If the position of the animal changed as a function of increased force, this was taken as indicating an internal command of non-zero value. Wolfgang and Helga Braemer (1958) did a series of these experiments on three species of head- and tail-standing fish. They found that:

Thayeria obliqua stands head up at approximately $25°$ regardless of the total force acting on its otoliths. In this fish, however, cleared preparations showed that the plane of the utricular otoliths was rotated forward through the same angle, so that in the fish's normal orientation the utricular otoliths are horizontal. Hence no bicomponent theory is needed here; the utricular anatomy suffices to explain the constant orientation in varying force fields.

Poeciliobrycon equus also has a rotated utricular statolith which changes its position with the ontogeny of the fish. In a centrifuge these "tail-standers" alter their position with increasing force fields, and the degree of alteration is explicable from the position of the otolith and the assumption of a simple additive command superimposed upon a shearing signal from the otolith organs. Since the bicomponent theory with equally weighted components would predict *no* alteration of the position of the fish with increasing force fields, the theory would decidedly *not* be operative here.

The behavior of the third fish species which the Braemers tested, *Chilodus punctatus*, appeared to be inexplicable on the basis of either otolith rotation or an additive command. This "head-stander" makes an angle of 45° to 55° with the horizontal in all force fields up to $2.2\,g$; in addition its otoliths are rotated forward (in the "wrong" sense!) by about 5°-10° out of the horizontal plane. This fish, then, is a candidate for the bicomponent theory of orientation in that it maintains approximately a fixed orientation with varying shearing forces on its utricular otoliths.

The Braemers noted it was likely that other otoliths were involved in the orientation because the unilateral removal of one utricular otolith hardly affected the orientation of the fish in the centrifuge.

Before we accept the bicomponent theory as *the* explanation of the behavior of this third fish, *Chilodus punctatus*, we should note that our alternatives have been somewhat widened by the fact that a theory which explains this behavior need not also model von Holst's dorsal light-reaction experiments. For von Holst's experiments were performed about the longitudinal or rolling axis of the fish, and we are talking here of an angular orientation about the pitching axis. Although my own experiments speak for it, no one knows for certain whether von Holst's constant shearing rule holds for the dorsal light reaction about the pitching axis of the fish with zero "command," because no one has ever performed the necessary experiments. Von Holst simply assumed that his rule held for all axes.

Mittelstaedt (chap. 26, this volume) introduced as evidence for his bicomponent theory the unpublished data of Wolfgang Braemer from a dorsal light experiment on *Chilodus punctatus*. The experiment was to place the fish in an otherwise optically homogeneous environment and rotate a light around the pitching axis of the fish through a full 360°. He then measured the angular position of the fish, α for each position of the light relative to the dorsoventral axis of the fish, β.

This experiment is particularly complicated to interpret for several reasons: first, no mention is made of the position of the eyes of the fish relative to the body, or whether it changes during the experiment. Second, even if the eye position of the fish were known to be constant throughout the experiment we would not know how the turning tendency varied with the angle of entry of the light into the eye. This is because there are no dorsal light experiments about this axis from which we can generalize. The biocomponent theory can be expressed as follows:

$$L \sin(\beta + W) + F \sin(\alpha + W) = 0 \qquad (1)$$

where L is a constant representing light intensity, F is the force acting on the otoliths, and W is the "commanded" angle. Even if the former objections are set aside and it is found that the fish obeys the prediction of the bicomponent theory, this does not imply that the fish must enforce this equality by the bicomponent algorithm. An alternative scheme is given in figure 1.

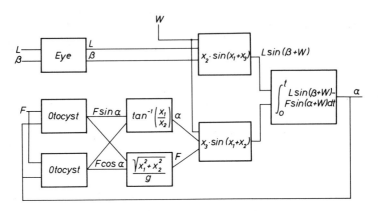

Fig. 1. An alternative to Mittelstaedt's bicomponent algorithm. It is assumed that the fish computes α from the outputs of two otocysts as well as F, the total force acting on its otolith. The angle, β, and the light intensity, L, are obtained directly from the eye. The angles α and β are added to the command value, the sine function of their sums are then weighted according to the strength of the light and the force, and the two interact to determine the output angle α.

To summarize my critique of the bicomponent theory as it applies to fish, I have argued that because they live in a patterned optical environment and normally swim belly down, few fish have a demonstrable need to overcome the difficulties which the bicomponent theory offers to solve. Of those three fish which have been investigated which do habitually make large angles with the surface of the water, two demonstrably do not use the bicomponent scheme to assume these positions. For the third fish which may employ the bicomponent theory, no evidence has been supplied which makes the theory anything more than a plausible alternative among a set of alternatives of unknown number.

The Role of the Semicircular Canals

So far we have discussed only the static aspects of fish orientation. The gravity receptors of fish, however, also play a dynamic role in continually returning the fish to its commanded orientation, and in so doing they come into play with the actions of the semicircular canals. My own researches concern the relative influence of these two organ systems in determining the dynamics of the orientation of fishes, and to this topic I shall now turn.

It was Steinhausen (1933) who first demonstrated experimentally that the cupula of the semicircular canal completely closes the canal like a spring-loaded, fluid-damped swinging door. He was the first to write the second-order differential equation describing its position in time.

Egmond, Groen, and Jongkees (1949) determined the important constants of this equation for humans, showing that the motions of the cupula and endolymph must be much more than critically damped. Lowenstein and Sand (1940) and later Groen, Lowenstein, and Vendrik (1952) demonstrated that the output of the ampullary nerve reflected the position of the cupula in the thornback ray, and that it, too, was highly damped. It is now generally recognized that semicircular canals are profitably viewed as velocity-sensing devices, in that over several decades of frequency of oscillation of the canal, the position of the cupula and the frequency of discharge of the ampullary nerve reflect the velocity of the canal itself (see, for example, Mayne 1965).

From the nerve transection studies of Lee (1894), the behavioral studies of Kreidl (1892) and the neurophysiological studies of Lowenstein and Sand (1940) we may infer that the semicircular canals of fishes function in continuously working closed feedback loops. These act to elicit motions of the fish which will return the cupulae to their normal resting positions—that is, normally to reduce their angular velocity, about any axis, to zero. Moreover, we can immediately see why such a response might be useful, in that this "semicircular canal reaction" would supply velocity proportional damping to the motions of the fish. It might be thought that the water in which the fish swam would be sufficient for this purpose, and indeed this might be so had the fish no forward motion of its own. But the damping of the water is inconsiderable in comparison with the other torques arising from the gravity response and the propulsive force of the fish.

The Semicircular Canal Reaction

How can we demonstrate the semicircular canal reaction in the free-swimming fish and measure its strength? One way is to follow the lead of von Holst and pit the semicircular canal reaction against that of the otolithic organs, measuring the strength of the one against that of the other. This may be accomplished by centrifuging a blind, free-swimming fish in a spherical flask and observing the plane on which it swims. The work of Harden Jones (1957) and our own experiments (Howland and Howland 1962) have shown that fish placed in such a situation tend to counteract the imposed rotation, swimming around and around their spherical flask continuously in an opposite sense to that in which they are being rotated. The basis of this reaction to absolute rotation is not completely understood, but I shall have more to say about it later. In any event, were there no semicircular canal reaction the fish would swim on a plane approximately parallel to the surface of the water in the centrifuge. But because the axis about which the fish swims is not parallel to the axis of the centrifuge, the fish can never completely compensate for the rotation of the centrifuge (fig. 2). This resultant, residual velocity vector remains to stimulate the semicircular canals and to draw the fish out of the plane of the apparent horizontal (fig. 3) until the shearing forces on the otoliths just balance the input to the semicircular canals.

We have performed this experiment many times on a special-purpose centrifuge which we have described elsewhere (Howland et al. 1966), and these results are given in figure 4. It may be seen that there is a linear relationship between the shearing forces acting on the otocysts, which would tend to return the fish to the plane of the apparent horizontal, and the magnitude of the velocity vector. The latter is stimulating the vertical semicircular canals of the fish and causing the plane on which it swims to tilt away from the apparent horizontal. It should be noted that this reaction demonstrates the gravity response simultaneously about both the pitch and the roll axes.

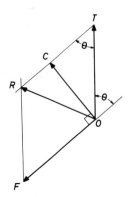

Fig. 2
Vector diagram of angular velocities. OT is the instantaneous angular velocity vector of rotation of the fish. The angle θ represents the tilt of the apparent horizontal away from the actual horizontal owing to the centrifugal force at that particular centrifuge radius. (The fish tends to swim in the plane of the apparent horizontal.) OR is the resultant of OT and OF. OC is the component of OR in plane of the fish's pitching and rolling axes. If the angular velocity of the fish is equal and opposite to the velocity of the centrifuge, $V = |OT|$, then the amplitude of the sinusoidal velocity signal to the semicircular canals is $V \sin \theta$ and its frequency is V.

Fig. 3
Geometry of the centrifuge experiment. A blind fish is allowed to swim freely in a 2-ℓ spherical florence flask. The flask is placed in a water-filled plastic box to eliminate refraction and the whole apparatus is mounted on the arm of a centrifuge of axis OT. In reaction to the rotation of the centrifuge the fish swims in a contrarotary sense in the plane XYZ, which makes an angle ρ with the apparent horizontal. The plane $ABCD$ is the water surface in the plastic box during centrifugation and indicates the apparent horizontal. The plane XYM is perpendicular to ABC and MY is parallel to ABC. $\rho = \angle XMY$.

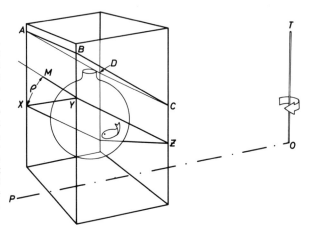

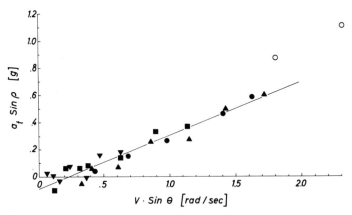

Fig. 4. Balance between otolithic organ and semicircular canal inputs. $V \sin \theta$ represents the amplitude of the velocity signal to the semicircular canals as the fish swims in the centrifuge on a plane which makes an angle ρ with the apparent horizontal. $a_t \sin \theta$ is the total shearing acceleration acting on the otoliths and a_t is the vector sum of the acceleration due to gravity and that due to centripetal acceleration. The four types of symbols represent data taken at the following centrifuge radii from 1.05 to 2.3 radians per sec: Circles, 4.075 m, triangles (apex up), 2.09 m, squares, 1.05 m, triangles (apex down), 0.515 m. Each point represents the averaged data of three fish centrifuged in both directions. Angles were measured from cinema films of television monitor screen. Regression line was fitted by method of least squares to all points except open circles. Its equation is $a_t \sin \theta = 0.3897 \, V \sin \theta - 0.08477$. The failure of the graph to intersect the origin is an artifact.

The Significance of the Reaction

How important is the semicircular canal reaction? To gain an idea of its significance we may assume for a minute that the angular orientation of the fish may be described by the simple second-order differential equation of a pendulum. The restoring force is the gravity reaction and the damping is provided by the canals. Calculations from the results of figure 4 show that the motions of the fish would be more than critically damped for any natural frequency of the fish greater than about 2/3 cycle per sec.

To put it another way, were a fish suddenly caught in a turbulence and rotated out of the horizontal with a velocity of about 1.5 radians per second (about 1.4 RPM), both the semicircular canal response and the gravity response would tend to restore the fish to its horizontal position. But the semicircular canal response would be the major contributor to the restoring force until the fish was rotated more than 45° out of the horizontal.

Without the semicircular canal reaction there is little doubt that the fish would go into prolonged underdamped oscillations each time it either commanded a new angular orientation or was significantly displaced from its old orientation by turbulence.

The Validity of the Theory of the Semicircular Canal Reaction

We must now fulfill the promise made in the introduction and cast a critical eye on the entire semicircular canal explanation of the reaction at hand. In fact we have no *direct* evidence that the canals are involved in this particular reaction. All we know is that some influence, proportional in strength to angular velocity, is opposing the normal orienting action of the otolithic organs. However, we have built a plausible argument on two main grounds; namely, previous knowledge of the role of the semicircular canals, which leads us to believe that they would function in this reaction, and a lack of alternative angular velocity receptors. Moreover, our theory is germane in that it suggests two possible lines of investigation which will further test it. The first involves determining if the semicircular canal and gravity reactions together will acount for the hitherto unexplained reaction to absolute rotation mentioned above. From the studies of a simplified model we have good evidence to believe that they will. In an attempt to answer this question, I am now employing a more complete numerical model which simulates the motions of a fish in three dimensions in a rotating environment. Second, although we have good reason to expect that the natural frequency of these goldfish semicircular canals may lie near 2/3 cycle per sec (as the assumption of critical damping of the motions of the fish would predict), there is a combination of behavioral and morphological measurements which will give us a more direct measure of this frequency. In this way I hope to strengthen or reject the theory in the future.

Since the above discussion was written more work has been done on the dynamic aspects of the angular orientation of fish (Howland 1968). It now appears that the natural frequency of the stabilization system of the fish about the pitching axis is approximately 1 radian per sec for a 15 g goldfish. In addition to hydrodynamic damping acting directly on the body of the fish, the phasic fibers of the otolithic organs also contribute to damping. I am in the process of attempting to verify, both neurophysiologically and behaviorally, a complete mathematical model for angular stabilization about the pitch and roll axes.

Meanwhile, attempts to show that the reaction of fish to absolute rotation were due to the interaction of the otolithic organ and semicircular canal reactions have failed, despite the "good" evidence mentioned. Both numerical and theoretical models (Howland 1968) showed that rotational torques arising from cross-coupled acceleration were simply too small to account for the reaction, whereas computations of the direct Coriolis force on the fish itself compared with the hydrodynamic drag studies on model fishes showed that these Coriolis forces were indeed significant. I now regard it as likely that the reaction to absolute rotation may be accounted for by Coriolis forces, acting on the entire body of the fish to push it sideways, combined with the "facing" reaction that such motion may cause via the lateral line.

Conclusion

We have discussed four different types of fish angular orientation behavior: the dorsal light reaction, the gravity reaction, the maintenance of an internally commanded position, and the semicircular canal reaction. We have seen how these behaviors may be plausibly, but not unambiguously, interpreted in terms of their underlying sensory structures. In criticizing Mittelstaedt's bicompetent theory I have advocated the search for alternative models and new tests for that theory. Likewise, in expounding my own experiments I have indicated in what directions we intend to test the theory in the future.

In my own experiments I have attempted to establish the semicircular canal response alongside the gravity response as an essential feedback loop in the orientation of fish. Although I am interested in explaining that response in terms of its underlying mechanism, it should not be forgotten that the *response itself* is of crucial importance to the fish, regardless of what "wiring diagram" of underlying structures we may ultimately accept.

Acknowledgments

This work was supported in major part by the Air Force Office of Scientific Research under grant AF EOAR 64-44 through the European Office of Aerospace Research (OAR), United States Air Force, and was performed at the Max Planck Institut für Verhaltensphysiologie at the kind invitation of Dr. Horst Mittelstaedt.

References

Braemer, W., and Braemer, H. 1958. *Z. vergleich Physiol.* 40: 529-42.

Egmond, A. A. van; Groen, J. J.; and Jongkees, P. 1949. *J. Physiol.* 110: 1-17.

Groen, J. J.; Lowenstein, O.; and Vendrik, A. J. H. 1952. *J. Physiol.* 117: 329-46.

Harden Jones, F. R. 1957. *J. Exp. Biol.* 34: 259-75.

Holst, E. von. 1950. *Z. vergleich Physiol.* 32: 60-120.

Howland, H. C. 1968. Semicircular canal and otolithic organ function in free-swimming fish. Ph.D. diss., Cornell University.

Howland, H. C., and Howland, B. 1962. *J. Exp. Biol.* 39: 491-502.

Howland, H. C.; Howland, B.; Ströbele, R.; and Jähde, J. 1966. *J. Appl. Physiol.* 21: 1938-42.

Kreidl, A. 1892. *Sitzb. Kais. Akad. Wissench. Wein.* Math-naturw. Klasse Bd CI, Abt. III (Nov. 1892)

Lee, F. 1894. *J. Physiol.* 15: 311 (part I); 17: 192 (part II).

Lowenstein, O., and Sand, A. 1940. *J. Physiol.* 99: 89-101.

Mayne, R. 1965. In *The role of vestibular organs in the exploration of space.* Washington D.C.: NASA.

Steinhausen, W. 1933. *Z. Hals-Nasen-Ohrenheilkunde* 34: 2-5.

Discussion

LOWENSTEIN: You put the blind fish in the bowl on the platform. How does that fit in with the Harden Jones experiment?

HOWLAND: It is the same experiment. I was a young graduate student watching my fish in the dark when he published his work. Here is the sequel of the story. I believe the gravity reaction functions all the time because I think the otolith organs work in closed feedback loops. The semicircular canals are probably running all the time because of the "tilting-out" behavior, and I think they are in closed feedback loops. There are three axes of semicircular canals and the otoliths work around two. They are fixed to the frame. The mathematics is rather complicated and I no longer trust my intuition. I've tried to solve the problem in a simple way and that's why I'm doing simulation experiments with a digital computer.

27
Central Nervous Responses to Gravitational Stimuli

W. R. Adey, *University of California at Los Angeles*

Responsiveness to gravitational stimuli in vertebrate tissues may arise in at least three ways. First, there may be a transducing of gravitational influences in specific neuronal receptors, such as those of the vestibular mechanism of the inner ear. Second, they may be sensed by their effects on the great physiological systems of vertebrates, including the cardiovascular, skeletal, and excretory systems. Third, central nervous tissue may respond through stretch and pressure receptors in the musculoskeletal system. Although their activation occurs primarily through muscular contraction, this response has a special importance in land-living vertebrates, where a terrestrial existence has fostered their development in groups of antigravity muscles. There is a fourth and as yet uncertain area of sensitivity in central nervous tissue, to be discussed below, involving the possibility of a direct sensing of certain accelerational stimuli, where these have a vibratory character.

All these factors must be considered in the substrates of behavioral processes. The significance of gravity may be profound and subtle in organisms whose entire evolutionary history has been shaped within its influence. Our inability to simulate its absence on earth emphasizes the uniqueness of space as a required environment for protracted investigations.

How may we investigate central nervous responses to gravitational stimuli? The past decade has seen remarkable new capabilities for directly recording electrical activity in central nervous and peripheral neuromuscular systems of performing subjects, so that it is no longer necessary to carry out such measurements within the constraints of the laboratory. Developments in aerospace electronic technology are probably the greatest single contribution to this dramatic new vista in environmental physiology. At the same time, and by the same methods, studies of the central nervous system have aided in understanding the brain as a window on the world without, and in comprehending the citadel of the neuron within.

Basic Cerebral Mechanisms in Sleep, Wakefulness, Arousal, Orienting, and Visual Discrimination: The Role of Gravitational Influences

We may take as a point of departure the postulate that gravitational influences may be a required element in maintaining cyclic aspects of central nervous activity. Their role would be quite different from transient gravitational influxes accompanying simple kinetic or vibratory stimuli. Although not necessarily having a *Zeitgeber* function itself, gravity may be one of a series of tonic influences that create an appropriate bias against which other factors, such as periodicity in light levels, may be effective in sustaining diurnal rhythms. Direct evidence on this point will be discussed after a brief review of neurophysiological mechanisms underlying sleep and wakeful states.

Mechanisms of Sleep and Wakefulness

In 1935, Bremer described effects of transection of the brainstem at two levels. Transection at the midbrain intercollicular level induced a persistent sleeplike state, with constricted pupils, slowed respiration, and recurrent bursts of high-amplitude EEG waves (spindles). Section at the spinomedullary junction produced a quite different picture, characterized by persistent wakefulness in EEG records and tracking of eye movements.

These studies stimulated much further work in elucidation of mechanisms of alerting and arousal. Attention has focused on the central zones of the pons and midbrain, and their ascending and descending connections (Moruzzi and Magoun 1949). This central zone contains a mixture of neurons and fibers (Scheibel and Scheibel 1967) and has been designated the reticular formation (fig. 1). It is accessible to peripheral influxes arising in special sense organs, including the eye and the vestibular mechanisms of the inner ear, as well as from somatic structures, including muscles and joints. In a simplistic sense, the reticular formation can be considered as an integrating way station for these peripheral influences, and the further ascent to the cerebral cortex occurs through thalamocortical paths that are separate from those mediating specific sensory functions. They constitute the "nonspecific" thalamocortical pathways and are essential elements in mechanisms of arousal from sleep. Electrical stimulation in the reticular formation can also evoke such an arousal (French, Verzeano, and Magoun 1952).

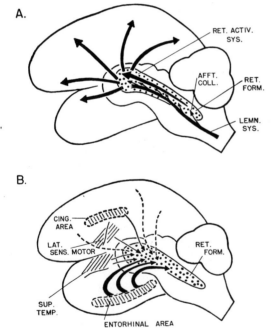

Fig. 1
a, Diagrammatic representation of ascending reticular system. Although the diagram shows inputs primarily from the lemnisci, activation also occurs substantially from spinothalamic and other non-lemniscal paths. b, Arrangement of main cortical areas (lateral sensorimotor, superior temporal gyrus, cingulate, and entorhinal) capable of modifying ascending reticular conduction (after Adey, Segundo, and Livingston 1957). Corticifugal projections from the entorhinal area exercised a major influence on ascending reticular conduction. (From Adey 1956.)

The reticular formation is also played upon by descending corticifugal influences, which arise principally in the "association areas" of the cerebral cortex, including temporal neocortex, medial frontal cortex, lateral sensorimotor areas, and allocortical regions of the medial temporal lobe (hippocampal gyrus) (Adey, Segundo, and Livingston 1957). These zones are also those from which behavioral arousal occurs at lowest thresholds to electrical stimulation (French, Hernandez-Peon, and Livingston 1955).

It appears that these corticoreticular and reticulocortical systems subserve finer aspects of consciousness than the mere determination of sleep and wakefulness. In particular, the evidence points strongly to their role in such functions as orienting based on visual and gravitational cues, and thus to their role in determining a sustained "set" of attention necessary to visual discrimination and decisions necessary in attainment of both short- and long-term behavioral and social goals. Two such corticosubcortical systems appear primarily involved in these functions. They arise in the frontal and the temporal lobes. In consideration of gravitational influences on the central nervous system, we shall limit our discussion to the part played by the temporal lobe and related subcortical structures. The latter constitute the limbic system, long considered to subserve a variety of viscerosomatic integrative processes (Herrick 1933).

Brain Electrical Concomitants of Orienting Reflexes

The orienting reflex was first described by Pavlov as a reaction to novel environmental stimuli. Its most typical aspects involve turning the head and eyes toward the source of the stimulus, but in a broader view, it includes a wide gamut of somatic and autonomic nervous components; and since reaction to novelty extends substantially beyond startle or alarm reactions to transient stimuli, orienting responses are of the highest significance in the organism's evaluation of its spatial environment. It is here that visual and gravitational cues are of the greatest importance. Modification or loss of gravitational cues in space flight may be reflected in the characteristics of orienting reflexes.

Brain Responses in Orienting Reflexes. In 1954, Green and Arduini described highly synchronized EEG wave activity in the hippocampus of the rabbit's temporal lobe during a wide variety of alerted behavior. They noted that this rhythmic hippocampal wave activity at 4-7 hertz (H), the "theta rhythms," bore an inverse relationship to simultaneous cortical records, which were asynchronous and low in amplitude (fig. 2). Subsequent studies by Grastyan et al. (1959) indicated that these hippocampal theta rhythms occurred during orienting responses in the cat. Further investigations have shown that these rhythms in the cat bear characteristic frequency signatures in relation to alerting, orienting, and discriminative responses (Raduoavacki and Adey 1965; Elazar and Adey 1967*a*). Alerted states are associated with rhythms at 4 H, orienting responses with 5-H rhythms, and visual discrimination with 6-H activity. Careful computer analysis has shown that these hippocampal concomitants are accompanied by less regular but still partially coherent theta activity in subcortical structures (subthalamus and midbrain reticular formation) and in visual cortex (Adey, Walter, and Hendrix 1961; Elazar and Adey 1967*b*). It was therefore decided to investigate the effects of sudden postural changes on EEG activity in limbic structures and at lower subcortical levels, including vestibular nuclei.

Effects of Posture on EEG Activity in Subcortical and Limbic Structures. With my colleagues Anatole Costin and Theodore Tarby, I have examined the effects of sudden postural changes on conscious cats with electrodes implanted in vestibular and cerebellar nuclei, midbrain reticular formation, anterior hypothalamus, septum, putamen, caudate nucleus, and dorsal and ventral hippocampi (Costin et al. 1967). The cats were lightly wrapped in a cloth restraint, to which they quickly adapted with repeated exposure. They were displaced from a prone position, regarded as a neutral or resting position, to either right or left side, or onto the back, and shifts in EEG patterns were sought for periods up to 60 sec.

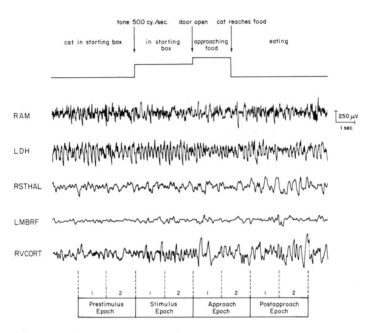

Fig. 2. Segments of EEG records during one trial of a cat, with the sequence of behavioral states shown by stepwise deflections of the event marker in the upper trace. Division of the record into successive segments for spectral analysis is shown below. Note the highly regular theta rhythms in the temporal lobe (hippocampal) lead (LDH) during the alerted behavior preceding discrimination. This rhythm becomes faster during the discriminative task performance. Simultaneous visual cortical records (RVCORT) are low in amplitude, but show fragments of theta rhythm in computer analyses. (From Elazar and Adey 1967a.)

EEG epochs 10 sec long were analyzed for autospectral density by digital computation, and for baseline data, variance was calculated and plotted as one standard deviation about the mean at each frequency. It was found that shifting to a lateral position produced a great enhancement of low-frequency activity at all implanted sites, particularly at frequencies from 1 to 4 H. These new spectral density peaks were largest immediately after the postural change, but subsequent decay was variable, and in septum and hippocampus (fig. 3) the peak often was sustained undiminished for more than a minute after the postural change. In other structures, a more rapid decay occurred or there were cyclic decays and recrudescences. Responses were frequently asymmetrical and were preponderant on the ipsilateral side.

These changes could arise from somatic, vestibular, or optic influxes, or from a combination of all three. Each component was separately evaluated. Blindfolding in no way modified the spectral shifts. Section of the dorsal columns of the spinal cord was followed by major alterations in hippocampal response patterns. Whereas preoperatively a lateral posture was associated with a major peak at 4.0 H, after dorsal column section this peak occurred at 5.0 H. Although not as large as the peak before operation, and although baseline variance was clearly greater in the spectrum from 1 to 5 H, the response persisted.

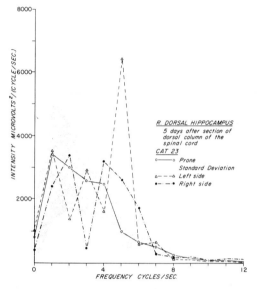

Fig. 3
Hippocampal EEG spectral density changes induced by shift in body position from prone to right and left lateral and to dorsal positions, after section of dorsal columns of spinal cord. Shaded band around graph of control data shows variance (one standard deviation). (From Costin et al. 1967.)

The vestibular nerves were therefore cut bilaterally at the internal auditory meatus, with and without prior section of the dorsal columns of the spinal cord. The evoked spectral peaks were abolished thereafter. It may therefore be concluded that vestibular influences predominate over spinal influxes in eliciting these widespread subcortical and limbic EEG responses by changes in posture. Nevertheless, as described below, spinal influxes may have considerable importance in man in maintaining normal sleep cycles.

Effects of Sinusoidal Whole-body Vibration on Cortical and Subcortical EEG Patterns. Transducing from surface and deep brain structures during whole-body vibration presents an imposing array of problems in detection and elimination of electromechanical artifacts associated with connecting lead displacement, possible shearing between electrodes and brain tissue, and field effects attributable to electromagnetic shakers. Despite these difficulties, and in the clear awareness that data gathered under these conditions cannot yet be fully elucidated, one may point to consistent changes in EEG activity that appear to have a physiological origin.

Whole-body vibration in the monkey over the range 5-40 H (Adey et al. 1963; Adey, Kado, and Walter 1967) induces EEG rhythmicity with the characteristics of a physiological "driving," and appears distinguishable from superficially similar phenomena of artifactual origin. The vibratory stimulus was given with a fixed double amplitude of 6.5 mm in the range 5-13 H, and at $2g$ peak-to-peak in the range 13-40 H. Spectral analyses showed little or no evidence of EEG driving below 9 H, despite powerful head movements. Driving at the shaking rate was frequency selective and maximal in the range 10-15 H (fig. 4). However, in many instances maximum energy peaks occurred at other than shaking frequencies, and without harmonic relationship to shaking frequencies.

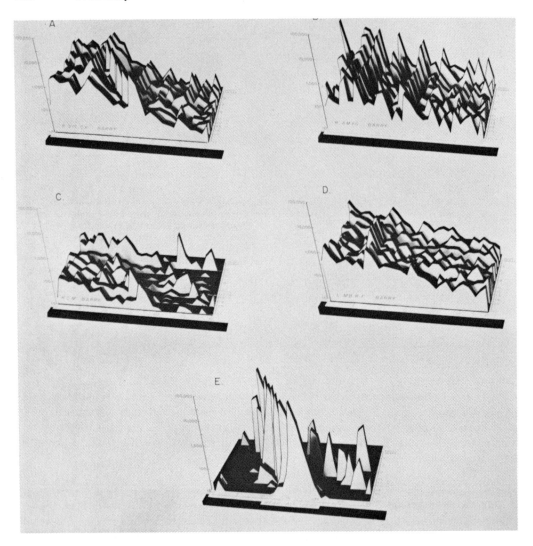

Fig. 4. Models of autospectral contours in normal monkey before and during shaking at decreasing frequencies from 17 to 5 H. EEG spectrum is depicted on abscissae, vibration spectrum on ordinates, and spectral power on Z-axis (in μV squared per cycle per sec) for visual cortex (A), amygdala (B), nucleus centrum medianum (C), midbrain reticular formation (D), and head accelerometer (E). (From Adey, Kado, and Walter 1967a.)

Coherence (linear predictability) was high between cortical and subcortical leads at EEG frequencies unrelated to concurrent shaking frequencies, and was absent from baseline records before and after shaking. This may imply aspects of cerebral system organization with ephemeral sharing of activity elicited by the vibratory volley. Coherence between head and table accelerometers and cortical and subcortical leads were below significant levels at fundamental driving frequencies below 11 H, although significant coherence peaks appeared at other EEG frequencies (fig. 5). Shaking in the range 11-17 H produced many coherent relationships at fundamental driving frequencies, and at harmonically related and unrelated EEG frequencies.

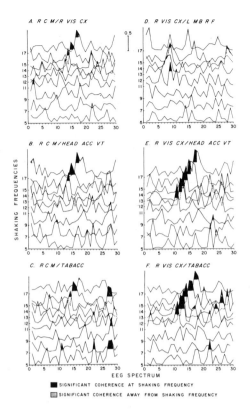

Fig. 5
Plots of coherence (linear predictability) between centrum medianum and visual cortex (A), vertical head accelerometer (B), and table accelerometer (C), during vibration. Similar plots are shown between visual cortex and midbrain reticular formation (D), head accelerometer (E), and table accelerometer (F), ordinates showing vibration spectrum abscissae the EEG spectrum, and z-axis the level of coherence. With 12 degrees of freedom, coherence levels were significant above 0.516. Significant coherence levels at the shaking frequency are shown in solid black and at points away from the shaking frequency in stipple. (From Adey, Kado, and Walter 1967a.)

It is noteworthy that harmonics appearing in the EEG were often of orders of magnitude larger than the harmonic content of the shaking table. For example, when the table fundamental was at 6 H, the amplitude of its 12-H harmonic would have to be multiplied by 1,200 to give the amplitude of the harmonic occurring in the EEG from the nucleus centrum medianum of the thalamus (Walter and Adey 1966).

This frequency-selective driving of the EEG was abolished in most cortical structures and much reduced in brainstem nuclei by anesthesia. It was not abolished by bilateral section of the vestibular nerves. Questions thus arise as to the physiological origin of this EEG driving. Evidence concerning its possible origin in spinal afferent volleys has been reviewed elsewhere (Adey, Kado, and Walter 1967a). There remains the possibility that direct mechanical excitation of cerebral tissue may occur during vibration, either transcranially or by hydraulic coupling through vascular or cerebrospinal fluid channels, although these remain matters for speculation. Weiss (1964) has drawn attention to the dynamics of membrane-bound incompressible bodies, and to the possible role of such incompressibility in neuronal mechanisms.

Role of Somatic Influences in Sleep and Wakefulness Cycles

Evolution of sleep patterns as a primary component of sleep-wakefulness cycles has been investigated from both phylogenetic (Klein 1963) and ontogenetic (Valatx 1963) viewpoints. In the premature human infant there is a preponderance of "activated" sleep (rhombencephalalic or rapid eye movement sleep) (Parmelee et al. 1967; Petre-Quadrens 1965). This phase of sleep in the adult has been strongly associated with dreaming (Dement and Kleitman 1957a).

The role that such factors as somatic sensory influxes might play in maintaining normal sleep patterns has not been extensively investigated. If the weightless state reduces tonic sensory influxes from deep somatic structures, and from graviception in vestibular and vascular systems, it might be anticipated that modification in sleep patterns would occur in ways resembling the effects of spinal transection. Transection of the cervical spinal cord in the cat is followed by decreased time in the wakeful state, at least on the basis of EEG patterns that show augmented time in spindling records (Ho et al. 1960; Hodes 1962). We have therefore examined EEG data from 14 patients with high cervical lesions, of whom the majority had suffered complete motor and sensory loss in the upper limbs and in the remainder of the body below this level (Adey, Bors, and Porter 1968).

In normal primates, from monkey to man, sleep is typically patterned about a 90-min cycle, with light sleep normally accounting for 30 to 40%, intermediate sleep for 20 to 30%, deep sleep 5 to 10%, and REM sleep 15 to 20%. In human subjects after high cervical lesions, sleep duration was reduced to about 5 hr and none showed a well-developed 90-min cycle. The proportion of light sleep was increased progressively following the lesion, so that in a series of patients with injuries of 2 to 14 yr duration, light sleep accounted for 78% of all sleep. Deep and REM sleep were markedly reduced (fig. 6). By contrast, even where total sleep duration was comparably reduced in patients with thoracic lesions, who retained partial or full function in the upper limbs, proportions of light, intermediate, deep, and REM sleep were similar to those in normal subjects.

We may extrapolate to the distorted sleep patterns seen here from models of circadian rhythms in normal subjects. Just as such factors as perturbations in temperature and steroid excretion exhibit

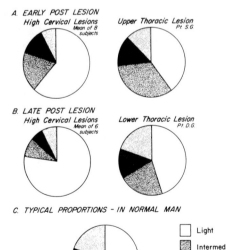

Fig. 6
Effects of high cervical lesions in man on proportions of light, intermediate, deep, and dream (rapid eye movement, REM) sleep, showing progressive decrease in deep and REM sleep with increasing time after high cervical lesions, but persistence of more normal proportions of these sleep phases in patients with thoracic lesions of comparable duration. (From Adey, Bors, and Porter 1968.)

oscillations about certain means, with maxima and minima that are temporally determined, so also can these be detected in relation to sleep and wakefulness, with maxima in neural and muscular activity in the waking state succeeded by significant amounts of deep and REM sleep. In the quadriplegic patient, these oscillations are both damped in amplitude and irregular in periodicity.

Citadels within the Central Nervous System: Excitability Thresholds in Receptors and Single Neurons

Recent studies have emphasized the very low energy levels capable of exciting sensory receptors. These energy levels are substantially below those at which metabolic systems would be triggered in the primary energy exchange. Odorous substances can be detected subjectively in concentrations of 10^{-13} molar (Beidler 1953). This is 10^4 times more dilute than the effective threshold concentration of any known enzyme system (Sumner 1953), so that the effective stimulus is apparently not dependent upon energy released by interaction with an enzyme. Similarly, auditory thresholds are only an order of magnitude above Brownian movements in fluid molecules of endolymph bathing the inner ear (Davis 1959). The eye can apparently detect as little as one photon of light falling upon a receptor (Hecht, Schlaer, and Pirenne 1941).

These findings may not be directly extrapolated to thresholds for neurons in the central nervous system, since they may represent specialized sensitivities in the peripheral transducer. Nevertheless, there is increasing evidence that central neurons do indeed respond to changes in their environment that arise in the concurrent activity of other neurons, but for which synaptic connections are not essential. This "whispering together," as it was described by Young (1951), on the one hand suggests a capacity in parallel processing in neurons of central nuclei that may characterize them uniquely, and on the other, is an immediate deterrent to simple analogies with nerve nets having binary elements that are activated solely by a "wiring diagram" of synaptic connections (Adey 1968).

At this time there is very little evidence that would directly relate gravity gradients across single central neuronal elements to basic aspects of their excitability. Nevertheless, the foregoing data indicate a susceptibility of either receptor end-organs or neurons, or both, to energy shifts substantially smaller than those appearing across single neurons in the course of movements of the head that are well within the physiological range of accelerations.

The question may be asked whether a central nervous system that has evolved in the earth's gravitational field may be directly sensitive to such low-level magnetic and gravitational fields, and whether such fields may provide a tonic driving influence to the *Zeitgeber* mechanisms of biological clocks. Obviously, emphasis has been and will continue to be placed on proved sensitivities of this area in specialized peripheral receptors, but there remains the possibility of direct sensibility in the specialized neuronal aggregates of cerebral tissue.

Although they are entirely speculative, serious consideration of such sensitivities demands a search for molecular mechanisms that might underlie their proved occurrence at the receptor level. Attention has focused on the occurrence at the cell surface and in the surrounding intercellular fluid of a hydrated network of macromolecules, mainly mucoproteins and mucopolysaccharides (Pease 1966; Rambourg and Leblond 1967). Schmitt and Davison (1965) have also suggested that these macromolecules may play a direct role in process of excitation and have emphasized the importance of recognition of an "electrogenic protein." The corollaries to such hypotheses have far-reaching consequences to classic models of the neuronal membrane.

Mucoproteins and mucopolysaccharides have two outstanding chemical characteristics. They bear numerous negative charges at fixed locations in the molecule (Elul 1966a, b). They are capable of binding water and ions reversibly, with substantial changes accompanying their attachment to, or removal from, the molecule (fig. 7). These fixed negative changes on the surface macromolecules bind selectively with cations, particularly divalent cations, and have a particular affinity for calcium ions (Katchalsky 1964). Recent studies have exemplified the fundamental role of calcium in neuronal excitability in ways suggesting interaction with surface macromolecules (Wang et al. 1966; Wang, Kado, and Adey 1968; Tasaki, Watanabe, and Lerman 1967). The classic phenomenon of ionic fluxes of sodium and potassium across the membrane, to which so much attention has been paid hitherto, may be secondary to processes involving calcium and its competitive binding with surface macromolecules (Bass and Moore 1968; Adey et al. 1969). It is at the level of such interactions, perhaps also involving the rupture of -SH bonds in macromolecules by electrons of low energy (Huneeus-Cox and Smith 1965), that we must seek answers to mechanisms of extreme neural sensitivity to environmental stimuli, including gravitational, electric, and magnetic fields.

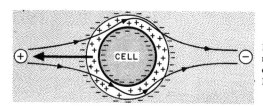

Fig. 7
Induction of movement in a body having net fixed surface charge in the presence of a nonuniform electric field. (From Elul 1966a.)

Central Nervous Data Relating to Effects of Weightlessness in Space Flight

The longest United States space flight in the Gemini program lasted 14 days. Astronauts Borman and Lovell were both monitored for cardiovascular and respiratory responses throughout the flight and, in addition, Borman's EEG was recorded in the first 55 hr of flight (Maulsby 1966). This tape-recorded flight data was extensively analyzed by auto- and cross-spectral methods and compared with laboratory and flight simulator records (Adey, Kado, and Walter 1967a). The laboratory task performances were identical with those presented to 200 astronaut candidates from a magnetic-tape control device. These data from the candidates have been extensively analyzed (Walter et al. 1967; Walter, Rhodes, and Adey 1967), and constitute a "normative library" (Kellaway and Maulsby 1968), with which Borman's flight data has been compared.

Borman's laboratory resting records were clearly normal, with a well-developed alpha rhythm at 9-11 H. At the same time, powers in the range 2-7 H were low. Visual discriminative tasks in the laboratory sequence sharply diminished alpha activity in all areas, and powers below 3 H and above 14 H were augmented considerably, (fig. 8). These findings agreed with those for 50 candidates in the normative library. In the Gemini simulator, with electrode placements, amplifiers, and recorders identical with flight equipment, EEG samples showed relatively low powers at all frequencies above 5 H (fig. 9). Both EEG channels showed a broad and ill-defined alpha peak at 9-13 H, but higher powers occurred in the theta band from 6-7 H.

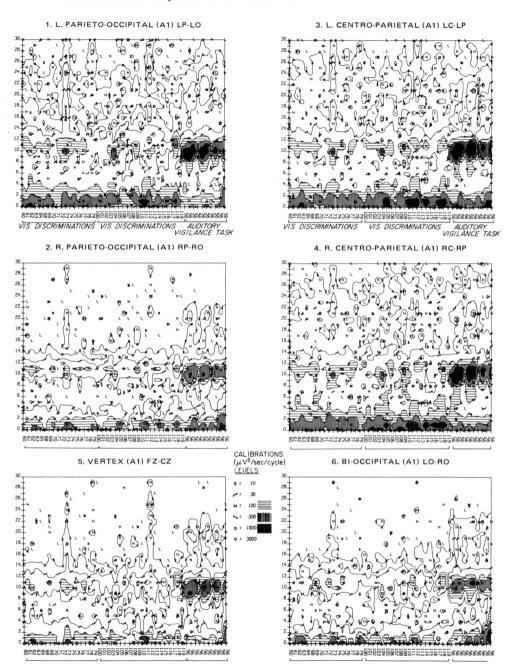

Fig. 8. Contour plots of EEG baseline records in laboratory visual and auditory task performances by astronaut F. Borman. Plots are from six EEG channels recorded simultaneously during performance of visual discrimination tasks in 3 sec (epochs 60-79), followed by more difficult visual discriminations each performed in 1 sec (epochs 100-119) and leading to an auditory vigilance task presented at 5-sec intervals (epoch 56). These plots are condensed to cover an elapsed time of many minutes. (From Adey, Kado, and Walter 1967b.)

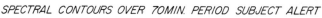
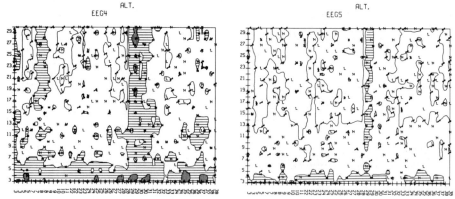
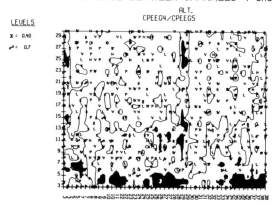
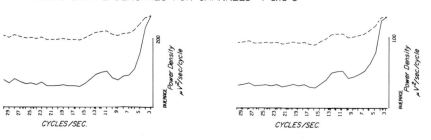

Fig. 9. EEG records from altitude chamber Gemini flight simulation, analyzed to show autospectral densities in two EEG channels and coherence between them. Electrode placements, amplifiers, and recording equipment were identical with actual flight systems. Averaged spectral densities for each channel (lower traces) showed enhanced theta activity (in range 3-7 H) by comparison with laboratory records (fig. 11). These averages were prepared from more than 40 epochs, each 20 sec in duration. Solid lines in lower traces, linear plot; dashed line, logarithmic plot. In the coherence plot, values above 0.7 (statistically significant level) are in black. (From Adey, Kado, and Walter 1967b.)

The decreased alpha and augmented theta activity in the altitude chamber test were clearly intermediate between laboratory tests and space flight records. Careful assessment of awake space flight records throughout the 55 hr after launch indicated increased power in the theta band compared with laboratory and flight simulator data (fig. 10). This augmented theta activity may arise in the physiological substrates of an orienting reflex, or reaction to the novelty of the space environment, and our findings suggest that it is indeed a response to weightlessness. Studies of Soviet cosmonauts reviewed elsewhere (Adey, Kado, and Walter 1967b) have shown similar augmentations in theta activity at frequencies below 8 H (fig. 11).

Speculatively, diminished levels of somatic sensory influxes and modification of vestibular inputs the weightless state may reduce sensory information normally used in orienting responses to alerting stimuli. Such sharp diminution in points of reference in the normal sensorium may also be responsible for the profound autonomic responses, including sweating, tachycardia, and hyperpnea that have frequently accompanied extravehicular activity. In particular, much more data are needed beyond the initial 50 to 60 hr of flight to establish the extent to which augmented theta activity may represent an adaptive phenomenon to early weightlessness, and the degree to which it may persist or recur in longer flights, especially in repeated episodic exposures to such environmental stresses as extravehicular activity.

Conclusion

Manifestly, orientation in the weightless state requires consideration not merely of vestibular mechanisms and closely related ocular coordination but of the whole hierarchy of functions in focusing of attention and visual discrimination. The former constitute the basic platform in a pyramid of increasingly complex central integration. The latter involve the interplay between cortical sensory systems and subcortical structures that are profoundly influenced by limbic activity. Limbic controls, particularly in the hippocampal system, appear essential to the fine focusing of attention necessary for laying down memory traces about spatially organized stimuli. Interference with such controls leads to degradation of spatial discriminative abilities in subtle but important ways that have particular relevance to problems of space flight, where gravitational cues no longer provide a key segment of environmental information and where compensatory mechanisms for this loss might evolve in prolonged space flight. To this time, we have no data on possible evolution of either partial or complete compensation.

The study of weightlessness and of its long-term central nervous effects demands the monitoring within a single subject, man or a subhuman primate, of the gamut of sensorimotor and higher nervous functions that relate to visual coordination, spatial orientation, recent memory, and discriminative ability in prolonged space flight.

Acknowledgments

Studies from our laboratory described here were supported by grants NB-01883 and NB-2503 from the National Institutes of Health, by contracts AF(49) 638-1387 with the United States Air Force Office of Scientific Research, by contract NONR 233-(91) with the Office of Naval Research, and by contracts NsG 237-62, NsG 502, NsG 505, and NsG 1970 with the National Aeronautics and Space Administration.

306 W. R. Adey

C. ALERT, BECOMING DROWSY.   3 HR. 4 MIN. TO 5 HR. 34 MIN.

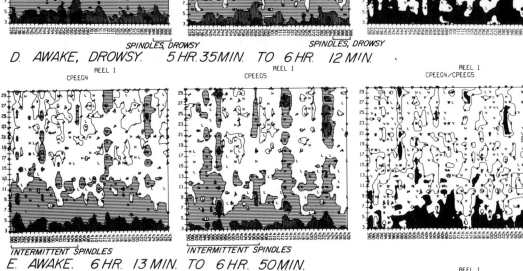

D. AWAKE, DROWSY.   5 HR. 35 MIN. TO 6 HR. 12 MIN.

E. AWAKE.   6 HR. 13 MIN. TO 6 HR. 50 MIN.

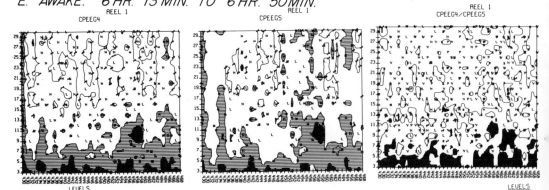

Fig. 10. Contour maps similar to those in fig. 8, covering the 4th through the 7th hr of space flight, for two EEG channels with spectral density (CPEEG4 and CPEEG5) and coherence (CPEEG4/5) plots, showing high theta powers (4-7 H) compared with subjects' ground-based records. (From Adey, Kado, and Walter 1967b.)

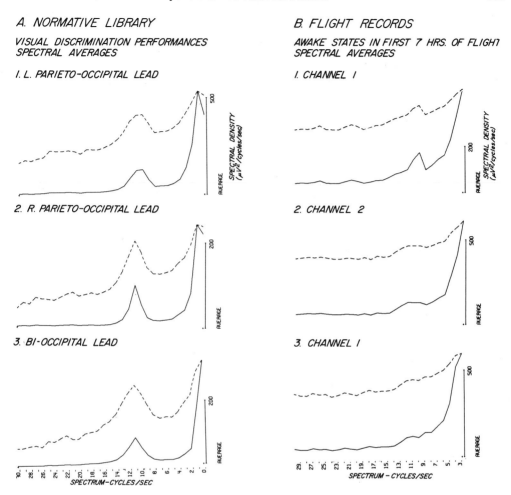

Fig. 11. Comparison of autospectral densities in similarly located leads on scalp of astronaut F. Borman in laboratory visual tasks (*left*) and in awake flight records (*right*), exemplifying diminished alpha peak (8-13 H) and augmented theta density (3-7 H) in flight records. For both laboratory and flight records, these averaged spectral densities were prepared from approximately 40 epochs, each of 20 sec. *Solid line*, linear density plot, *dashed line*, logarithmic plot. (From Adey, Kado, and Walter 1967b.)

References

Adey, W. R. 1956. *Austr. Ann. Med.* 5: 153-62.

Adey, W. R. 1968. In *International conference on cybernetics of central nervous system, Washington, D.C.*, ed. L. Proctor. Boston: Little, Brown.

Adey, W. R.; Bors, E.; and Porter, R. W. 1968 *A.M.A. Arch. Neurol.*, October.

Adey, W. R.; Bystrom, B.; Costin, A.; Kado, R. T.; and Tarby, T. J. 1969 *Exp. Neurol.* 23: 29-50.

Adey, W. R.; Kado, R. T.; and Walter, D. O. 1967a. *Electroenceph. Clin. Neurophysiol.* Suppl. 25: 227-45.

———. 1967b. *Aerospace Med.* 38: 346-59.

Adey, W. R.; Kado, R. T.; Winter, D. W.; and Delucchi, M. R. 1963. *Electroenceph. Clin. Neurophysiol.* 15: 305-20.

Adey, W. R.; Segundo, J. P.; and Livingston, R. B. 1957. *J. Neurophysiol.* 20: 1-16.

Adey, W. R.; Walter, D. O.; and Hendrix, C. E. 1961. *Exp. Neurol.* 3: 501-24.

Bass, L., and Moore, W. J. 1968. In *Structural chemistry and molecular biology*, ed. A. Rich and N. Davidson, pp. 356-69. San Francisco: Freeman.

Beidler, L. M. 1953. *Ann. Acad. Sci.* 58: 52-57.

Bremer, F. 1935. *C.R. Soc. Biol. (Paris)* 118: 1235-42.

Costin, A.; Elazar, Z.; Walter, D. O.; and Adey, W. R. 1967. *Physiologist* 10: 148.

Davis, H. 1959. In *Handbook of physiology*, Section 1, vol. 1, *Neurophysiology*, ed. J. Field, V. E. Hall, and H. W. Magoun, pp. 565-84. Washington: Amer. Physiol. Soc.

Dement, W. C., and Kleitman, N. 1957a. *Electroenceph. Clin. Neurophysiol.* 9: 673-90.

———. 1957b. *J. Exp. Psychol.* 53: 339-46.

Elazar, Z., and Adey, W. R. 1967a. *Electroenceph. Clin. Neurophysiol.* 23: 225-40.

———. 1967b. *Electroenceph. Clin. Neurophysiol.* 23: 306-19.

Elul, R. 1966a. *Trans. Faraday Soc.* 62: 3484-92.

———. 1966b. *J. Physiol.* 189: 351-65.

French, J. D.; Hernandez-Peon, R.; and Livingston, R. B. 1955. *J. Neurophysiol.* 18: 74-95.

French, J. D.; Verzeano, M.; and Magoun, H. W. 1952. *Arch. Neurol. Psychiat.* 69: 519-29.

Grastyan, E.; Lissak, K.; Madarasz, I.; and Donhoffer, H. 1959. *Electroenceph. Clin. Neurophysiol.* 11: 409-30.

Green, J. D., and Arduini, A. 1954. *J. Neurophysiol.* 17: 533-57.

Hecht, S.; Schlaer, S.; and Pirenne, M. H. 1941. *J. Gen. Physiol.* 25: 819-40.

Herrick, C. J. 1933. *Proc. Nat. Acad. Sci., Washington* 19: 7-14.

Ho, T.; Wang, Y. R.; Lin, T. A. N.; and Cheng, Y. E. 1960. *Physiol. Bohemoslav.* 9: 85-92.

Hodes, R. 1962. *Electroenceph. Clin. Neurophysiol.* 14: 220-32.

Huneeus-Cox, F., and Smith, B. H. 1965. *Biol. Bull.* 129: 408.

Katchalsky, A. 1964. In *Connective tissue: Intercellular macromolecules*, pp. 9-42. Boston: Little, Brown.

Kellaway, P., and Maulsby, R. L. 1968. Final report to National Aeronautics and Space Administration, Washington, D.C., under contract NAS9-1200.

Klein, M. 1963. Etude polygraphique et phylogenique des Etats de Sommeil. Thesis, L'institute de physiologie de faculté de Lyon.

Maulsby, R. L. 1966. *Aerospace Med.* 37: 1022-26.

Moruzzi, G., and Magoun, H. W. 1949. *Electroenceph. Clin. Neurophysiol.* 11: 455-73.

Parmalee, A. H.; Wenner, W. H.; Akiyama, Y.; Schults, M.; and Stern, E. 1967. *Develop. Med. Child Neurol.* 9: 70-77.

Pease, D. C. 1966. *J. Ultrastruc. Res.* 15: 555-83.

Petre-Quadrens, O. 1965. *J. Neurol. Sci.* 3: 151-61.

Radulovacki, M., and Adey, W. R. 1965. *Exp. Neurol.* 12: 68-83.

Rambourg, A., and Leblond, C. P. 1967. *J. Cell. Biol.* 32: 27-53.

Scheibel, M. E., and Scheibel, A. B. 1967. In *The neurosciences*, ed. G. C. Quarton, T. Melnechuk, and F. O. Schmitt, pp. 577-601. New York: Rockefeller University Press.

Schmitt, F. O., and Davison, P. F. 1965. MIT Neurosciences Research Program, *Bulletin*, no. 3, part 6: 55-76.

Sumner, J. B. 1953. *Ann. N. Y. Acad. Sci.* 58: 68-72.

Tasaki, I.; Watanabe, A.; and Lerman, L. 1967. *Amer. J. Physiol.* 213: 1465-74.

Valatx, J. L. 1963. Ontogenese des differents etats de sommeil. L'institut de physiologie de la faculté de Lyon.

Walter, D. O., and Adey, W. R. 1966. *Ann. N. Y. Acad. Sci.* 128: 721-1116.

Walter, D. O.; Kado, R. T.; Rhodes, J. M.; and Adey, W. R. 1967. *Aerospace Med.* 38: 371-79.

Walter, D. O.; Rhodes, J. M.; and Adey, W. R. 1967. *Electroenceph. Clin. Neurophysiol.* 22: 22-29.

Wang, H. H.; Kado, R. T.; and Adey, W. R. 1968. *Fed. Proc.* 27: 749.

Wang, H. H.; Tarby, T. J.; Kado, R. T.; and Adey, W. R. 1966. *Science* 154: 1183-84.

Weiss, P. 1964. *Proc. Nat. Acad. Sci.* U.S. 52: 1024-9.

Young, J. Z. 1951. *Doubt and certainty in science: A biologist's reflections on the brain.* New York: Oxford University Press.

28
Vestibular Influences on the Brain Stem

Hiroshi Shimazu, *University of Tokyo*

Dr. Adey (chap. 27, this volume) has given us considerable information regarding functional changes in the mammalian central nervous system which may be associated with space flight. This research field may be related not only to the fine vestibular function controlling the oculomotor, postural, or autonomic activities, but also to states of consciousness, alertness, or emotional perturbation. I would like to focus my discussion on some aspects of his recent research, particularly on function of the vestibular apparatus and related neuronal activities in the central nervous system.

The first point considers the influences from the vestibular apparatus on the core of the brain such as the reticular formation. Adey, Kado, and Walter (1967) found EEG "driving" occurring at certain frequencies of whole-body vibration. They stressed that the EEG change had the characteristics of a physiological driving and appeared distinguishable from similar phenomena of artifactual origin. Lowenstein and Roberts (1951) showed that the impulse activities synchronous to table vibration were recorded from the nerve branch of the saccule in the vestibular labyrinth, and that the synchronism was observed even with low-frequency vibration such as 20 cycles/sec. These synchronous discharges may evoke an oscillatory change in brainstem activities which might be reflected in the rhythmic driving of cerebral electrical activity. However, Adey, Kado, and Walter (1967) have shown that vestibular influxes are not essential and, indeed, that driving in midbrain and cortical structures may increase after vestibular denervation. Thus, they suggested that an origin of the EEG driving may be in mechanoreceptors of thoracoabdominal structures and that the vestibular influxes may normally inhibit brainstem responsiveness to spinal ascending inflow.

The possibility of vestibular inhibition of brainstem activities is an interesting idea, since there have been many studies on convergence of vestibular and spinal influences on the vestibular nuclei. However there appears to be no evidence in the literature of vestibular afferents exerting an inhibitory influence on the somatic nervous activities in the brainstem. Gernandt and Gilman (1960) have shown that the conditioning stimulation of the vestibular nerve enhances the cortically induced potentials in the brainstem. Shimazu and Precht (1966) suggested that crossed influences of the vestibular input on the secondary neurons in the contralateral vestibular nuclei through the pontobulbar reticular formation are also excitatory. On the other hand, vestibular neurons excited from the ipsilateral horizontal canal (type I neurons) are inhibited from the contralateral vestibular nerve (Shimazu and Precht 1966). This appears to be an example of inhibitory action of vestibular influxes. However, this inhibitory effect is mediated specifically through the commissural pathway interconnecting the bilateral vestibular nuclei and may not function as a modulator of general responsiveness of the brainstem to various sensory inputs. It was possible to record usual type I responses of vestibular neurons to horizontal rotation in a chronic preparation where the ipsilateral labyrinth was destroyed; the animal was in the stage of so-called central compensation (Precht, Shimazu, and Markham 1966). Since the vestibular input comes only from the contralateral side in this condition, the response obtained on the deafferented side must be related to the crossed inhibitory influence from the intact side. Thus, the commissural inhibition may act as a compensatory mechanism for vestibular-induced oculomotor or postural adjustment.

The nature of the commissural inhibition was studied by Mano, Oshima, and Shimazu (1968) with intracellular recording techniques, and was found to be postsynaptic. By measuring the latencies of the onset of the inhibitory postsynaptic potentials (IPSPs) in the vestibular neurons after stimulation of the contralateral vestibular nerve or nuclei, it was suggested that there are two kinds of crossed inhibitory pathways. First, the primary afferent fibers excite monosynaptically the inhibitory commissural neurons, which in turn cross the midline and produce the monosynaptic IPSP in the medial nucleus neurons. Second, the commissural inhibition is mediated through inhibitory interneurons on the homolateral side, these interneurons being excited from the contralateral nuclei. The latter pathway may correspond to the mechanism suggested by Shimazu and Precht (1966) for the commissural inhibition in the horizontal canal system.

In contrast to the vestibular influences on the somatic activity in the brainstem, there may be both inhibitory and excitatory effects on the autonomic functions mediated through the medulla. Spiegel (1946) showed that caloric, galvanic or rotational stimulation of the labyrinth produced a decrease in blood pressure together with a slowing of the cerebral circulation. Recently, Uchino et al. (1970) have found that weak stimulation of the vestibular nerve inhibits spontaneous discharges of the postganglionic sympathetic nerve, although strong stimulation produces concomitant excitation as described by Megirian and Manning (1967). These latter investigators suggested that the excitation of sympathetic activities may be attributable to small fibers in the vestibular afferents. According to Uchino et al. (1970), sympathetic inhibition could be induced by the vestibular afferents of middle and large size which are similar to those related to the somatic function. If we could assume that the stronger gravitational force increases afferent impulses along the vestibular nerve, these increased inputs would depress the blood pressure. Adey (1964) has reported that during head-to-tail accelerations of increasing intensity alterations of patterns of brain-wave activity such as flattening or epileptiform discharges are observed. Although cardiovascular changes of vestibular origin alone appear to be too weak to cause functional changes in cerebral activities, they might be partly relevant, in addition to depletion of the cerebral circulation caused by strong centrifugation.

References

Adey, W. R. 1964. *Life Sci. Space Res.* 2: 267-86.

Adey, W. R.; Kado, R. T.; and Walter, D. O. 1967. *EEG Clin. Neurophysiol.* Suppl. 25: 227-45.

Gernandt, B. E., and Gilman, S. 1960. *J. Neurophysiol.* 23: 516-33.

Lowenstein, O., and Roberts, T. D. M. 1951. *J. Physiol.* 114: 471-89.

Mano, N.; Oshima, T.; and Shimazu, H. 1968. *Brain Res.* 8: 378-82.

Megirian, D., and Manning, J. W. 1967. *Arch. Ital. Biol.* 105: 15-30.

Precht, W.; Shimazu, H.; and Markham, C. H. 1966. *J. Neurophysiol.* 29: 996-1010.

Shimazu, H., and Precht, W. 1966. *J. Neurophysiol.* 29: 467-92.

Spiegel, E. A. 1946. *Arch. Otolaryng. Chicago* 44: 61-72.

Uchino, Y.; Kudo, N.; Tsuda, K.; and Iwamura, Y. 1970. *Brain Res.* 22: 195-206.

VI Plant and Animal Gravimorphism

29
Gravimorphism in Higher Plants

Leszek S. Jankiewicz, *Research Institute of Pomology, Skierniewice*

The term "gravimorphism" was introduced by Wareing and Nasr (1958) to define the effects of gravity on plant morphogenesis. This term is now widely used. It has, however, no generally used analogies in botanical sciences. It would be advisable to introduce the terms "photomorphism," "thermomorphism" and similar terms, or to replace the term "gravimorphism" by "gravimorphogenesis" (see Borthwick and Hendricks 1961, Klein 1965, Wagner and Mohr 1966). Dostal (1962) used the terms "gravimorphose," or "barymorphose," "photomorphose," "mechanomorphose." Rawitscher (1932) wrote about "geomorphotic phenomena." The nomenclature in this field should be revised and made uniform.

Wareing and Nasr (1958-61) include in gravimorphism several phenomena: the effect of gravity on the growth of nonramified plants, the effect on the growth of lateral buds and shoots and on the correlations among them, and also the effect on flower bud-formation. These authors do not include in gravimorphism the orientation of plant organs relative to gravity, that is, geotropism, geoepinasty, and related changes in form. In my opinion these phenomena should be included, since the shape or habit of a plant depends largely on the growth *direction* of its parts. It is sufficient to compare creeping versus erect types of growth.

Straight Growth of Nonramified Plants

Horticulturists and plant physiologists long ago made the observation that when a shoot was inclined from vertical to a horizontal or downward position it grew weakly (Voechting 1884, Brzeezinski 1910, Champagnat 1954). This phenomenon was investigated recently by Wareing and Nasr (1958), and Tromp (1967). They found that in a horizontal position apple, plum, cherry and black currant plants restricted to a single shoot grow less than in a vertical position. The difference was accentuated when horizontal plants were rotated around their axis. Weaker growth of these horizontally rotated plants resulted from shorter internodes and, in some cases, from a smaller number of internodes.

The effect of gravity on growth intensity is referred to in older and in some recent papers as a "geogrowth reaction" or as a "geotonic effect on growth." Almost all of these papers show that the inclination of the stem from its normal vertical position to a horizontal position causes growth inhibition. Mature stems of grasses, however, as well as the stems of some other plants which show intercalary growth, behaved differently. Their meristems, situated at the nodes, renewed or increased their growth after inclination to the horizontal position (Rawitscher 1932, Anker 1962, Larsen 1962*a, b*).

The data concerning the growth intensity of plants rotated on a horizontal clinostat are often contradictory. Several authors observed growth inhibition in horizontally rotated plants, whereas others noted growth stimulation (Rawitscher 1932, Anker 1962, Larsen 1962a, b, Bara 1957). This discrepancy might be caused by differences in the rates of rotation used. Larsen (1953) observed that the growth inhibition of Artemisia roots depended on the rate of rotation of the clinostat. Secondary effects of clinostat rotation, for example, vibration (Lyon 1961), also might affect the growth in some experiments. In a recent carefully done experiment of Dedolph et al. (1965), it was shown that oat plants rotated on a horizontal clinostat develop larger roots, but this treatment had no influence on the growth of the shoot. Also, Veen (1964) has shown stimulation of *Vicia faba* roots grown on a horizontal clinostat.

Lateral Bud and Shoot Development

This phenomenon has been investigated chiefly in fruit and forest trees. When the young tree is placed in a horizontal position, the buds on its upper side generally develop into long shoots, whereas those on the underside develop only slightly (fig. 1) or not at all (Wareing and Nasr 1958, 1961; Jankiewicz 1964; Smith and Wareing 1964a, b). In poplar trees, the buds on the lower side often begin to develop, but after a few days become yellow and die (Jankiewicz et al. 1967).

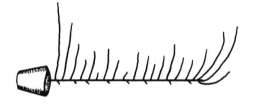

Fig. 1
Young apple tree placed for one vegetative season in a horizontal position.

The geoinduction time for differential bud growth varies among species. One-year-old poplars showed differential bud growth in response to only a 24-hr induction (fig. 2). One-year-old apple trees can be induced by a 5-7 day horizontal treatment (Borkowska 1966).

Apple trees are most sensitive to gravimorphic induction at the time the buds are activated and during the beginning of growth. The buds may be geoinduced, however, even during deep dormancy in the autumn. After being bent so that they were transverse to the direction of gravity for two months (October 17 to December 17), they showed, in the spring, marked inhibition of growth on what was formerly the underside (Borkowska 1966, and in press). This result may be better understood in the light of Champagnat's observation (1962) that considerable change takes place in the correlations between buds of apical and basal regions of vertical Sambucus shoots during the dormancy period. Another interesting fact about the apple trees geoinduced in the fall was that the upper and lower buds did not differ in their fresh weight up to the time they began to swell (Borkowska 1968).

The cause of differential growth of buds on the two sides of horizontally placed shoots has been investigated by several workers. The angle that a bud makes in relation to gravity appears not to be important, since the buds on the upper and underside of the poplar or apple tree form very similar, or even the same, angle (fig. 3), whereas their development is strikingly different (Smith and Wareing 1964a, Jankiewicz 1964).

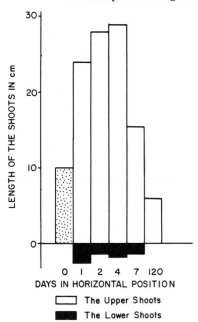

Fig. 2
The length of the shoots on the upper side (white columns) and on the underside (black columns) of a poplar tree placed for varied times in a horizontal position. "O" represents the lengths of the vertical controls.

Fig. 3
The orientation of buds on the apple tree stem.

The effect of differential illumination can be excluded as a casual factor, as willow trees (Smith and Wareing 1964a) and apple trees (Jankiewicz, unpublished) show similar gravimorphic differences between the upper and lower buds in darkness and in light. Also, bending of the stem during horizontal treatment had little effect on the differences in bud growth (Smith and Wareing 1964a). However, the arched shoots of the apple possess a lower water conductivity than those that are straight (Cristopheri and Giachi 1964).

There is the likelihood that the buds located on the upper side of a woody stem correlatively inhibit those on the underside. Splitting a one-year-old apple tree longitudinally in a horizontal plane, and inserting an impermeable barrier between the halves, caused marked alleviation of bud inhibition on the underside (Mullins 1965a). A similar effect was obtained by Borkowska (1968, and Borkowska et al. 1967) by removing a 2 mm strip of bark along flanks of horizontal trees. In this case it was shown that the inhibitory factor moves from the upper half of the tree to the lower in the bark. Removing all the buds of an apple tree except the lower ones markedly alleviated the inhibition of those buds remaining.

In willow trees, on the other hand, the inhibition of bud activation on the underside is apparently not caused by influences coming from the upper buds (Smith and Wareing 1964a). Also, the buds on the underside did not break if they were separated from factors coming from upper buds by an impermeable barrier inserted between two halves of a longitudinally split tree. Smith and Wareing (1964a) conclude that bud break in the willow tree "is determined by some mechanism, acting between the bud and the immediately adjacent tissue, which is affected by the orientation of the adjacent tissues relative to gravity."

It seems clear that the gravity-mediated bud inhibition on the underside of a tree occurs in several steps. The "early step" is probably the accumulation of a growth stimulator, and/or growth inhibitor, on the underside of the stem. Leach and Wareing (1967) found that labeled IAA accumulates on the lower side of horizontal poplar stems in a manner similar to that found for coleoptiles and stems of herbaceous plants (Gillespie and Thimann 1963, Goldsmith and Wilkins 1964, Lyon 1965). Nečesany (1958) and Leach and Wareing (1967) also found that the natural growth stimulators accumulate in a larger amount on the lower side of woody stems. There exists, however, a discrepancy of results concerning growth inhibitors—Nečesany found more on the upper side of poplar stems, whereas Leach and Wareing found more on the lower side. However, these workers used different plant materials and different methods. Even single buds with a stem segment cultured in vitro show differences in growth depending on orientation (Borkowska 1969). Substances inhibiting bud growth are probably present in effective amounts even in short sections of the stem. It is also possible that gravity acts directly on the buds.

The reactions involved in this "early step" probably caused some undefined but durable changes in the lower buds, so that a reduction in flow of nutrient to them occurred. In horizontal poplar trees fed with ^{32}P-labeled *ortho*-phosphate through the roots for three days, the lower buds absorbed less phosphorus per bud and per unit of dry weight (Borkowska 1968). Similar results were obtained earlier by Smith and Wareing (1966). This might be associated with a lower auxin production by the lower buds (Booth et al. 1962), but the auxin contents of the upper and lower buds have not yet been investigated. An interesting idea of Sachs and Thimann (1967) might explain the two seemingly contradictory facts—accumulation of growth stimulators on the lower side of the stem and low auxin production by the lower buds. They suggest that an organ which receives the auxin from elsewhere in the plant curtails its own production of this hormone.

The next step consists of correlative inhibition between the shoots and buds. In vertical as well as in horizontal apple and poplar trees only a portion of the buds develop into long shoots. The others grow into short shoots of different vigor, or do not develop at all. In vertical apple trees practically every bud in the middle and upper part of the whip can form a long shoot if the competition from other buds is limited. If a shoot has even a small disadvantage during its early stage of development, it becomes the victim of competition. The mechanism of fast augmentation of differences between the shoots has not been explained as yet. MacIntyre's (1964) hypothesis may be helpful in this respect—the stronger bud produces more auxin, which stimulates development of vascular strands that supply it with nutrients, and this causes still better growth and auxin production. This cycle of interdependence is a typical positive feedback (Jankiewicz 1967). In this fashion the stronger bud progressively gains in competitive efficiency. In horizontally oriented trees all buds located on the underside have a lower growth potential at the beginning of the growth period, so they are soon dominated. The hypothesis of MacIntyre does not exclude the possibility that the bud which forms a stronger shoot directly inhibits the other buds by production of a transported growth inhibitor. Goodwin and Gansfield (1967) suggest the existence of a very labile inhibitor in the sprouting potato tuber which disappears quickly from the tuber after removing the dominant uppermost shoots. The sprouts are inhibited only if they are below a certain vigor limit. Dörffling (1966) and Arney and Mitchell (1969) have also found evidence that growth inhibitors may play a role in correlative bud inhibition.

Gravity and Apical Dominance

The effect of gravity on apical dominance is readily demonstrated in one-year-old decapitated stems of several woody plants. When the stems are growing in an upright vertical position, the apical laterals are most vigorous, and dominate those located below. The buds near the base of the stem usually do not develop at all. Wareing and Nasr (1961) and Jankiewicz et al. (1967) have shown that in cherry and poplar trees those relations can change completely when a tree is placed in a horizontal position. The basal buds then produce the strongest shoots, whereas the apical buds grow weakly (fig. 4). In one-year-old poplar trees kept in a horizontal position for 2-7 days at bud bursting time, the ratio between the apical and lower buds changed markedly as compared with the vertical controls (table 1). The buds in the basal part of the tree which otherwise would be inhibited produced numerous shoots in response to the 7-day horizontal treatment (Jankiewicz et al. 1967).

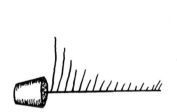

Fig. 4
Apical dominance in the poplar tree, and its inversion by horizontal orientation of the plant.

TABLE 1. The Development of Shoots in One-year-old Poplars (Populus hybrid No. 277) after Placing the Seedlings in Horizontal Position at Bud Opening Time

Days in Horizontal Position	Ratio of Length of Apical Shoots to Length of Middle Shoots	Shoot Growth in the Basal Part of the Tree
0	5.3	-
1	3.1	-
2	2.5	-
4	1.8	-
7	1.2	++
120	0.6	+++

Apical dominance in the apple tree is less dependent on gravity than in the cherry and poplar trees. Mullins (1965b) has found, in vertical as well as in horizontal and inverted trees, two centers of vigorous growth—apical and basal. The apical center is more active in vertical trees, the basal in horizontal or inverted trees. The basal center was more active in all trees during the second part of the vegetative season. In the cassava, inverted shoots show weak apical dominance, whereas basal dominance occurs in those horizontally oriented (Longman 1968). Poplars, on the other hand, show marked basal dominance in both orientations (Jankiewicz et al. 1967).

The domination of one branch over the others in vertical trees depends on their angles in relation to gravity. When an uppermost branch was tied in a horizontal position, and the second branch kept vertical, the second branch became dominant (Wareing and Nasr 1961).

When a branch of a woody plant is arched or bent into a loop, the most vigorous laterals appear at the highest point of the arch, or proximate to it (Wareing and Nasr 1961, Fiorino 1968, and Crabbé 1969). The cause of this phenonenon was investigated in the willow tree by Smith and Wareing (1964b). They showed that a growth factor necessary for shoot development is generated in the roots and is monopolized by the uppermost shoots of the arc or loop. It was also shown that the other shoots were limited by the root factor; they developed better when an additional section of the stem was rooted. The root factor passed easily across a girdle, indicating that it was probably transported in the xylem. This growth factor produced in the roots was not a mineral nutrient since the roots continued to produce it even when submerged in distilled water. Longman's work (1968) with cassava shoots, rooted basally, apically, or at both ends, also suggests that root factors play an important role in dominance. These results find support in more recent investigations. Sitton et al. (1967), Itai and Vaadia (1965), Nitsch and Nitsch (1965), and Luckwill and Whyte (1968) have found cytokinin and gibberellin activity in root exudates. Šebanek (1966a) has presented data indicating that in pea plants the root system is a site of gibberellin synthesis and export.

The uppermost shoots of the arc also are able to "monopolize" nutrients. If a branched birch seedling was bent into a loop or arc, the branch in the uppermost position on the stem accumulated ^{32}P most efficiently. This difference was clearly visible 2-4 days after reorientation (Smith and Wareing 1966).

Why the uppermost buds or shoots in the arc become the most powerful "sinks" for nutrients and growth substances supplied by the roots remains unanswered. Smith and Wareing (1966) suggest that looping induces profound changes in auxin metabolism in the buds, resulting in abundant production of auxin in the uppermost buds of the loop or arc. Several recent papers show the powerful effect of auxin on translocation within the plant (Booth et al. 1962, Šebanek 1965, Davies and Wareing 1965, Hew et al. 1967). Gibberellin and cytokinin stimulate the action of auxin in this respect (Šebanek 1966b, Wareing and Seth 1964, Seth and Wareing 1967). Šebanek and Hink (1967) has shown that if auxin is applied to one part of the stem and gibberellin to another, the gibberellin is diverted to the site of auxin application.

The change from apical into basal dominance when a plant is placed in a horizontal or an arched position indicates that the key role in these phenomena is played by a substance whose transport is polar and dependent on gravity. We know at present only one natural factor which exhibits these properties—the auxin indoleacetic acid. We are therefore prone to ascribe to it this key role. It is possible, however, that other endogenous substances with similar properties exist.

The Effect of Gravity on Flower Bud-set

Training the branches in a horizontal or downward position in order to enhance flowering and fruiting has long been applied by fruit growers (Brzezinski 1910, Champagnat 1954). Only recently, however, has this phenomenon become the object of scientific studies. Inversion of soybean plants accelerates their flowering (Fisher 1957). Van Overbeek and Crusado (1948) induced flowering in pineapple plants by tilting them to the horizontal. Most of the subsequent investigations have been with woody plants.

Since flower bud-set depends largely on growth vigor of the plant or branch, we may expect that gravity influences flower bud-differentiation by affecting vegetative growth. However, there probably exists a more direct effect of gravity on flower bud-formation. The formation of male flower buds on horizontal branches of the Japanese larch (Longman and Wareing 1958, Melchior 1960, 1961) may be an example of an effect of gravity that is probably not mediated by growth vigor. The buds of the larch were found to differentiate into male flowers only if they were pointing horizontally or downward. The buds which were pointing upward were exclusively vegetative. If the horizontal branch was reverted 180° during flower differentiation, the flower buds were formed on the new lower side. Thus it was demonstrated that flower buds were formed on the physically lower side of the stem irrespective of its original morphological orientation (Longman and Wareing 1958, Longman et al. 1965), though the causal physiological change is as yet unknown.

The competition between shoot growth and flower formation in fruit plants is stressed by many authors (Kobel 1954, Champagnat 1954, Chandler 1957). The nature of this connection, however, has been little studied. Differentiation of apple flower buds takes place in the shoot apices which already have ceased active growth (Swarbrick 1928, Fulga 1966). Therefore, shoots growing for a long time have little chance to become reproductive. It was also found in the black currant that the long shoots which have not finished their extension growth inhibit the formation of flower initials (Nasr and Wareing 1961). There are several indications that the same is true for apple trees (Grochowska 1966). Horticultural practices which affect growth (pruning, use of dwarfing rootstocks, spraying with growth regulators) as a rule also influence flowering (Tromp 1967).

Since bending the branches strongly influences their growth, it can be expected that it will also influence flower initiation. There are reports of a positive effect of branch inclination on flower bud-formation. Heavier flowering in horizontal trees, rotated or nonrotated, than in vertical ones was observed by Wareing and Nasr (1958), Longman et al. (1965), and Tromp (1967). Concomitant decrease of shoot length and earlier cessation of extension was also noted in horizontal trees. Also, Neumann (1962) observed a higher number of flower buds and earlier flower bud-differentiation accompanied by reduced growth in apple branches tied in a horizontal position.

On the other hand, Mullins (1965*b*) could not produce more flowering in horizontal and inverted trees as compared to vertical ones. In his experiment, however, the branches of both horizontal and vertical trees were, more or less upright, and there was no difference in total shoot growth between the differently treated trees. Other workers have not found an increase of flower bud-initiation after the inclination of branches of apple trees and of black currant (Jonkers 1962, Longman et al. 1965, Husabo 1965, Mika 1969*a*, Mitov 1969).

It would be an oversimplification to think that changes in the position of a fruit tree or branch in relation to gravity can affect flowering only by controlling the vegetative growth. Mika (1969*b*) found that labeled assimilates are transported differently in vertical and arched shoots. In actively growing shoots radioactive metabolites were transported predominantly toward the apex, but several times greater amounts were found in the apices of arched shoots than in the vertical. In the shoots which terminated their growth, more labeled assimilates were retained in the arched shoot, and less was transported toward the trunk and roots. On the other hand, ^{32}P, applied through the roots, was found in smaller amounts in the arched than in the vertical shoots of the same tree. Kato and Ito (1962) obtained results analogous to those of Mika (1969b), that is, higher concentrations of carbohydrate were found in horizontally-trained and sagging apple shoots. Kato and Ito observed a close negative correlation between the carbohydrate content in shoot apices and the auxin content. They suggest that carbohydrates control the auxin metabolism. Our present knowledge suggests, rather, that the reverse is true. Shternberg and Kulikova (1957) obtained different results—less carbohydrates in the apical part of horizontal and arched branches; this is a discrepancy that remains to be resolved. Fiorino (1968) observed flower bud-formation to occur preferentially in the apical portion of arched pear shoots, which corresponds with the preferential transport of assimilates to that region (Mika 1967, 1969b).

Orientation of Plant Organs

Aboveground plant organs orient themselves primarily by their sensitivity toward gravity. Phototropism, at least at high light levels, plays a less important role. The lateral shoots and roots and sometimes the main axes of many plants grow obliquely or transversely to the direction of gravity (plagiogeotropism). In most instances plagiotropic growth can be explained as a result of the interaction of two factors: negative geotropism and epinasty in shoots, and positive geotropism and hyponasty in roots (Kaldewey 1962, Rufelt 1962). In the actively growing lateral shoot, the epinasty counteracts the negative geotropism, thus preventing or diminishing the upward curvature. Lyon (1963) has shown that the epinastic tendency is caused by the lateral transport of auxin toward the upper side of the oblique shoot. The apex of the main shoot induces epinasty in the laterals (Münch 1938). Since auxin paste applied to a decapitated axis gives the same effect, it is postulated that the active apex induces epinasty in the laterals by production of auxin (Kaldewey 1962). Older, lignified, woody axes orient themselves by the formation of reaction wood (see Westing 1965, 1968).

The position of a branch is often the result of the interaction of several factors. Consider the pine branch as an example (Jankiewicz 1966). As the pine shoot emerges from the bud it grows vertically upward for about two weeks (fig. 5). Then its basal part begins to deviate toward the horizontal. This movement is probably caused by the mechanical pressure exerted on a branch base by the intensively enlarging tissue located in the crotch. This downward movement is counteracted by strong geotropic curvature of the distal portion of the branch. As a result, the arc-like form appears (fig. 6). During the second and third year of branch growth the basal part no longer changes its angle with the vertical, but the middle and apical parts move gradually downward during the time of most intensive shoot and needle growth (fig. 7). There are good indications that the increasing weight of new shoots and needles causes this movement. Compression wood, which is formed on the lower side of the branches, usually is not able to counteract fully this change of branch position. The weight of snow frequently causes additional irreversible change in branch position (Jankiewicz 1966). This example shows that geotropism is only one of the factors that define branch position.

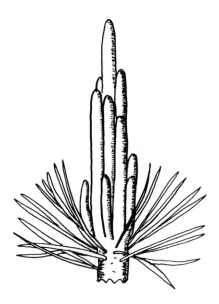

Fig. 5. Strong negative geotropism in the initial growth of pine tree shoots.

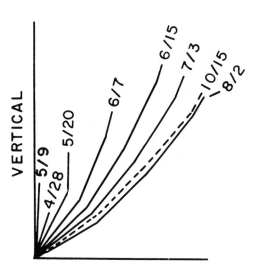

Fig. 6. Changes in the position of a pine branch during the first year of its growth.

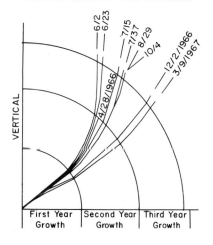

Fig. 7
Changes in the position of a pine branch during the third year of its growth. The change during the autumn (compare Oct 4 and Dec 2) was caused by the weight of snow.

Flower Morphology

Flower shape in several species is influenced by gravity (see reviews of Rawitscher 1932, and of Halbsguth 1965). Simple examples are the geotropic curvatures of styles and filaments. Often the proximal part of a filament or style is negatively geotropic and the distal part is positively geotropic, which results in S-shaped curvature.

In some flowers which are inserted laterally on the axis, for example, *Epilobium angustifolium* (Voechting 1886), the arrangement of petals is altered by the continuous reorientation on the clinostat. Normally most of the petals are grouped on the upper side of the receptacle, whereas in the flowers that develop on a clinostat, the petals are arranged radially.

The influence of gravity on flower morphology in wild gladiolus (*Gladiolus paluster Gaud.*) is greater. The lowermost leaves of the perigonium are much larger than those on its opposite (upper) side. This is a gravity effect, since the flowers developing on the clinostat have the leaves of the perigonium of equal size, and display radial symmetry (Haeckel 1931).

Summary

The influence of gravity on plant morphogenesis has only recently come under intensive investigation, and the nomenclature in this field is still uncertain. This paper has described, and to a limited extent interpreted, research on the influence of gravity on linear growth, lateral bud and shoot development, apical dominance, the initiation of flower buds, orientation of plant organs, and flower morphology. Several generalizations in these domains can be made:

1. Unramified shoots, coleoptiles, or roots inclined from the vertical toward the horizontal grow at a slower rate. In some species this effect is accentuated when a plant is rotated on a clinostat, in others the reverse is true.

2. Buds on a horizontally oriented shoot develop asymmetrically. Those located on the upper side form vigorous laterals, whereas those on the underside form only short shoots or do not develop at all. There are evidences that auxin, accompanied by an inhibitor, accumulates on the underside of the stem and evokes the changes resulting in inhibition.

3. Apical dominance depends on a position of a shoot in relation to gravity. When a shoot is placed in a horizontal or downward position, usually the basal laterals become dominant. This throws new light on several hypotheses concerning apical dominance.

4. Fruit bud-set is often enhanced by inclination of a tree or branch toward the horizontal position. In most cases this effect seems to be mediated by a slower rate of growth of inclined shoots.

5. There is a direct effect of gravity on the initiation of flower bud-formation.

6. The form (habit) of a plant depends on the orientation of its parts in relation to gravity. Data are presented on the mechanism of vertical branch-angle formation in pine. The shape of some flowers, as well as the differential growth of flower parts, is also determined by orientation with respect to the direction of gravity.

References

Anker, L. 1962. In *Encyclopedia of Plant Physiology*, vol. 17 (2). W. Ruhland and E. Bünning eds., Berlin: Springer, pp. 103-52.

Arney, S. E., and Mitchell, D. L. 1969. *New Phytol.* 68: 1001-1015.

Bara, M. 1957. *Rev. Fac. Sci. Univ. Istanbul* B.22: 209-38.

Booth, A.; Moorby, J.; Davies, C. R.; Jones, H., and Wareing, P. F. 1962. *Nature* (London) 194: 204-5.

Borkowska, B. 1966. *Bull. Acad. Polon. Sci.* Cl.V. 14: 563-67. 1968. Studies on gravimorphism of apple trees (in Polish). Doctoral Thesis, Central College of Agriculture, Warsaw.

———. 1969. *Bull. Acad. Polon. Sci.* Cl.V. 17: 503-5.

Borkowska, B.; Jankiewicz, L. S.; and Srzednicka, W. 1967. *Wiss. Zeitschr. Univ. Rostock, Math. Naturwiss.* 16: 561-62.

Borthwick, H. A., and Hendricks, S. B. 1961. In *Encyclopedia of Plant Physiology*, vol. 16, W. Ruhland and E. Bünning eds., Berlin: Springer, pp. 299-30.

Brzezinski, J. 1910. *Hodowla drzew i krzewów owocowych*. Warsaw: Gebethner and Wolf.

Chandler, W. H. 1957. *Deciduous orchards*. 3rd ed. Lea and Febiger, Philadelphia.

Champagnat, P. 1954. *The pruning of fruit trees*. London: Crosby Lockwood and Son.

———. 1962. *Ann. Gembloux* 68: 137-52.

Crabbé, J. 1969. *Bull. Rech. Agronom. Gembloux.* 4: 220-39.

Cristoferi, G., and Giachi, M. 1964. *Effetti della curvatura sulla conductibilita idrica e sulla structura anatomica dei rami di melo*. Rome: Cons. Nat. Ric.

Davies, C. R., and Wareing, P. F. 1965. *Planta* (Berl.) 65: 139-56.

Dedolph, R. R.; Naqvi, S. M.; and Gordon, S. A. 1965. *Plant Physiol.* 40: 961-65.

Dörffling, K. 1966. *Planta* (Berl.) 70: 257-74.

Dostál, R. 1962. *Zemedelska Botanika–Fysiologie rostlin* CSAZV (Praha).

Fiorino, P. 1968. *Riv. Ortoflorofruttic. Ital.* 52: 603-16.

Fisher, J. E. 1957. *Science* 125: 396.

Fulga, I. G. 1966. *Trudy Mold. nauč. iss. inst. sadov. vinogr. vinod.* 10: 109-122.

Gillespie, B., and Thimann, K. V. 1963. *Plant Physiol.* 38: 214-25.

Goldsmith, M. H. M., and Wilkins, M. B. 1964. *Plant Physiol.* 39: 151-62.

Goodwin, P. B., and Gansfield, P. E. 1967. *J. Exp. Bot.* 8(55): 297-307.

Grochowska, M. J. 1966. *Prace Inst. Sad.* (Warsaw) 10: 51-92.

Haeckel, J. 1931. *Flora.* 125: 1-82.

Halbsguth, W. 1965. In *Encyclopedia of Plant Physiology.* vol. 15(1). W. Ruhland and E. Bünning, eds., Berlin: Springer, pp. 331-82.

Hew, C. S.; Nelson, C. D.; and Krotkov, G. 1967. *Amer. J. Bot.* 54: 252-56.

Husabo, P. 1965. *Forsk. forsok landb.* 1965: 227-37.

Itai, C. Y., and Vaadia, Y. 1965. *Physiol. Plant.* 18: 941-44.

Jankiewicz, L. S. 1964. *Acta Agrobot.* (Warsaw) 15: 21-50.

---. 1966. *Acta Agrobot.* (Warsaw) 19: 129-42.

---. 1967. In *Zarys Fizjologii Sosny Zwyczajnej* [The physiology of Scots pine]. Warsaw: PWN, pp. 223-46.

Jankiewicz, L. S.; Srzednicka, W.; and Borkowska, B. 1967. *Bull. Acad. Polon. Sic.* Cl. V, 15: 111-13.

Jonkers, H. 1962. *XVI-th International Congr. Brussels.* pp. 441-43.

Kaldewey, H. 1962. In *Encyclopedia of Plant Physiology*, vol. 17(2) W. Ruhland and E. Bünning eds., Berlin: Springer, pp. 200-321.

Kato, T., and Ito, H. 1962. *Tohoku J. Agr. Res.* 13: 1-21.

Klein, R. M. 1965. *Physiol. Plantarum* 18: 1026-33.

Kobel, F. 1954. *Lehrbuch des Obstbaues auf physiologischer Grundlage,* Berlin: Springer.

Larsen, P. 1953. *Physiol. Plantarum* 6: 735-74.

---. 1962a. In *Encyclopedia of Plant Physiol.*, vol. 17(2), W. Ruhland and E. Bünning eds., Berlin: Springer, pp. 34-73.

---. 1962b. In *Encyclopedia of Plant Physiol.*, vol. 17(2), W. Ruhland and E. Bünning eds., Berlin: Springer, pp. 153-99.

Leach, R. W. A., and Wareing, P. F. 1967. *Nature* 214: 1025-27.

Longman, K, A. 1968. *Ann. Bot.* 32: 553-66.

Longman, K. A., and Wareing, P. F. 1958. *Nature* 182: 379-81.

Longman, K. A.; Nasr, T. A. A.; and Wareing, P. F. 1965. *Ann. Bot.* 29: 459-73.

Luckwill, L. C., and Whyte, P. 1968. In *Plant growth regulators.* Monograph 31. London: Soc. Chem. Industry, pp. 87-101.

Lyon, C. J. 1961. *Science* 133: 194-95.

———. 1963. *Plant Physiol.* 38: 145-52.

———. 1965. *Plant Physiol.* 40: 953-61.

MacIntyre, G. I. 1964. *Nature* (Lond.) 203: 1190-91.

Melchior, G. H. 1960. *Naturwiss.* 47: 502.

———. 1961. *Silvae Genet.* 10: 20-26.

Mika, A. L., 1967. On the growth inhibition and the stimulation of flower bud set due to bending of branches. Doctoral Thesis, Central College of Agriculture, Warsaw.

———. 1969a. *Hort. Res.* (Edinburgh) 9: 93-102.

———. 1969b. *Biol. Plant.* (Prague) 11: 175-82.

Mitov, P. 1969. In "Giornate di Studia sulla Potatura degli alberi da frutto." *Soc. Orticola Ital.* (Florence), pp. 249-55.

Mullins, M. G. 1965a *Ann. Bot.* 29: 73-78.

———. 1965b *J. Hort. Sci.* 40: 237-47.

Münch, E. 1938. *Jb. Wiss. Bot.* 86: 581-673.

Nasr, T. A. A., and Wareing, P. F. 1961. *J. Hort. Sci.* 36: 1-10, 11-17.

Nečesany, V. 1958. *Phyton.* (Buenos Aires) 11: 117-27.

Neuman, D. 1962. In *Physiologische Probleme im Obstbau*, pp. 87-100. Berlin: Dtsch. Akad. Landwirt.

Nitsch, J. P., and Nitsch, C. 1965. *Bull. Soc. Botan.* (France) 112: 1-10.

Overbeek, J. van, and Crusado, H. J. 1948. *Amer. J. Bot.* 35: 410-12.

Palmer, J. H. 1964. *Planta* 61: 283-97.

Rawitscher, F. 1932. *Der Geotropismus der Pflanzen.* Jena: Gustav Fischer.

Rufelt, H. 1962. In *Encyclopedia of Plant Physiology*, vol. 17(2), W. Ruhland and E. Bünning eds. Berlin: Springer, pp. 322-342.

Sachs, T., and Thimann, K. V. 1967. *Amer. J. Bot.* 54: 136-44.

Šebanek, J. 1965. *Biol. Plant.* (Prague) 7: 380-86.

———. 1966a. *Biol. Plant.* (Prague) 8: 470-75.

———. 1966b. *Biol. Plant* (Prague) 8: 213-19.

Šebanek, J., and Hink, J. 1967. *Planta* (Berlin) 76: 124-28.

Seth, A. K., and Wareing, P. F. 1967. *J. Exp. Bot.* 18: 65-77.

Shternberg, M. B., and Kulikova, R. F. 1957. *Bot. Zhurn.* 42: 1079-87.

Sitton, D.; Itai, C.; and Kende, H. 1967. *Planta* 73: 296-300.

Smith, H., and Wareing, P. F. 1964a. *Ann. Bot.* 28: 283-95.

———. 1964b. *Ann. Bot.* 28: 297-309.

———. 1966. *Planta* 70: 87-94.

Swarbrick, T. 1928. *J. Pom. Hort. Sci.* 7: 100-129.

Tromp, J. 1967. *Naturwiss.* 54: 95.

Veen, B. W. 1964. *Acta Botan. Neerl.* 13: 91-96.

Voechting, H. 1878/84. *Ueber Organbildung im Pflanzenreich.* Bonn.

———. 1886. *Jahrb. wiss. Bot.* 17: 297-346.

Wagner, E., and Mohr, H. 1966. *Planta* 71: 204-21.

Wareing, P. F., and Nasr, T. A. A. 1958. *Nature* 182: 379-81.

———. 1961. *Ann. Bot.* 25: 321-40.

Wareing, P. F., and Seth, A. 1964. *Life Sci.* 3: 1483-86.

Westing, A. R. 1965. *Bot. Rev.* 31: 381-480.

———. 1968. *Bot. Rev.* 34: 51-78.

Discussion

LEOPOLD: I think that some of these gravimorphisms might be related to ethylene production in the tissues. We know from the work of the California group, under Harlan Pratt, that when tension or pressure is exerted on the pea seedling ethylene is produced. Gravitationally induced growth changes might conceivably be related to the production of ethylene by physical stress imposed on the tissue.

GALSTON: In support of Dr. Leopold's suggestion, certainly the case of the induction of flowering in pineapple by prostration of the plant can be attributed to ethylene. We know that under the conditions of auxin accumulation in the lower part of the growing point, auxin concentrations are sufficiently high to induce ethylene formation, and it has been shown that these quantities of ethylene are sufficient to induce flower bud-formation in the pineapple. In this one case I think it is fairly clear that the prostration effect and the gravity-induced changes are, in fact, due to ethylene.

30
Geoepinasty, an Example of Gravimorphism

Harald Kaldewey, *University of Saarland*

It is clear from Dr. Jankiewicz' paper (chap. 29, this volume) that gravimorphisms are induced by a long-lasting directionally oblique influence of gravity on a plant organ (see also Wareing and Nasr 1958). As a result of this stimulus, physiologically upper and lower sides typically develop. They coincide with the physical upper and lower sides during induction but are independent of the original morphological organization of the organ. The growth differences observed may be attributed to a redistribution of growth-regulating substances. Does plagiotropism have these characteristics of gravimorphism?

Characteristic of plagiogeotropism is a long-lasting induction time, obtained either by reorienting an organ experimentally (cf. Kaldewey 1962) or, under natural conditions, by an inherent, morphologically fixed growth reaction, called "adjusting movement" (*Einstellbewegung*) in the German literature (Zimmerman 1932). The plagiogeoinduced organ shows physiologically upper and lower sides which are independent of initial morphological organization, with different growth characteristics that lead to the nonorthogeotropic orientation. Moreover, it has been shown that asymmetric distribution of auxin is involved in the altered patterns of growth (Leike and von Guttenberg 1961, 1962; Lyon 1963*a, b*; 1965*a, b*). Thus, plagiogeotropism appears to be a characteristic example of gravimorphism. In this context we will analyze the geoepinastic bending of Fritillaria axes as a prototype of geoinduced plagiotropic growth.

The Geoepinastic Movement of *Fritillaria Meleagris* L.

The development of the axis of *Fritillaria meleagris* is characterized by an early erect stage, followed by a prefloral bending movement and then by a postfloral straightening of the stem. During the prefloral and the floral bent phases, the curvature begins at the base of the stem, moves progressively acropetally, together with the acropetally moving elongation zone, which is always found below the peak of curvature (fig. 1).

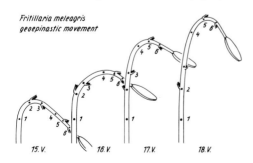

Fig. 1
Growth movements of the stem of *Fritillaria meleagris* L. during the prefloral bent phase. The numbers show that the peak of curvature is moving acropetally during the 3 days of development, and their changing positions indicate the zone of main stem elongation at the base and below the bent region.

Thus the bent stage, seemingly static in contrast to the phase of prefloral bending and postfloral straightening, is actually a continuous movement when observed more closely. We find within the curved region a straightening zone below a bending one, both zones moving acropetally.

The subepidermal cells of the physiologically upper and lower sides have been found to be convenient for observing the growth patterns involved in the movement. Measurements and counts of these cells have shown that the cell number is fixed in an early stage of the bent phase and does not vary in the two sides. Hence, the process of bending and straightening is composed only of cell elongation within and below the curvature zone.

Since the cell number is equal in both sides, and also in the corresponding opposite sides of each stem zone, the cell length yields a measure of length of the organ. The ratio between cell lengths in the upper and lower sides (length quotient) thus indicates the ratio between the lengths of the two sides, and this ratio is a measure of the degree of curvature. The peak of the curve, obtained by plotting the length quotients calculated along the shoot axis, coincides with the peak of curvature (fig. 2).

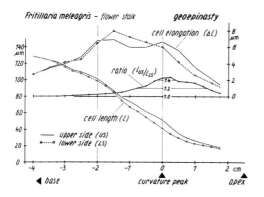

Fig. 2
Length, elongation, and ratio of the length of the subepidermal cells of the physiologically upper (US) and lower sides (LS) in the axis of Fritillaria meleagris L. during the floral bent phase. The elongation (Δl) has been calculated from the differences of the average cell lengths between successive 2.5-mm stem sections. The quotients between the cell lengths of the opposite sides (l_{US}/l_{LS}), a measure of the degree of curvature, were calculated in successive 5-mm sections. Mean of 5 plants.

The cell lengths along the stem show a continuous basipetal increase on both sides of the axis, the cells having an initial length of about 15 μ at the apex and about 100 μ in the erect region just below the curvature, here still elongating up to about 300 μ when mature. Since there is an age gradient in the cells from the apex to the base, and since aging is associated with elongation, one may say that the difference in length between cells of the successive stem sections is a measure of growth rate of these sections along the axis toward the base. The differences between the average cell lengths of successive 2.5 mm sections of the upper and lower sides are also plotted in figure 2. The curves show the lower growth rate of the apical portion and increasing rates toward the base, reaching a maximum in the basal region of the curvature. Thus, the peak of curvature lies above the main zone of elongation.

Comparing the curves for the upper and lower sides, it is evident that the growth curve of the upper side differs from the "grand period curve," the cells elongating in the form of a double-peaked curve. In the apical region of curvature the cells of the upper side elongate more quickly than the cells of the lower side, whereas the contrary is true within and immediately below the peak of curvature, with both the curves coinciding at the base of the curvature. The accelerated growth of the young apical cells of the upper side appears to be the reason for the initial bending of the young erect axis as well as for the maintenance of curvature during further development. Since the young but still erect plants do not bend, and bent plants straighten when grown on a clinostat, it is also evident that the bending movement is geoinduced, that is, is a geoepinasty.

Geoepinasty and Auxin Distribution

It has been shown (Kaldewey 1957, 1962, 1965) that the geoepinastic movement depends on the presence of the ovary as the auxin source for the stem. However, normal growth and bending could also be observed when the flower bud was replaced by a lanolin paste containing indoleacetic acid (IAA) (10^{-3} g/ml). These findings point to the involvement of auxin in geoepinasty, and the question arises whether the asymmetric growth acceleration of the young cells could be traced back to differences in auxin distribution.

From the foregoing it is clear that geoepinasty is caused by growth differences along the plant axis. Therefore it was decided to study the transport and distribution of auxin in the upper and lower halves in short sections along the stem. The low amounts of endogenous auxin made it difficult to determine its distribution with precision. Therefore, plants in the prefloral bent stage were supplied with exogenous auxin. A paste of IAA-2-^{14}C ($2 \cdot 10^{-4}$ g/ml with 4% of the IAA labeled) was applied to the decapitated apical end of the axis and left for at least 7 hr before it was analyzed for the distribution of radioactivity. The procedure is outlined in figure 3.

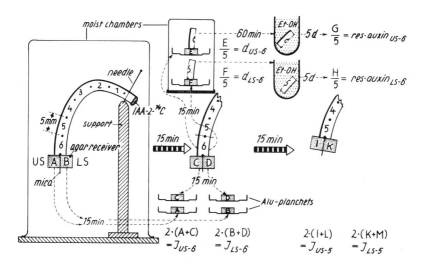

Fig. 3. A sketch of methods used for determining the transport and distribution of ^{14}C in the bent axis of *Fritillaria meleagris* L., which has been decapitated and apically supplied with a paste of IAA-2-^{14}C ($2 \cdot 10^{-4}$ g/ml containing 4% of labeled IAA) at least 7 hr before the beginning of the test. From the activity found in the receivers of the two 15-min diffusion tests the transport intensity (J) in the otherwise intact plant was calculated. The transport density (d) could be calculated from the 60-min diffusate of the 5-mm stem sections, and the content of residual auxin or its radioactive metabolites by exhaustive extraction (res-auxin).

To begin with, the axis thus treated was cut off in the erect region below the curvature and fixed on a support in its normal position by a thin needle. The upper and lower sides of the axis were separated with a mica plate. Agar receivers were applied to the cut basal end for 15 min, then replaced by fresh receivers for 15 min more. From the activity found in these receivers the auxin efflux per unit time, that is, the transport intensity, was calculated. After the second 15 min period a basal 5-mm stem section was cut off and bisected longitudinally, and the upper and lower halves were put upon agar receiver blocks for 60 min to determine their contents of diffusible auxin. These values divided by the length of the section give the transport capacity in the terms of van der Weij (1932), here proposed to be called "transport density" (*Transportdichte*) (cf. Kaldewey 1967 *a*, *b*). Finally, the bisected sections were extracted for their remaining non-diffusible auxin. These measurements were made for successive 5-mm sections up to the apex of the stem.

The curves drawn from the values thus obtained (fig. 4) show that in both the upper side and the lower side the transport intensity as well as the transport density increases basipetally and then decreases, the maxima lying in the region of the curvature peak. Since maximum growth appears below the curvature it is evident that the growth rate does not directly depend on the transportable auxin available. This proves true also in the basal part of the bent region, which is characterized by a slower growth of the upper side even though more diffusible auxin is found here than in the lower side.

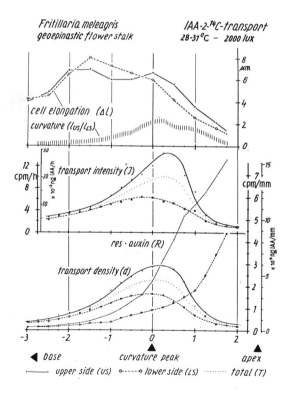

Fig. 4
Transport and distribution of ^{14}C, originating from IAA-2-^{14}C, applied to the decapitated apexes of bent axes of *Fritillaria meleagris* L. The transport intensity (I) and the transport density (d) have been evaluated by diffusion tests, the residual auxin (R) by exhaustive ethanol extraction (see fig. 3). The cell elongation (Δl) and the curvature (l_{US}/l_{LS}) were calculated from measurements of the subepidermal cells (see fig. 2). Mean of 5 plants.

In the apical part, however—and this is the region where the curvature is initiated—the differences in growth rate between the upper side and the lower side correspond to the differences observed in the content and transport of diffusible auxin. More extractable auxin is to be found in the upper side as well.

Conclusions

We suggest that an asymmetry in the auxin transport system is involved in the geoepinasty. The geoinduction itself appears to consist of an asymmetric alteration of the young apical cells, leading to physiologically upper and lower sides differentiated on the basis of their ability to transport auxin.

In accordance with the findings of Leike and von Guttenberg (1961, 1962) the geoinduction is independent of the presence of auxin. The georeaction, however, does depend on auxin. Under the influence of the larger amounts of auxin, either diffusible or extractable, the young cells of the physiologically upper side are induced to accelerated elongation, making this side convex, which gives rise to the geoepinastic bending movement. However, the ability of the cells to elongate—in spite of the presence of the high amount of auxin—seems to be limited. After the initial positive deviation from the normal growth, seen in the lower side, the elongation rate of the cells of the upper side falls below the "normal" rate. During this phase of retarded growth of the upper side, the lower side overcomes the lag of the initial elongation phase, so that finally the lengths of the cells in both sides become equal. Consequently the axis straightens.

It may be mentioned again that the growth pattern described is the result of the long-lasting oblique direction of gravity, and that the growth reaction is seen in that region of the stem where the cells just enter their auxin-sensitive elongation phase.

In contrast to the gravimorphotic process just described, we can distinguish the geotropic reaction induced by a brief geoexposure. The geoepinastic Fritillaria axes also react geotropically (Kaldewey 1962, 1963). The sensitive region for this geoinduction, however, lies below the region of epinastic curvature. In the geotropic region the cells of both sides, still elongating, have achieved equal length. They grow rapidly and, perhaps because of their low auxin supply, seem to be sensitive to the quickly appearing geoinduced differences in auxin distribution.

Summary

Plagiogeotropism possesses the typical characteristics of a gravimorphism. As a prototype of geoinduced plagiotropic growth, the geoepinastic bending of *Fritillaria meleagris* has been analyzed.

Within the curved region of the axis a straightening zone is found below a zone of bending. Both of these zones move acropetally, along with the acropetal displacement of the zone of main elongation, during the development of the axis. The process of bending and straightening has been traced back to temporary lateral differences in elongation rate, brought about in young apical cells of the two physiologically different sides by the prolonged effect of gravity acting obliquely.

The normal course of elongation and movements depends on the presence of the ovary as the natural source of auxin for the stalk. Further, since this function of the ovary may be replaced by IAA, auxin must be involved in these growth processes. Therefore the transport and distribution of ^{14}C, originating from an IAA-2-^{14}C paste applied apically to decapitated plants, has been investigated. It has been found that the transport intensity and the transport density, when traced basipetally, show an increase followed by a decrease, the maximum appearing in the region of the peak of curvature. Furthermore, the transport intensity as well as the transport density could be shown to be higher in the physiologically upper tissues than in the lower. It has been suggested that these differences are the result of a geoinduced asymmetric alteration of the young apical cells in their ability to transport auxin, and thereby cause the enhanced growth of the upper-side cells just entering the auxin-sensitive elongation phase.

Finally, this gravimorphotic process should be distinguished from the negative geotropic reaction induced by a brief geostimulation, which appears in a region below the curved part of the axis in older but still elongating cells.

References

Kaldewey, H. 1957. *Planta (Berl.)* 49: 300-44.

―――. 1962. In *Handbook of plant physiology*, ed. W. Ruhland and E. Bünning, XVII/2: 200-321. Berlin: Springer Verlag.

―――. 1963. *Planta (Berl.)* 60: 178-204.

―――. 1965. *Planta (Berl.)* 67: 55-74.

―――. 1967a. *Ber. Dtsch. Bot. Ges.* 80: 238-51.

―――. 1967b. *Wiss. Z. Univ. Rostock, Math. Nat. Reihe* 16: 485-92.

Leike, H., and Guttenberg, H. von. 1961. *Naturwissenschaften* 48: 604-5.

―――. 1962. *Planta (Berl.)* 58: 453-70.

Lyon, C. J. 1963a. *Plant Physiol.* 38: 145-52.

―――. 1963b. *Plant Physiol.* 38: 567-74.

―――. 1965a. *Plant Physiol.* 40: 18-24.

―――. 1965b. *Plant Physiol.* 40: 953-61.

Wareing, P. F., and Nasr, T. A. A. 1958. *Nature (Lond.)* 182: 379-81.

Weij, H. G. van der. 1932. *Rec. trav. bot. neerl.* 29: 379-496.

Zimmermann, W. 1932. *Jahrb. wiss. Bot.* 77: 393-506.

Discussion

GALSTON: Could you comment further on the participation of the ovary? You have told us that the geoepinasty and the geotropism are auxin dependent in the sense that you must have either the ovary or an exogenous supply of auxin. I was wondering whether the ovary was involved as the georeceptor or solely as the auxin supplier.

KALDEWEY: The geonegative reaction begins within 2 hr in a region 70 or 80 mm below the ovary. However, if differences in the auxin delivery from the ovary to the two sides of the axis are involved, as a consequence of the supposed geoperception, the reaction should start at the earliest after 7 or 8 hr, since the transport velocity of the endogenous auxin of Fritillaria has been shown to be about 10 mm/hr. To me this suggests that the main role of the ovary here is not as a georeceptor, particularly since the georesponse occurs with exogenous IAA in place of the ovary.

31
Plant Responses to Chronic Acceleration

Stephen W. Gray and Betty F. Edwards, *Emory University*

Plants have been placed in all possible positions with respect to the earth's gravitational field in order to study their responses. They have been subjected less frequently to forces higher than that of normal gravity by centrifugation. Only rarely have such forces been applied for more than a few minutes. Our own work considers wheat seedling responses to chronic acceleration between one and 500 times gravity, and the factors which influence these responses.

Review of the Literature

Centrifugation has been employed in the study of the geotropic response since 1806, when Knight demonstrated that growing seedlings assume the position of the resultant between the force of gravity and the force of centrifugation. Curiously, two recent studies have been undertaken to confirm this observation which has been fundamental to geotropic studies for over a century and a half (Westing 1964, Krasochkin and Moshkov 1966).

Since Knight's work, most studies using the centrifuge have been devoted to controlling the strength as well as the direction of the stimulus for the geotropic response. The effects produced in these experiments, however, must be considered the result of acute, rather than chronic, acceleration. Larsen (1962) has admirably summarized this work.

The time factor was introduced when Rutten-Pekelharing (1910) used centrifugal acceleration at right angles to Lepidium roots to determine the relation between the force and the time over which it was applied. This work resulted in the "Reciprocity Rule" which states that force multiplied by time produces a constant geotropic response. This rule has been shown by Lundegårdh (1926) and Johnsson (1965) to hold for a small range of accelerative forces and at some temperatures only.

Recently, Dennison (1961) subjected sporangiophores of phycomycetes to accelerations of up to $4.35\,g$ and proposed a dual geotropic response mechanism consisting of a rapid, transient component and the better known, slow, long-term response. The details of this work are to be found in chapter 6 of this volume.

Differential transport of substances by the force of gravity was implied by the work of Schecter (1935), who altered the polarity of the alga Griffithsia by a force of $150\,g$. Ootaki (1963) also reversed the developmental axis of young fern protonema by reversing their position in a field of $5,000\,g$. Similar alteration of polarity of pollen grains of Vinca was effective only at a force of $20,000\,g$ (Beams and King 1944).

Transport of root-promoting substances in segments of Salix stems was enhanced by centrifugation in the normal position at 640 g for 90 min or 2,540 g for 60 min, but inhibited by similar centrifugation in the reversed position (Kawase 1964). Rooting, enhanced by centrifugation, was further stimulated by subsequent etiolation or by treatment with Indole Acetic Acid (IAA) (Kawase 1965). In our own experience (Edwards 1951), intact wheat seedlings produced more than the normal number of lateral roots when subjected to 500 g.

Certain plant cells are capable of recovering from the effects of very high g forces. Zell (1953) found that radish seedlings survived centrifugation up to 176,000 g for 20 hr with recovery time between two 10-hr exposures. Subsequent growth was retarded and recovery was faster in meristematic tissue than in nonmeristematic tissue. Zell concluded that, like X-radiation, centrifugation has a cumulative effect. Bouck (1963) stratified the contents of pea root cells by centrifugation at 20,000 g and observed recovery after exposures up to 24 hr. His electron micrographs of stratified cell contents agree with the earlier description by Beams and King (1935) of bean root tips centrifuged at 400,000 g. Beams (1949) was able to stop protoplasmic streaming in Elodea leaves at a force of 135,000 g applied for 30 min. Streaming was resumed following the exposure. Saez (1941) produced stratification in Lathyrus seeds at forces as low as 3,000-6,000 g.

The effects of acceleration on chromosomes were first observed by Kostoff in 1938. Seedlings of various plants were centrifuged for a few hours each day at forces between 7,000 and 15,000 g with resulting monosomies, trisomies, and tetraploid cells due to failure of chromosome separation during division. Hilbe (1942) produced polyploidy, giant cells, and binucleate cells in seedlings centrifuged at 50,000 g and above; in the same year Camara showed that chromosome structure could be altered by centrifugal force. Yu and Dodson (1960) found decreased germination and increased chromosome aberrations in barley seedlings centrifuged between 200 and 2,500 g for various periods up to 20 hr.

That centrifugation following irradiation produces more chromosomal aberrations than irradiation alone, was first demonstrated by Sax in 1943. This was confirmed by Wolff and von Borstel in 1954 and by Kumar and Natarajan in 1966. The effect of postirradiation centrifugation is to interfere with the reconstitution of chromosomes broken by the irradiation. The experiments on Tradescantia microspores by Yeargers (1964) suggest that the relation between irradiation and acceleration cannot be so simply explained when the two stimuli are presented simultaneously. As the exposure lasted only eight min, these experiments must be considered acute rather than chronic.

Montgomery et al. (1963) inhibited growth of *E. coli* cultures by centrifugation for 24 hrs at 110,000 g. Cells so treated were of larger size and some of their intracellular components were larger (Montgomery et al. 1964a). Similar effects could be obtained by exposure to Cobalt 60 radiation at 55.8 R per hour for periods up to 30 hours (Montgomery et al. 1964b). This is one of the few observations of morphological change, other than stratification of cell contents, resulting from chronic acceleration.

Experimental Work

Our own work on chronic acceleration has been done with seedlings of winter wheat, *Triticum vulgare* (variety Sanford unless otherwise specified). Seeds were selected, sterilized for 30 sec in

0.05 percent mercuric chloride and soaked for 30 min to initiate germination. They were planted in 50 ml centrifuge tubes on sterile sand or vermiculite saturated with boiled tap water.

Acceleration was produced by centrifugation, usually at a radius of 22 cm; the tubes were free-swinging, allowing the seedlings to assume the line of resultant force without touching the walls. The tubes were closed to prevent air movement, but were not sealed.

Several external factors which influence the response of an organism to gravity or increased gravity were also considered. These are: light, temperature, and radiation which affect root and coleoptile responses separately. To avoid differences which might be related to light variation while growing, all the seedlings were kept in the dark. Unless otherwise specified, they were grown at 25-26° (Gray and Edwards 1955). Radiation experiments were also conducted in conjunction with centrifugation of seeds and seedlings (Edwards 1963).

Growth under Chronic Acceleration

Coleoptile Height. The wheat seedling grows straight up from one end of its seed and thus its height is easily measured. The straight, vertical character of the coleoptile is unaltered by forces up to $100 g$. At $150 g$, $300 g$ and $500 g$ coleoptiles over 10 mm are remarkably curved (fig. 1). The curvature bears no constant relation to the position of the plant in the centrifuge or the direction of revolution. S-shaped curves are frequent. Occasionally the coleoptile lies flat on the substrate or it may even be broken in two by its own weight after achieving a certain height.

The height of the control coleoptile increases in a typical sigmoid growth curve. The linear phase begins at about 48 hr (9.2 mm) during which growth is about one mm per hr. After emergence of the first true leaf from the tip of the coleoptile at about 80 hr, growth of the coleoptile declines. After 96 hr, when the average height is about 40 mm (fig. 2), little further growth takes place.

Seedlings subjected to centrifugation follow a similar growth curve with reduced slope and final height. Plants subjected to a force of $10 g$ show a slight but significantly accelerated growth rate up to 72 hr, after which they level off and do not quite attain the final height of control plants. Seedlings subjected to higher forces show a decreased velocity of growth and a decrease in total height attained. The decrease in height at any given age is linear with respect to the force of gravity applied up to $150 g$ (fig. 3).

The effects of centrifugation are not constant over the whole of the growth period of the coleoptile. Centrifugation at $25 g$ during the first 12 hr produces no significant change in growth rate. Similar centrifugation during the second 12-hr period produces a marked stimulation of growth for the next two days, followed by a retardation of growth on the fourth day. Centrifugation on the third or fourth day results only in retardation of growth (Edwards and Gray 1956).

The growth stimulation phase has been further studied by growing excised segments from control and centrifuged coleoptiles (var. Georgia 1123) in White's plant culture medium with 3 percent coconut milk. Segments 4.5 mm long were cut 2 mm below the coleoptile tip. After 48 hr of culture, segments from 72 hour seedlings that were 6 to 10 mm tall elongate to 199 percent of their initial length, while segments from seedlings of the same age and height that have been centrifuged at 150 g elongate 218 percent (fig. 4). Segments of coleoptiles over 10 mm tall elongate less, and the difference between control and centrifuged seedling segments is not significant (Gray and Edwards 1965a).

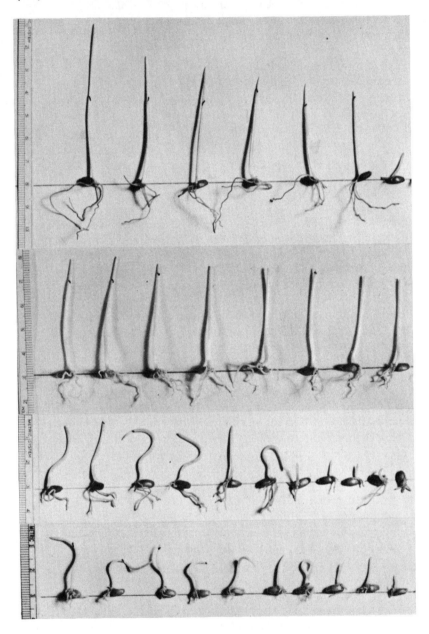

Fig. 1. Typical wheat seedlings, uprooted at 96 hr of age. Top row: control seedlings grown at 1 g. Second row: seedlings grown at 50 g. Third row: seedlings grown at 150 g. Bottom row: seedlings grown at 500 g. The dots indicate the tip of the coleoptile through which the first leaf has emerged.

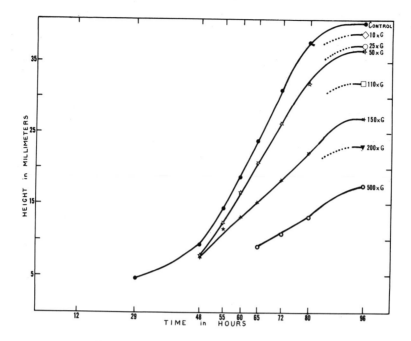

Fig. 2. Curves of coleoptile growth at various gravitational forces.

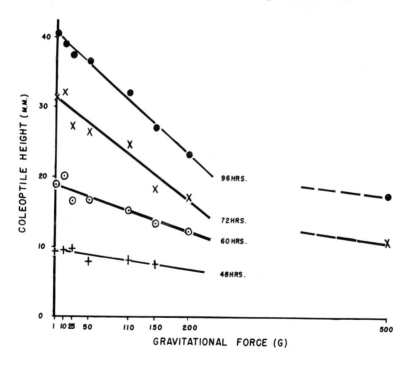

Fig. 3. Coleoptile height attained at various ages under varying forces of gravity.

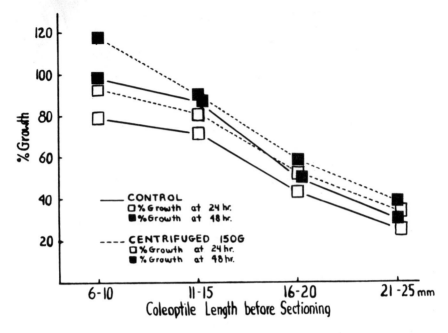

Fig. 4. Growth of 2 mm sections of control and centrifuged coleoptiles cultured for 24 and 48 hr.

Coleoptile Diameter. Continuous centrifugation affects the diameter of the coleoptile as well as its height. Wheat coleoptiles are elliptical in cross section, having a vascular bundle on each side in the longer diameter, which is about 1.6 mm in control plants; the ratio of diameters is 0.72 and the cross-sectional area is 1.41 mm^2. Both diameters increase equally rather than proportionately when the seedling is centrifuged, so that the coleoptile becomes more nearly circular in cross section. The area of cross section increases by 38 percent at 150 g, and by 54 percent at 500 g. A curious parallel to this change has been observed in the femurs of mice subjected to centrifugation by Wunder (1960).

Changes in the diameters of the centrifuged coleoptiles were also effectively adaptive under conditions of chronic acceleration. Calculations based on the shape of the cross section of the coleoptile show that the moment of inertia increased with increased accelerative force, and tensiometer experiments prove that the moment of bending increased by 15.4 percent at 150 g. Thus seedlings grown at forces higher than earth's gravity become more resistant to deformation.

Coleoptile Cell Size. To determine the basis for the reduced height and increased diameter of centrifuged coleoptiles, parenchyma cells were measured. Parenchyma cell length proved to be a function of coleoptile age and is not affected by accelerative forces up to 500 g. Cell width, on the other hand, is increased by continuous centrifugation. Exposure to 150 g for 96 hr results in a 73 percent increase in cell diameter (fig. 5). We conclude that, as parenchyma cell length remains unchanged, in coleoptiles shortened by exposure to centrifugal force, cell division rather than cell elongation has been affected by the acceleration.

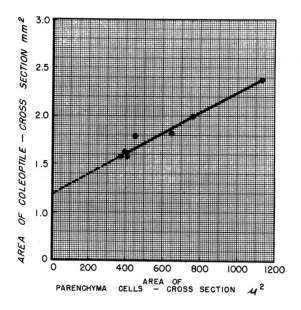

Fig. 5. Relation of parenchyma cell cross-section area to coleoptile cross-section area. 150 g for varying lengths of time.

Root Length. The effect of accelerative forces on the roots of wheat seedlings differs from the effect on the coleoptile. Under the conditions of our experiments, the roots, although growing in the direction of the accelerative force, must overcome the resistance of the substrate. For each cubic millimeter of space the roots occupy, an equal amount of water and substrate must be displaced upward toward the free surface. Under centrifugation this is accomplished with increasing difficulty as the effective weight of the overlying substrate increases.

Centrifugal force delays the appearance and early growth of the roots in relation to the growth of the coleoptile, hence roots are both absolutely and relatively shorter in centrifuged plants. During the linear portion of the growth curve of the coleoptile, the increase in total root length remains constantly 2.4 times the increase in coleoptile height at all of the accelerative forces employed (fig. 6).

Roots of centrifuged plants grown in sand appear grossly different in texture and color from those of control plants, and there is an increase in the number of lateral roots at the higher forces. This is probably the result of increased transport of root-promoting substances (Kawase 1964).

Work Expended under Chronic Acceleration

The increase in diameter and decrease in height of the coleoptile suggest that these changes are adaptive in effect. If the accelerative force is increased, the amount of work required for vertical growth is also increased. The work required to erect a coleoptile against gravity may be expressed as:

$$\text{Work} = \text{weight} \times g \times \frac{\text{height}}{2}$$

or

$$\text{Work} = 0.04726 \text{ gm} \times 1\,g \times \frac{4.046 \text{ cm}}{2} = 0.095607 \text{ gm cm}$$

(for an average control seedling at 96 hr at earth's gravity).

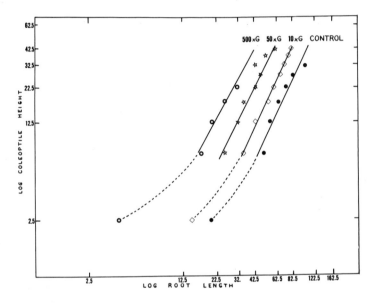

Fig. 6. Relative growth of coleoptiles and roots of seedlings subjected to various gravitational forces.

If this value for work done is calculated for coleoptiles subjected to various accelerative forces above $1\,g$, it is found that the work accomplished increased to 38.2 times that of the control at $50\,g$, and 132 times at $500\,g$. At $50\,g$ the seedling is still achieving 76.4 percent of its normal accomplishment, measured by material lifted and height attained, at a cost of 38.2 times the effort expended by a control plant! Above about $100\,g$ the work done increases more slowly and the growth is no longer compensatory (fig. 7).

To determine whether the increased effective weight of the coleoptile on its basal cells influenced subsequent growth, 48 hr coleoptiles were loaded with 200 mg lead cones. This added a structural load equivalent to $20\,g$ at the start and to $5\,g$ at 96 hr. Growth of such coleoptiles was compared with the growth of those subjected to a centrifugal force of $20\,g$ at 48 hr, decreasing to $5\,g$ at 96 hr. The results were strikingly different. Coleoptiles bearing the lead weights grow to a final height of 4.06 mm less than unweighted control plants, while those centrifuged to a comparable load became 8.28 mm taller than the controls. Obviously increased weight alone provides no stimulus to growth. A recent paper by Dennison and Roth (1967) reports that phycomycete sporangiophores grow faster under compression and more slowly under extension. The failure of wheat seedlings to grow faster under weight loading may be caused by the greater rigidity of their cell walls in comparison of those of sporangiophores.

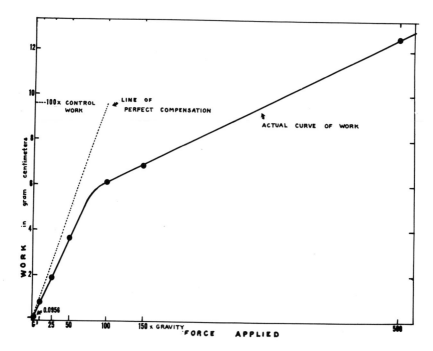

Fig. 7. Work done by growth of coleoptile against increasing gravitational force.

Temperature Effects on Responses to Chronic Acceleration

Temperatures above and below 25° C produce less than optimal growth of wheat seedlings (var. Georgia 1123), and alter the response to chronic acceleration. At 21°C, continuous centrifugation at 150 g results in faster growth of coleoptile and secondary roots for the first three days, and a final size at 96 hr identical with that of control plants grown at the same temperature.

At 29°C centrifugation at 150 g results in slower growth of organs over the entire period. The growth retardation of the coleoptile and secondary roots is similar to that seen in centrifuged plants grown at 25°C, but the growth of the primary root is much more retarded (fig. 8). There was no period during the 96 hr in which growth of centrifuged plants exceeded that of control plants. Seedlings appear to be less sensitive to chronic acceleration when grown at lower than optimum temperatures (Gray and Edwards 1965b).

X-Radiation and Chronic Acceleration

Wheat seeds are very resistant to low doses of X-radiation. In our laboratory we failed to establish an LD-50 even with a total dose of 160,000 r. Our results are corroborated by several other investigators working with wheat (Moutschen 1958, Natarajan et al. 1958, Saric 1958).

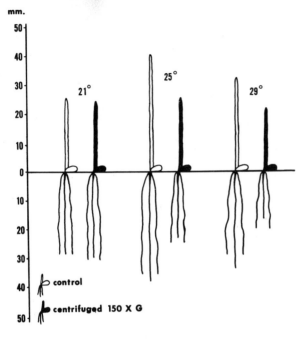

Fig. 8
Response of coleoptiles and roots of 96 hr wheat seedlings to centrifugation at 150 g at different temperatures.

In all of our experiments a single dose of X-radiation was administered at 50 r per minute following the initial soaking period. An initial dose of 600 r produced coleoptiles slightly taller than those of control seedlings. The final attained height of the coleoptile decreases when the dose reaches 2,400 r. The seedlings subjected to 20,000 r are about one-half normal height. Seedlings grown continuously at 150 g also attain about one-half their normal height. When irradiation is followed by centrifugation, there is no further decrease in attained coleoptile height until the dose reaches 20,000 r (Edwards 1963) (figs. 9, 10).

Roots of non-centrifuged seedlings become shorter than normal following a single radiation dose of 4,800 r. Roots of continuously centrifuged seedlings are reduced to 25 percent of their normal length, but a radiation dose of 10,000 r is required to further depress their growth. Thus the threshold of sensitivity to radiation as measured by growth retardation appears to be raised by subsequent continuous centrifugation.

While acceleration of 150 g and radiation doses above 20,000 r each produce a 50 percent reduction in final coleoptile height, the two together are not simply additive, but inhibit growth proportionately. Interestingly, similar curves of response were found in fibroblast tissue cultures at much lower doses of X-radiation and similar centrifugal forces (Edwards 1963). Thus, centrifugation, or increased gravity, and radiation are definitely not synergistic and may possibly show some "protection." Sax (1943) had shown that centrifugation following irradiation alters chromosome movements yet causes no increase in the number of aberrations produced by a given dose of X-radiation. These effects of increased gravity should be considered in conjunction with new data from Russian and American experiments in orbiting satellites on the effects of less than normal gravity.

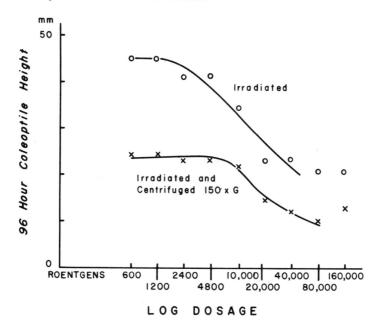

Fig. 9. Height attained by coleoptiles of seedlings subjected to various single doses of X-radiation and subsequently grown at 1 g or 150 g.

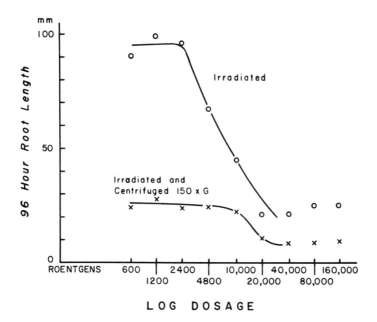

Fig. 10. Length attained by roots of seedlings subjected to various single doses of X-radiation and subsequently grown at 1 g or 150 g.

Biosatellite II experimenters reported various effects on plant species. Sparrow and his associates (1969) found an increased number of disturbed spindles in cells of roots and microspores of Tradescantia plantlets grown in weightlessness, with or without gamma radiation. They also found increased pollen abortion and stunting of root hairs in plantlets irradiated while in orbit, although they found somewhat fewer somatic mutations for stamen hair color. Russian scientist Delone and her colleagues reported similar spindle disruption and microspore aberrations in Tradescantia grown on Cosmos 110 and Vostok 5 and 6 (Delone 1964, 1966, refer to Sparrow 1969).

Possibly due to random cell distribution, two strains of lysogenic bacteria (*E. coli* and *Salmonella typhimurium*), showed higher mean density, whether irradiated or not, following growth in a liquid medium in weightlessness. These strains also showed a relatively lower induction of bacteriophage P-22 under reduced gravity conditions (Mattoni 1968). However, Neurospora conidia showed no difference in induction of mutations when irradiated in flight (deSerres and Webber 1968). For discussion of the biosatellite experiments, see chapter 37 below.

Histology and Histochemistry of Seedling Tissues

Tissues from seedlings grown in a centrifuge at greater than normal gravitational forces show definite differences from seedlings of the same age and size grown at normal gravity. For histological studies, entire seedlings, roots and shoots, were fixed in formalin, acetic acid, and alcohol (FAA), followed by chromic acid, acetic acid, and formalin (CRAF); they were then embedded in paraffin and serially sectioned at 5 to 7 micra. For acid hematein reaction for phospholipid distribution, tissues were fixed in Lewitsky's fluid. The histochemistry sections were cut at 16 to 20 micra on the cryostat; control and centrifuged tissues were mounted on the same cover slip in order to eliminate possible differences due to involuntary variations of technique. We feel that our methods show that the tissues reflect the gross morphological differences previously described for control and centrifuged seedlings. The changes observed in centrifuged seedlings may be seen in those grown at $150\,g$, but they are best shown in seedlings grown at $300\,g$. Tissues from $300\,g$ seedlings have been selected for demonstration in the accompanying plates. plate I shows the general relationship of the seed and the seedling organs in very young seedlings.

The paraffin sections were stained with hematoxylin and fast green, and for periodic acid-Schiff (PAS) reaction, or by the Feulgen technique for deoxyribonucleic acid (DNA) localization. Every fifth slide was withheld for special staining procedures. The PAS technique, which localizes carbohydrates, demonstrates the characteristic positions of starch grains and amyloplasts in centrifuged and control seedlings. It is more difficult to interpret differences in response to nuclear stains. While the hematoxylin slides show only a few differences in nuclear location, they also show normal mitotic figures in centrifuged plants. There is no evidence that the mitotic process is disturbed by centrifugation, although the number of mitotic figures may be fewer. The Feulgen technique shows no differences in amount or distribution of DNA in the nucleus of either type seedling. The nucleus is usually located close to one of the cell walls (plate IV, no. 10). This location is only occasionally altered.

The PAS technique offers much more revealing differences between our seedling types. The ability of this technique to localize all types of carbohydrates makes it possible to evaluate the reactivity of root tips and vascular tissues, as well as the position of starch grains and amyloplasts in cells throughout the tissues. Because the relationship of these granules within specific cells may be

Plant Responses to Chronic Acceleration

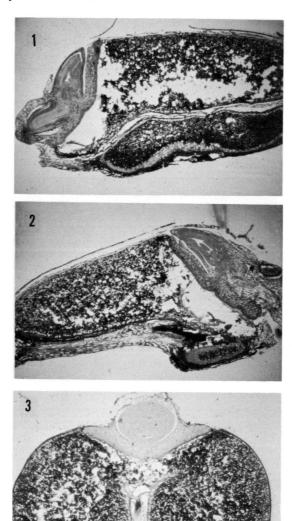

PLATE I

WHEAT SEEDLINGS WITH SEEDS. Sectioned 8 hr after soaking to show relationship of seedling parts to endosperm. Periodic Acid-Schiff reaction, magnification approximately 3.25X.

1. *Control seedling* grown erect with relation to normal 1 g. Sectioned parallel to long axis of seed and showing erect shoot and young primary root still within its coleorhiza at an angle to base of seedling.

2. *Centrifuged seedling* grown at 300 g. Sectioned parallel to long axis of seed, providing a longitudinal section of young seedling not yet emerged from seed coat.

3. *Centrifuged seedling* grown at 300 g. Sectioned at right angle to long axis of seed and providing cross section of seedling parts. This section shows base of coleoptile and scutellum applied to endosperm.

related to the perception of gravity, the technique is of particular importance in assessing the reaction of our wheat seedlings to different forces of gravity. Audus (1962) correlates the sedimentation of starch-containing amyloplasts in root tips with changes in position with respect to gravity. His "statolith" theory is discussed in chapter 13 above, and it seems quite probable that similar plastids in the coleoptile are also receptors for the gravity stimulus. Plates II through VI compare the localization of amyloplasts in centrifuged and control seedling tissues.

Since the areas of response to gravity have been shown to be the coleoptile tip and the root cap, these have been selected for demonstration. The internode area of the wheat seedling, or meristem, located between the origin of the root vascular bundle and the vascular bundles of the coleoptile, is a narrow region at the base of the shoot; it is not shown because photographs of comparable magnification were not available. Similar increase in sedimentation of the starch granules has been observed in these areas of centrifuged seedlings, however. The scutellar region, because of its general reactivity, has also been illustrated, although it has no demonstrable response to changes in the gravity stimulus.

In centrifuged seedlings, many more cells in all areas show sedimentation of starch amyloplasts. The amyloplasts are more closely packed into the basal areas of the cells, whether the plants are 8 hr old, 24 hr old, or older. This is clearly illustrated in the accompanying plates.

We found the distribution of other cell components very similar in both centrifuged and control seedlings. For example, there is little difference between the two when tissues are fixed in Lewitsky's fluid and stained for phospholipids by the acid hematein method (see plate VII).

When the frozen sections used to demonstrate histochemical localization, we have usually been able to make three enzyme tests on each seedling. Peroxidase, succinic dehydrogenase, and acid phosphatase have all given consistent results. Glucose-6-phosphatase and 5-nucleotidase have given less constant results.

Peroxidase activity is consistently greater in centrifuged seedlings, whether sections are taken through the meristematic area, through the midsection of coleoptile and leaf, or through the tip of the coleoptile. The typical peroxidase reaction is associated with the vascular tissues and with the cell walls of other tissues. This is in agreement with the suggestion of Van Fleet (1959) that peroxidase is associated with the deposition of lignic substances, although others, including Galston at this conference, correlate peroxidase reaction with the site of IAA metabolism and the oxidation of IAA by the peroxidase (Jensen 1962). If translocation of IAA is the important factor in the response of seedling organs to gravity, the localization of peroxidase should provide a clue to its movement within the seedling. We are continuing our work on this problem, comparing centrifuged seedlings with those grown in Biosatellite II as well as with normally grown erect seedling tissues. The quantitative evaluation of enzyme concentration has been made possible by the use of disc gel electrophoresis. The results for one peroxidase experiment may be seen in plate VIII, no. 24.

Succinic dehydrogenase shows greater variability. As seen in plate VIII, no. 23, there is a markedly greater reactivity at the base of the centrifuged seedling. In the sections treated histochemically we saw this reactivity concentrated largely in the scutellar epithelium, which shows many more of the dark blue granules characteristic of this reaction than any other tissue examined, regardless of the

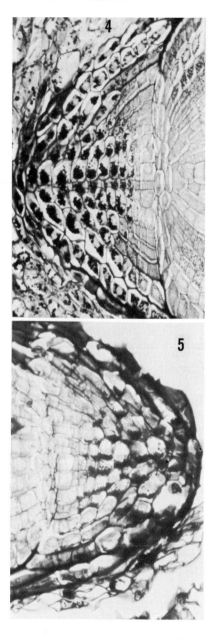

PLATE II

ROOT CAPS OF YOUNG WHEAT SEEDLINGS. Shows distribution of statolith starch granules under different gravity conditions. Periodic Acid-Schiff reaction, magnification 72X.

4. *Control seedling* root cap similar to that seen in (1), with starch granules sedimented toward lower surface of cells (young roots not yet emerged are at an angle to the direction of gravity).

5. *Centrifuged seedling* (300 g), root cap showing similar granules tightly packed at lower surface of cells.

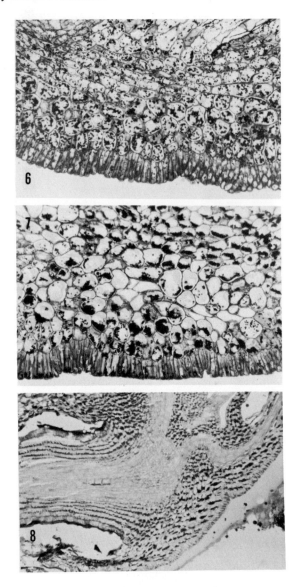

PLATE III

SCUTELLAR AREA OF WHEAT SEEDLINGS. Periodic Acid-Schiff reaction.

6. *Control seedling*, epidermis and subepidermal cells of scutellum showing normal distribution of starch granules throughout the cells. 48hr seedling, approximately 33X.

7. *Centrifuged seedling* scutellum, showing accumulation of starch granules in lower portion of epidermal and most subscutellar cells. 48 hr seedling grown at 300 g, approximately 33X.

8. *Centrifuged seedling* scutellum, primary root and first internode of meristematic area of 8 hr seedling centrifuged 24 hr at 300 g, showing at lower magnification that most of the cells have layering of starch plastids except within vascular bundles, which appear light, 16.25X.

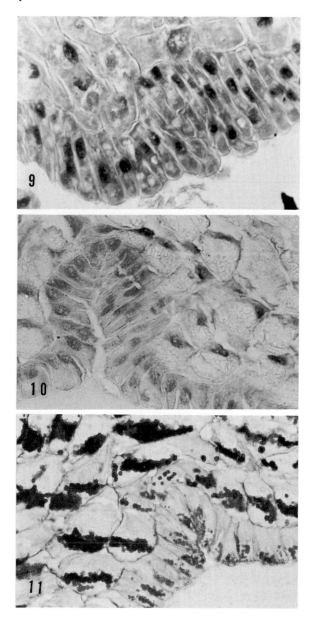

PLATE IV

SCUTELLAR AREA OF WHEAT SEEDLINGS. Different stains, magnification 142X.

9. *Control seedling* scutellar area showing normal random distribution of nuclei within the cells. Hematoxylin and Fast Green stain, showing few plastids which are not in any one area of cells.

10, 11. *Centrifuged seedling* scutellar area. Sections from adjacent slides. No. 10, stained with Hematoxylin and Fast Green, shows nuclei tend to lie against upper cell wall. No. 11 shows the accumulation of plastids in lower portion of cells in both epidermal and subepidermal cells.

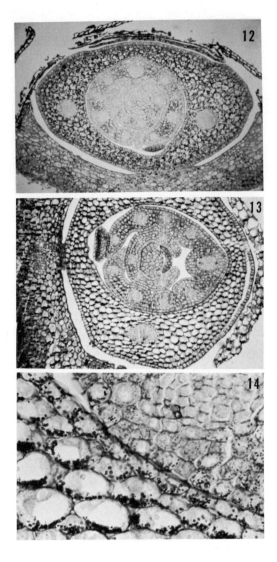

PLATE V

BASE OF LOWER COLEOPTILE AND LEAF of 24 hr wheat seedlings. Periodic Acid-Schiff reaction.

12. *Control seedling*, part of scutellum and epiblast and base of coleoptile enclosing first two leaves within it, showing normal Periodic Acid-Schiff granule distribution. There are no statolith cells normally apparent in this area. 15.5X.

13. *Centrifuged seedling*, base of coleoptile, enclosed leaves, part of scutellum and epiblast. Note that most cells in all tissues show the layering of plastids, except in areas of vascular tissue. 15.5X.

14. *Centrifuged seedling*, inner edge of coleoptile and part of leaf. Area enlarged from portion of area shown in no. 13 above to show detail of cells. Approximately 137.25X.

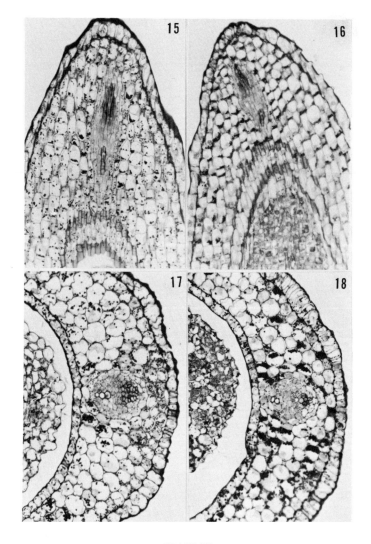

PLATE VI

WHEAT SEEDLINGS, TIP OF COLEOPTILE AND LEAF. Magnification 38.5X.

15. *Control seedling*, long section of coleoptile tip slightly to one side and through coleoptile vascular bundle. Note the layering of plastids in the cells adjacent to the bundle only. This is the normal site of statolith containing cells. Surrounding cells maintain a random distribution of plastids.

16. *Centrifuged seedlings* (300 g), section similar to no. 15, but showing layering of plastids in many more cells of the tip of the coleoptile. Note that at this gravitational force there is no layering of plastids in the leaf.

17. *Control seedling*, coleoptile vascular bundle and parenchyma with small area of leaf. The plastids immediately around the vascular bundle show some layering. The xylem elements appear more numerous than in the adjacent section of a centrifuged seedling.

18. *Centrifuged seedling* shows a "checkerboard" effect due to the layering of plastids in many cells of the parenchyma, but not within the cells of inner or outer epidermis where the granules are always smaller.

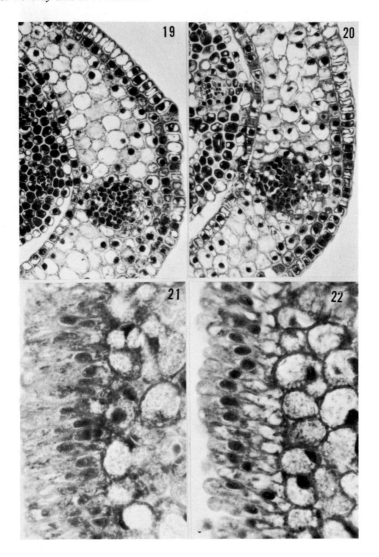

PLATE VII

PHOSPHOLIPID DISTRIBUTION IN WHEAT SEEDLING TISSUES. Acid Hematein method, magnification 136.5X.

19. *Control seedling*, edge of coleoptile and leaf coleoptile through vascular bundle and edge of leaf in a section similar to no. 17, but stained with acid hematein for localization of phospholipids. The reaction is slightly heavier in the control leaf here than in no. 20.

20. *Centrifuged seedling*, same area as nos. 18 and 19. Shows that distribution of phospholipids is very similar in both control and centrifuged seedling shoot tissues.

21. *Control seedling*, epidermis and underlying cells of scutellum.

22. *Centrifuged seedling*, epidermis and underlying cells of scutellum showing a heavier phospholipid concentration around the cell membranes of subepidermal cells than seen in No. 21.

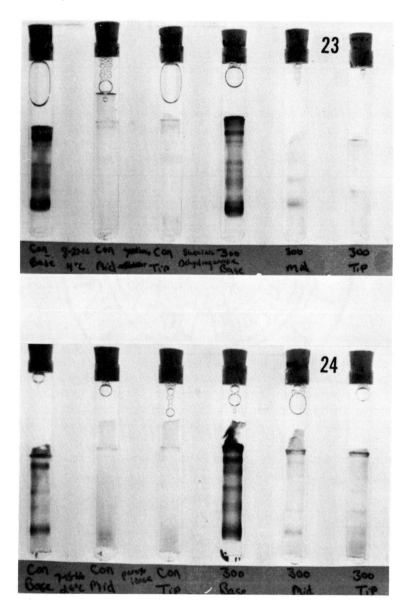

PLATE VIII

DISC ELECTROPHORESIS GELS of wheat seedling tissues.

23. *Succinic dehydrogenase* reaction in base of seedling, midsection, and tip of coleoptile of control seedlings and of centrifuged seedlings (300 g), showing the greater reactivity and greater number of bands of reactivity in centrifuged seedling tissues as compared with control tissues. The coleoptile tip of the control seedling appears to be more active with this enzyme.

24. *Peroxidase* reaction in similar tissues. With this enzyme, all tissues of the centrifuged seedlings are more reactive than those of the control seedlings.

strength or direction of gravity (clinostat, erect control, centrifuged, or Biosatellite). The least concentration of this enzyme appears in erect control plants. The leaf within the coleoptile of control seedlings shows less reactivity than centrifuged leaves or seedlings, and probably accounts for heavier bands from the midsection of $300\,g$ plants shown in plate VIII, no. 23. The significance of these differences in reactivity are not yet understood.

Acid phosphatase produces a granular reaction most marked in the cells immediately below the scutellar epidermis. In addition, the cortex of the apical portion of the root, the xylem elements of its central cylinder, and the root cap are highly reactive, as was reported by Avers (1963). Acid phosphatase reactions are somewhat weaker in centrifuged than in control seedlings.

Our results from wheat seedlings grown in orbital weightlessness ($10^{-5}\,g$) aboard Biosatellite II show changes in the coleoptile-root length ratio; root growth is retarded in free fall, and the coleoptile shows a smaller diameter (Edwards and Gray 1968). There were fewer cells in mitosis in the root tips of our flight seedlings, and significantly larger nuclei in both primary and lateral roots. Different, random starch grain distribution in coleoptile and root tip was the most obvious cytological change in flight seedlings (see plate IX, Edwards 1969). Histochemical reactions were more intense in sections of these seedlings than in ground controls, suggesting increased metabolic activity. No permanent deleterious effects of lack of gravity were observed, and seedlings recovered from space grew to apparent normal maturity.

We hope that with further study, and with the correlation of histochemical localization with disc electrophoretic studies evaluated by densitometer tracings, we may begin to explain the differences we have observed in wheat seedlings grown under various gravitational conditions. Since gravity has a qualitative effect on plant orientation in addition to the quantitative results which we have reported, we feel that the basic biological response of tissues to gravity may be elucidated by the comparison of seedling growth at near weightless conditions with seedling growth in the simulated hypergravity of chronic acceleration.

Summary

Chronic acceleration between 10 and $500\,g$ produces the following effects on wheat seedlings grown in the dark at $25 \pm 1°C$.

1. The total height attained by the coleoptile in the four days of its development decreases proportionally to the accelerative force employed. The time required to reach maximum height is unchanged. Accelerative force applied only during the second day of coleoptile development produces a transient increase in growth rate, followed by a decrease.

2. With higher accelerative force, the coleoptile diameter increases and the cross section becomes more circular. There is an accompanying increase in the resistance to bending of the coleoptile. Parenchyma cell diameter increases, but cell length remains unchanged.

3. As the accelerative force increases, total root length decreases both absolutely and relatively in relation to coleoptile height.

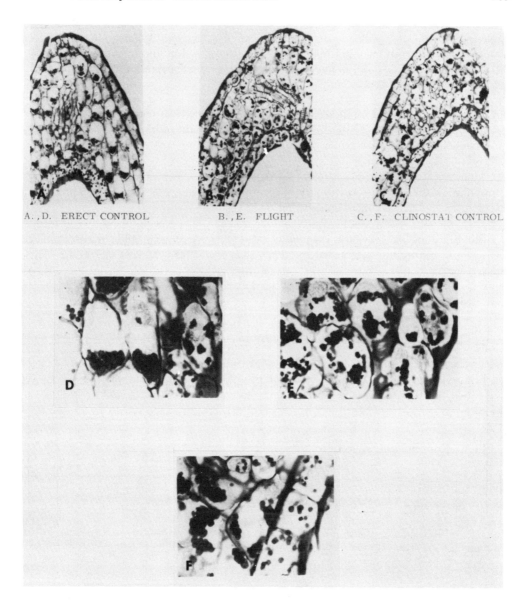

PLATE IX

WHEAT SEEDLING COLEOPTILE TIPS. PAS and Fast Green stain at 100X and 400X magnification.

A and D. Erect control showing characteristic starch grain distribution at 1 g.

B and E. Flight seedling coleoptile tips showing random distribution of starch grains from similar area as above.

C and F. Clinostat control seedling tip showing disturbed starch grain distribution characteristic of this simulated low gravity environment.

4. The work done by the seedling in erecting its shoot increases proportionally to the increased accelerative force, up to about $100\,g$, beyond which the work increases but is no longer compensatory. That this increased work output is not governed by the increased effective weight of the coleoptile on its base is shown by the absence of compensatory growth produced by weight-loading uncentrifuged coleoptiles.

5. Continuous accelerative forces retard growth more effectively at the temperature optimal for gravity than at lower temperatures. Transient growth stimulation is observed more clearly at temperatures a few degrees below the optimum.

6. Continuous accelerative force of $150\,g$ following single-dose X-radiation reduces the sensitivity of seedlings to the growth retardation produced by the radiation. The retardation produced by centrifugation and that produced by radiation are not additive.

7. Histological and histochemical changes in the seedlings accompany the gross morphological changes. The configurations assumed by starch granules and amyloplasts ("Statolith starch") support the view that these organelles are gravity receptors. Differences in distribution of peroxidase, succinic dehydrogenase, and acid phosphatase in control and centrifuged seedlings are illustrated.

Acknowledgments

This investigation was supported in part by research grants from the Division of Research Grants and Fellowships of The National Institutes of Health, U.S. Public Health Service, The Carnegie Foundation, and the National Aeronautics and Space Administration, Grants No. NsG-521 and NsG-529.

References

Audus, L. J. 1962. In *Symposia of the Society of Experimental Biology*, no. 16, "Biological receptor mechanisms," pp. 197-226.

Avers, C. J. 1963. *Annals of Histochem.* 8: 115-20.

Beams, H. W. 1949. *Biol. Bull.* 96: 246-56.

Beams, H. W., and King, R. L. 1935. *Proc. Roy. Soc. London*, Sec. B 118: 264-76.

———. 1944. *J. Cell. and Comp. Physiol.* 24: 109-14.

Bouck, G. B. 1963. *J. Cell Biol.* 18: 441-57.

Camara, A. 1942. *Agronomia Lusitana.* 4: 199-211.

Dennison, D. S. 1961. *J. Gen. Physiol.* 45: 23-38.

Dennison, D. S., and Roth, C. C. 1967. *Science* 156: 1386-88.

deSerres, F. J., and Webber, B. B. 1968. *BioScience* 18: 590-95.

Edwards, B. F. Changes in morphology and growth of wheat coleoptiles subjected to centrifugation. 1951. Thesis, Emory University.

———. Effects of radiation and supragravitational forces on growth. 1963. Dissertation, Emory University. Diss. Abstr. 24, no. 64-175.

———. 1969. Weightlessness experiments on Biosatellite II. In *COSPAR Life Sciences and Space Research VII*, pp. 84-92. North Holland Publishing Co., Amsterdam.

Edwards, B. F., and Gray, S. W. 1956. *J. Cell. & Comp. Physiol.* 48: 405-20.

Gray, S. W., and Edwards, B. F. 1955. *J. Cell. & Comp. Physiol.* 46: 97-126.

———. 1965a. National Aeronautics and Space Administration Contract NAS 2-1556, Third Quarterly Report. Washington, D.C.

———. 1965b. National Aeronautics and Space Administration Contractor Report. NASA CR-303, Washington, D.C.

———. 1968. *BioScience* 18: 632-633, 638-45.

Hilbe, J. J. 1942. *Proc. Iowa Acad. Sci.* 48: 457-66.

Jensen, W. A. *Botanical Histochemistry. Principles and practice.* 1962. San Francisco: W. H. Freeman and Co.

Johnsson, A. 1965. *Physiol. Plant.* 18: 945-67.

Kawase, M. 1964. *Physiol. Plant.* 17: 855-65.

———. 1965. *Physiol. Plant.* 18: 1066-76.

Knight, T. A. 1806. *Phil. Trans. Roy. Soc.* (London) 96: 99-108.

Kostoff, D. 1938. *Cytologia* 8: 420-42.

Krasochkin, R. V. and Moshkov, B. S. 1966. *Fiziol. Rast.* 13: 177-83.

Kumar, S., and Natarajan, A. T. 1966. *Genetics* 53: 1065-69.

Larsen, P. Geotropism. An introduction. 1962. In *Handbook of Physiology*. vol. 17(2), pp. 34-73. W. Ruhland and E. Bünning, eds. Berlin: Springer-Verlag.

Lundegardh, H. 1926. *Planta* 2: 152-240.

Mattoni, R. H. T. 1968. *BioScience* 18: 602-8.

Montgomery, P. O'B.; Van Orden, F.; and Rosenblum, E. 1963. *Aerospace Medicine* 34: 352-54.

Montgomery, P. O'B.; Neumeyer, B.; and Rosenblum, E. 1964*a*. *Aerospace Medicine* 35: 360-61.

Montgomery, P. O'B.; Rosenblum, E.; and Stapp, B. 1964*b*. *Aerospace Medicine* 35: 731-33.

Moutschen, J. 1958. UN International Conference for Peaceful uses of Atomic Energy. Proc. II 27: 217-22.

Natarajan, A. T.; Sikka, S. M.; and Swaninathan, M. S. 1958. Ibid. 27: 321-31.

Ootaki, T. 1963. *Cytologia* 28: 21-29.

Rutten-Pekelharing, C. J. 1910. *Rec. Tran. bot. neerl.* 7: 241-346.

Saez, F. A. 1941. *An. Soc. Cient* (Argentina) 132: 139-50.

Saric, M. R. 1958. UN International Conference for Peaceful uses of Atomic Energy. Proc. II 27: 233-48.

Sax, K. 1943. *Proc. Natl. Acad. Sci.* 29: 18-21.

Schecter, V. 1935. *Biol. Bull.* 68: 172-79.

Sparrow, A. H.; Schairer, L. A.; and Marimuthu, K. M. 1969. Final Report for NASA Project P-1123 (unpublished).

Van Fleet, D. S. 1959. *Canad. J. Bot.* 37: 449-58.

Westing, A. H. 1964. *Science* 144: 1342-44.

Wolff, S., and von Borstel, R. C. 1954. *Proc. Natl. Acad. Sci. U.S.* 40: 1138-41.

Wunder, C. C. 1960. *Nature* 188: 151-52.

Yeargers, E. 1964. *Radiation Botany* 4: 101-6.

Yu, C. K., and Dodson, E. O. 1960. *Genetics* 46: 1411-23.

Zell, L. W. Effects of Ultracentrifugation on seed germination and seedling development. 1953. Dissertation: Univ. Mich. Diss. Abstr. 13: 293.

Discussion

HOWLAND: I didn't quite understand what the point was in calculating the total energy involved in growth. Compared to the chemical energy in the plant, that's absolutely trivial stuff.

GRAY: It's work that has to be put out against gravity. While it may be relatively trivial, the budgeting of the available energy might very well affect this particular aspect; there is plenty of energy but it isn't necessarily available in nonlimiting quantities for all activities. You're quite right, the total amount of the energy used for upward growth is small, but it is increased by several orders of magnitude during centrifugation.

SMITH: I think that this partition of energy in the accelerated organism is quite important in understanding the nature of gravity effects. We have found no difference in the partial efficiency of growth (the increment of tissue energy produced per unit of feed energy utilized) in chronically accelerated chickens. Similarly, Bjurstedt (of Sweden) has observed no change in the partial efficiency for muscular work in centrifuged humans. Consequently, the greater energy requirement of animals in increased acceleration fields probably represents the imposition of additional work, rather than a change in the energetics of physiological processes. In animals this additional work most likely represents maintenance of muscle tonus, locomotion, circulation, and the like. It is difficult to identify analogues of these processes in plants.

GORDON: Is my impression correct that you find hyper-g does not affect the length of the parenchyma cells?

GRAY: It seems not to have.

GORDON: Then I don't understand how the length of the coleoptile is shortened at the higher accelerations without decreasing the length of the parenchymatous cells, which constitute a major function of the coleoptile.

WILKINS: What happened to those cells?

AUDUS: There must be fewer cells!

ETHERTON: There are two possibilities. One is fewer cells, the other is that the lengths of the cells are measured later than the lengths of the tissues. During this time interval the cells may change shape, becoming less flattened and more elongated.

GRAY: There were fewer cells. When you say flattened, do you mean they were under compression? It is possible that there was some compression but it would be very small. The reduction in coleoptile height produced by centrifugation is far too great to be the result of elastic compression of tissues.

LARSEN: In connection with the time studies, consider the effect of light as a parallel phenomenon. Light stimulates coleoptile growth, but elongation will *cease* sooner, and a similar effect might come in here. With gravity or acceleration you might have a stimulation, perhaps temporary, and then the coleoptiles might stop growing sooner.

EDWARDS: One of our slides showed this, after the seedlings were centrifuged for the first 24 hr.

AUDUS: Have you made these studies for a wide range of g? I ask because there seems to be some evidence that the starch which doesn't normally sediment, that is, the metabolic starch in the parenchyma as distinguished from the "statolith" starch, in fact did sediment at the high g.

EDWARDS: We have made slides from seedlings grown at 150, 425, and 300 $\times$ g. What you call "metabolic starch" sometimes sediments but not always. We do not know the critical g value for their displacement.

PICKARD: It would be interesting to compare the amounts and activities of various enzyme systems in coleoptiles which have been treated with different auxin concentrations. It might be that centrifugation changes hormonal control of growth, and that simply varying the hormone level independent of centrifugation would create enzyme changes resulting in some of the effects you measure.

GALSTON: I would like to ask about those interesting cells at the base of the coleoptile. A few days ago Dr. Wilkins reminded us that grass nodes show a very particular response to gravity in that there is an actual stimulation of mitotic activity. I am also reminded that in such things as Avena coleoptiles the response to light can be differentiated into a tip and a base response. The base shows a very sharp tipping over toward high intensities of light. I wonder whether you have taken a look at the growth patterns on the base of the coleoptile?

EDWARDS: Yes, we have, staining alternate sections with iron-haemotoxylin and then PAS. With the nuclear stains, we have not seen any difference in the pattern of division or of growth in this basal area between the clinostat and the control, or the centrifuge and the control.

KALDEWEY: Did you look at the distribution of applied auxin in the centrifuged coleoptiles?

EDWARDS: No, we have not run any experiments with applied auxin.

WEIS-FOGH: Is there a chance that ethylene is the active substance, that it can, at least, mimic the action of auxin? We've been considering coleoptiles that are in the millimeter dimension. If ethylene, which is volatile, is actually the active substance, it would be next to impossible to put up concentration gradients across dimensions as small as these, unless we could be sure that it all occurs in the solid or liquid phase. If there is an air/liquid interface we would be in trouble. More simply, the order of magnitude of the diffusion coefficient in air is 10^6 times greater than that in liquid. This means that even if we have an area filled with air, which is only of the order of, say, one one-thousandth of the total cross section of area—a very small area indeed—you will find that the role of diffusion inside the tissue will be one thousand times greater than had it been occupied by liquid. These dimensions would make it next to impossible to postulate a system working on gradients of ethylene in an air-filled space. This may not be relevant. It may be relevant that at the growing point the intercellular space is apparently free of air. Or is this not the case?

EDWARDS: There are intercellular spaces along the whole length of the coleoptile, but I don't know whether they are with air or liquid. These spaces are seen not only on fixed slides but also in hand-cut sections.

GALSTON: Dr. Weis-Fogh, I think your analysis is correct if you imagine that ethylene is just produced once and then quickly diffuses. What probably actually happens is the induction by auxin of an enzymatic system for the constant production of ethylene. The enzyme, by the way, may very well be a peroxidase, which is relevant to Dr. Edwards' statement. Peroxidase can act on methionine and release carbon atoms 3 and 4 as ethylene. If you imagine a constant ethylene production at one locale, then even with rapid diffusion one could still have a gradient attributable to the auxin gradient.

WEIS-FOGH: Have you ever worked out how extraordinarily small this gradient would be? I have worked on such diffusion systems. This is what worries me.

GORDON: I wonder whether the endogenous concentrations of auxin in these phenomena are high enough to stimulate the production of ethylene?

GALSTON: Stanley Burg has shown them to be so. More to the point, whatever they are, if you tip an organ horizontally, a pea root for example, you do get ethylene production on the lower side.

LEOPOLD: Avena is a particularly insensitive material to ethylene, so there is little likelihood that ethylene is involved in the georesponses of Avena seedlings. But in some of the plant materials Dr. Jankiewicz was speaking about, the organ dimensions are larger and the sensitivity to ethylene is greater. So it might be that gravimorphism would be a better place for ethylene to play a role.

BANBURY: Where the starch grains are displaced by centrifugation, have you observed any physical changes in the grains—alterations in their surface, their structure, or their size?

EDWARDS: We can always distinguish the sections taken from a clinostated plant, as long as fixation was done immediately after the clinostat was stopped, by the distribution of the amyloplasts. But even at 1000X magnification we have not observed any cytological differences.

32
Chronic Acceleration of Animals

A. H. Smith and R. R. Burton, *University of California at Davis*

Terrestrial animals generally exist in physical environments characterized by variability—the variation being recognized as "climate" and "weather." It is well understood that these animals are endowed, in differing degrees, with systems for homeostasis and physiological adaptation which tend to minimize the influence of environmental variation upon their physiological function. These environmental factors not only have been variable but also have existed in extreme states for long (geological) periods of time—and it is considered likely that the internal regulatory mechanisms were developed by evolutionary processes in response to these changing conditions.

But there has always been one invariant in the environment of terrestrial animals—the accelerative force, gravity. Consequently, it might be anticipated that these animals would be intolerant of prolonged changes in the ambient accelerative force. However, various investigators—principally at The University of Iowa, the University of California (Davis) and NASA's Ames Research Center—have shown that a variety of terrestrial animals can become adapted to greater acceleration fields (i.e., tolerate a simulated increase in gravity). For technical reasons, these research programs have been based on the long-term exposure of animals to centrifugal forces. Although this treatment involves turning as well as the production of accelerative forces, only the latter appears to be of physiological significance. Wunder, Milojevic, and Eberly (1966) have shown that labyrinthectomized and intact hamsters do not differ significantly in rate of growth or feed intake, either at normal gravity or in fields of 5-$6\,g$. Also, intact chickens subjected to rotations of $60°$/sec, which at the particular radius of rotation does not produce a significant accelerative force, have not been found to differ physiologically or behaviorally from static controls.

Consequently, the principal effect of accelerative forces, as produced by centrifugation, is an increase in the weight-to-mass ratio. By this treatment, an animal of 1 kg mass may come to weigh 2 or 3 kg, so that more work becomes necessary for equivalent locomotion, maintenance of tonus, etc. Thus many of the effects of chronic acceleration may resemble those of increased exercise at normal gravity. However, there are other effects peculiar to hyperdynamic environments, and these generally are related to the specific weight (wt/vol). But there also are limitations on the effects of accelerative forces on organisms. Over the range tolerated by terrestrial organisms (generally $<10\,g$) and at normal temperatures, accelerative forces are much too weak, compared with thermal energies, to directly affect thermochemical reactions. For example, to move usual cell organelles selectively requires fields on the order of $1,000\,g$, and to separate large molecules (e.g., proteins) requires fields on the order of $100,000\,g$ (Davson 1966).

However, even weak forces can become effective if their influence is "amplified" through action on a large mass. The moon's gravitational force at the earth's surface is measured in "micro-g"; but when this small force is amplified through the volume of the seas, highly energetic tides result. The situation in chronically accelerated animals may be analogous. For example, a moderate g-field acting on the whole organism may produce forces on the antigravity muscles that approach 1,000 times its direct effect on those tissues. Also, where vascular columns are long and parallel to the field, the hydrostatic aspect of blood pressure is greatly enhanced. These hemodynamic effects may be complex. In the great vessels of the body cavity the effect is maximal, perhaps grossly affecting circulation. In the muscles, the effect tends to be compensated for by tissue pressures—which attenuate to zero at the body surface.

Large organs, such as the brain, which exist in a density gradient may be particularly susceptible to chronic acceleration. The forces on the restraining elements of such organs will be proportional to the acceleration field, and these may be attenuated through a relatively small part of the organ. It has been known for some time that the application of mechanical forces to some parts of the brain leads to whole-body metabolic changes (e.g., Claude Bernard's "piqûre" and the metabolic sequellae in some persons recovering from brain concussions).

Consequently, a wide variety of physiological responses to chronic acceleration may be anticipated in terrestrial animals—some of a general nature (resembling exercise, etc.) and others that are specific to the treatment.

Systematic Aspects of Animal Response and Physiological Adaptation to Chronic Acceleration

Although the numbers and species of terrestrial animals exposed to chronic acceleration have been rather few, some systematic factors are apparent in the response to this treatment.

Size

The effect of size upon the properties of systems, both physical and biological, has been known for some time (Galileo's "Principle of Similitude," 1638), and it has been discussed in some detail for biological systems by D'Arcy Thompson (1917). At normal gravity conditions, the effect (load) increases in proportion to the cube of some dimension, whereas the capability of support (strength) increases only with its square. The apparent result of this arrangement is a relative increase in the load-bearing structures of terrestrial animals of increasing size (Thompson 1917).

Animal	Mass	Skeleton (% of body mass)
Mouse, wren	20-30 g	8%
Dog, goose	5 kg	13-14%
Man	75 kg	17-18%

The role of gravity in inducing these anatomical changes is indicated by the lesser skeletal component and lack of similar variation among marine mammals of different size (e.g., porpoises and whales).

This "inverse-size" relationship also applies to the ability of animals to tolerate increased accelerative forces—for example, among homeotherms tolerance appears to be inversely and exponentially related to body size:

Animal	Approximate Mass	Approximate g-Tolerance
Mouse	20 g	7 g (Wunder 1962)
Rat	200 g	5 g (Oyama and Platt 1965b)
Chicken	2,000 g	3 g (Burton and Smith 1965)

Similar size-effect relationships have been observed within species—Wunder, Herrin, and Cogswell (1959) found that among Drosophila larvae, the severity of growth repression is proportional to initial size. However, in a group of chickens exposed to fields up to 4 g, Smith, Winget, and Kelly (1959a) observed that survival was correlated with the mean size for the group (fig. 1). This is somewhat different from the earth-gravity situation, where maximum longevity is found in the "moderately large" individuals of a species. The lethal aspects of the extreme sizes in centrifuged chickens were explained as: larger animals are more affected by "similitude limitations," and the smaller are "metabolically deficient."

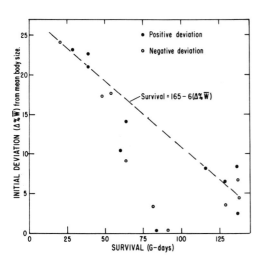

Fig. 1
Initial body size and survival during chronic acceleration. The survival time of chicks (starting at 6 wk of age) to an acceleration field that increased, incrementally, to a maximum of 4 g, is indicated. Very large and very small individuals appear to be equally susceptible to the treatment, and it appears that those deviating 28% from the mean size would be nonviable.

Physiological Adaptation and Phenotypic Adaptation

The ability of animals to tolerate accelerative forces above some injurious threshold depends upon the development of special adaptive processes (fig. 2). For example, if a group of chickens is introduced rather gradually to a 3-g environment (at 1/4 to 1/2 g/wk), after 3 mo, they may have suffered a 30% mortality. However, if hatch-mates are introduced abruptly to that environment, all will die within 48-72 hr. Of course, the general nature of this physiological adaptation, let alone the details, is currently unknown.

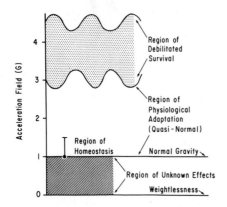

Fig. 2
Schema of the chicken's response to chronic acceleration. There appear to be several categories of response of chickens, and presumably homeotherms generally, to long-term esposure to increased accelerative force. Where the increment is small, little if any change is evident, homeostatic mechanisms apparently accommodating any internally induced changes. However, with more intense fields, some initial debility may be apparent, which attenuates as the animal becomes physiologically adapted. But there are limits to this adaptational capacity, and beyond these the animal survives for a time in a progressively debilitated state.

When groups of chickens are introduced to a 3-g environment at different rates (varying from 0.1-0.5 g per wk increments) the mortality incurred is proportional to the rate of increase. However, the exact relationship is more complex. At moderate fields (up to 1.5 g for chickens) the rate of g-increase has little effect; but its influence becomes progressively greater as field strength increases. So the adaptational capacity curve appears to be a hyperbolic function.

The durability of this adapted state is rather paradoxical. Even after many months of exposure, a few individuals become debilitated and die (Burton and Smith 1965). This situation, commonly encountered with other stressors, has been referred to by Selye (1950) as "fatigue," implying that the adaptation is an active process which must be maintained. However, in other individuals the adaptation appears to be very persistent. When well-adapted animals are removed from the centrifuge, physiological and anatomical changes generally disappear within a few weeks. But these animals can be reintroduced, abruptly, to rather great acceleration fields without debility—apparently having retained the adaptation for several months at normal gravity (Smith and Burton 1965).

Other Systematic Factors in Chronic Acceleration Tolerance

Several investigators have shown that young animals do not survive chronic acceleration as well as older ones (Burton and Smith 1965; Steele 1962; Wunder 1962), although they are smaller and have a greater metabolic intensity. In chicks, exposure to increased accelerative forces at ages substantially less than skeletal maturity (which occurs at 90 days) leads to gross skeletal deformation and visceral displacement, which appears to be at least a contributory factor in chronic acceleration death (Burton and Smith 1965).

Young rats (less than 14-18 days of age) fail to exhibit metabolic changes induced in older animals by acute exposure to a field of 4.5 g, such as increases in lipid synthesis in the liver (Feller and Neville 1965) and increases in liver glycogen synthetase (Oyama, Medina, and Platt 1966). Also, unweaned mice fail to respond to acute acceleration with the increases in blood glucose, blood corticosterone, and liver glycogen that are found in more mature animals (Oyama and Platt 1965a). Thus, very young animals, which are essentially poikilothermic, respond differently from homeothermic adults and are much less tolerant of chronic acceleration. The mature metabolic system and its regulatory mechanisms would appear to be very important in physiological adaptation to hyperdynamic environments.

According to Selye (1950), females generally tolerate environmental stressors better than do males. However, Smith, Winget, and Kelly (1959a) observed no difference in survival between sexes in chronically accelerated chicks. But with the onset of egg production, a marked increase in female mortality results from oviduct prolapse (Burton and Smith 1965), and this can be eliminated by preventing laying (by treatment with androgens or progesterone), after which the mortalities of females and males become comparable.

Posture also affects resistance to acceleration. Britton, Corey, and Stewart (1946) found that bipeds were much more tolerant of centrifugation than were quadrupeds. This difference is attributed to the visceral vasomotor apparatus of bipeds, which is necessary because their major blood vessels are parallel to the field of gravity. Burton and Smith (1965) observed that chickens treated with reserpine, limiting sympathetic accommodation and largely inactivating visceral vasomotor response, became much more sensitive to chronic acceleration.

Heritability of Adaptation

A particularly important aspect of physiological adaptation to chronic acceleration is the rather great heritability of that capacity, which has been demonstrated in the fowl (Smith and Kelly 1961) but presumably applies generally. When the survivors of centrifugation experiments are allowed to reproduce for a few generations, the mortality resulting from such treatment is drastically reduced in the offspring (perhaps to 10% of the mortality of unselected stock for an equivalent treatment). The nature of this selection, like all artificial selections, is a simple increase in gene frequency—or, expressed physiologically, no new adaptations appear to be developed; those present in a population are merely "collected" into individuals. When animals of the selected strain fail to tolerate chronic acceleration, they go through the same syndrome as susceptibles from unselected stocks. Conversely, acceleration-adapted individuals of selected and unselected stocks are indistinguishable. Without chronic exposure to acceleration, no distinguishing properties have been found in the selected strain. Although the acceleration-selected strain is quantitatively more adaptable, its maximum tolerance ($3\,g$) does not appear to have been increased. Short-term selection to higher fields—e.g., by several weeks' exposure to $4\,g$—was found to be of no protective value to progeny in long-term exposure to lesser fields.

However, the selection progress in developing the acceleration-tolerant strain was rather rapid—a condition which geneticists generally associate with a metabolic basis, because of the rather direct process-enzyme-gene sequence (Wagner and Mitchell 1955). Consequently, this aspect of chronic acceleration tolerance indicates that it has a metabolic basis.

Pathology: The Nature of Systems Failure during Chronic Acceleration

Exposure of animals to accelerative forces above some limit is a very lethal treatment. Where the field is very intense, Britton, Corey, and Stewart (1946) found that survival may be measured in minutes, and that limiting organic lesions may be evident. In early chick embryos, for example, Besch, Smith, and Goren (1965) and Besch, Smith, and Walker (1965) were able to correlate acceleration-induced death with the erosion of cells from the blastoderm.

At lesser fields and longer exposures a more variable mortality is encountered. In chickens, the debilities of "chronic acceleration sickness" follow a restricted pattern, including one syndrome which is reversible and another (specifically involving a leg paralysis) which is uniformly fatal. At autopsy, a variety of findings are encountered—however, none is indicative of a limiting organic lesion, either grossly or microscopically. Similar results have been reported for autopsies on acute and chronic acceleration deaths in rats (Casey et al. 1967), baboons (Menninger, Murray, and Robinson 1967) and dogs (Murray, Prine, and Menninger 1965). Consequently, the lesion must lie at the submicroscopic, and perhaps the molecular level, and so the pathological aspects of susceptibility to chronic acceleration indicate that it also may have a metabolic basis.

Physiological and Anatomic Alterations in Chronically Accelerated Terrestrial Animals

When terrestrial animals are exposed chronically to increased acceleration fields, a variety of changes in anatomic proportion and physiological function are encountered. With the more severe treatments, approaching the tolerance limit, our results for repeated experiments with chickens tend to become rather variable. Presumably, this variation arises from the presence of a spectrum of physiological conditions relative to stress and adaptation under these marginal conditions. It is quite important, for the rationalization of such data, that criteria be developed to distinguish the animal's condition. Hematological parameters, particularly lymphocyte frequency, appear to be promising for evaluating centrifuging chickens (Burton and Smith 1967a) (fig. 3). As indicated, chickens exposed to moderate fields (<1.75 g) do not appear to be stressed. Feller et al. (1965) observed a similar (but quantitatively greater) threshold field for enhanced lipid synthesis in livers of chronically accelerated rats, changes not being evident in animals exposed to 3.6 g or less. At greater (marginal) fields, hematological criteria for chickens correlate well with exercise capacity (Burton and Smith 1967b) (fig. 4). Lymphocyte frequencies also are correlated with longevity—when lymphocytes are much less than 30%, the outcome is fatal.

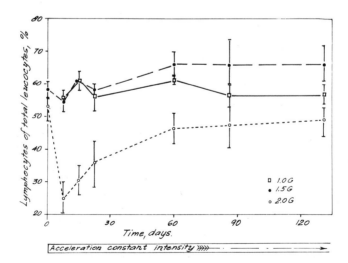

Fig. 3. Mean relative lymphocytic response to chronic increased acceleration in three groups of male chickens. Stress response in chickens is accompanied by a drastic reduction in lymphocytes. With this, and other experiments, it appears that acceleration stress does not develop in fields less than 1.75 g. Also, recovery from the stress (physiological adaptation) requires about 2 mo in a 2-g field. In nonadapting birds, the lymphocytopenia persists until death.

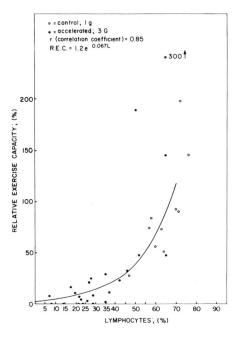

Fig. 4
Correlation of lymphocyte frequency and exercise capacity. The "relative exercise capacity," REC, is the ratio (as %) between preacceleration and acceleration exercise time, to exhaustion. A reduction in this performance (i.e., REC < 100) is associated with a lymphocytopenia, both in controls and in centrifuged birds.

However, a fairly discrete series of changes has been reported by various investigators as occurring in animals after fairly long-term exposure to accelerative forces. These changes tend to be similar among several species of terrestrial animals, and they are generally related in degree to field intensity. From the duration of exposure, it would appear that these changes represent physiological adaptation to the hyperdynamic environment.

Growth

Growth (as indicated by increase in body mass) is repressed by exposure to chronic acceleration. This is particularly evident soon after exposure, when feed intake is repressed (Oyama and Platt 1965b; Wunder 1961). After some weeks—depending upon the organism and acceleration field—growth rate and feed intake increase and usually exceed control values. Several investigators have reported greater body sizes, as well as growth rates, in some experiments involving exposure to moderate fields (e.g., 1.5 g), and in turtles (Dodge and Wunder 1963), a markedly enhanced growth was obtained in fields as strong as 5 g. This departure is of particular interest since, as has been noted by the investigators, turtles exist equally well in an aquatic (weightless) environment and at normal gravity. These comparative studies, being pursued at Iowa, should be particularly informative in understanding the influence of accelerative forces on growth.

With longer centrifugation, animals continue to grow, and a mature body mass becomes established which is less than that of animals at normal gravity—the decrement being proportional to the acceleration field. Over the tolerable-acceleration range in white leghorn chickens (1-3 g) Smith and Burton (1967) have reported the decrement in mature size (A, kg) to be rectilinearly related to field strength (g):

$$A = 2.29 - 0.23 g \quad [r = 0.95; p < 0.01]$$

The kinetics of late growth also are altered in hyperdynamic environments. In this aspect of determinate animal growth, body mass (M) approaches a limit (A) hyperbolically:

$$M = A - Be^{-kt}$$

Where:

M is the body mass (kg) at time t (days);

A is the "mature size" (kg), the asymptote which body mass approaches exponentially;

B is an integration constant; and,

$-k$ is the growth rate constant, which is negative, since the increments are decreasing.

With increases in the ambient accelerative force, the growth rate constant ($-k$) becomes numerically greater, and rectilinearly with field strength:

$$-k = 0.85 + 0.24\,g$$

Since the greater acceleration decreases ultimate size (A) and increases the growth constant ($-k$), maturity (as indicated by cessation of growth) is attained at an earlier age. However, in centrifuging chickens, sexual maturity is delayed by approximately a month.

The smaller body mass attained by chickens in a hyperdynamic environment has been shown by Smith and Burton (1967) to be closely regulated. Alterations in body mass, induced by brief (3-day) fasts are rapidly regained upon realimentation (fig. 5), and at least as rapidly as by normal-gravity control birds. This clearly indicates that the smaller body mass in chronically accelerated animals is not merely a matter of feed restriction or of reduction in synthetic capacity.

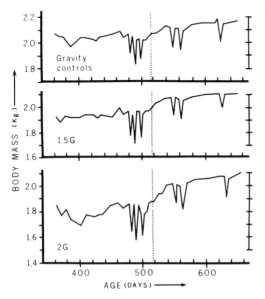

Fig. 5
Maintenance of mature size in chronically accelerated male chickens. Mean body masses are indicated for treatment groups at various times during and after exposure to chronic acceleration (starting at 110-180 days, and ending at 516 days). Standard errors were rather uniform: controls, 0.09-0.11; 1.5 g, 0.05-0.08; and 2 g, 0.06-0.10. A series of six weighings between 300 and 360 days of age (before fasting trials) indicated individual coefficients of variation to be less than 4%.

After return to normal gravity, body masses increased (and this included another environmental component which was shared by the controls, but was distinguishable with a rate analysis). The body mass differential between control and previously centrifuged animals ceased to change after 40 days for the 1.5-g group (at -1.9% Δ body mass), and after 60 days for the 2-g group (at -3.2% Δ body mass).

Body mass losses resulting from six fasting periods (3 days each) are obvious. These losses in body substance are regained upon realimentation, and as readily in centrifuged birds as in controls.

The acceleration-induced size decrements are not permanent, and even after a year's treatment, return to normal gravity results in essentially regaining the usual mature size (there are residual differences, but these are not statistically significant with groups of a dozen or so). During the late growth, a reduction in the acceleration field is followed very rapidly by a change in the growth pattern (Smith and Burton 1967) (fig. 6). Since these animals are skeletally mature, the increased body mass represents a plasticity of soft tissue.

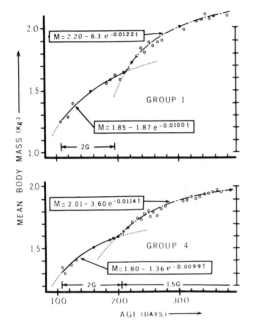

Fig. 6
Body mass response to a change in the acceleration field. Mean body masses are shown for two groups of male chickens before and after a change in the acceleration field. Standard errors for group 1 varied from 0.05 around 100 days of age to 0.12 around 350 days of age; and, for group 4, 0.04 and 0.08, respectively.

Equations for the growth periods were calculated on a "least squares" basis. From these, it is apparent that changes in the acceleration lead to a rapid growth response, which, from the onset, is characteristic of the secondary field. The relationships between estimated and observed mature sizes can be compared in groups of continuously treated animals:

	mean mature size (kg)	
	observed	estimated
1 g	2.06 ± 0.03	2.16 ± 0.10
1.5 g	1.93 ± 0.08	1.98 ± 0.05
2 g	1.83 ± 0.01	1.88 ± 0.03

A comparison between various groups for each treatment indicates that the "overestimate" is relative to the ultimate size (S), rather than acceleration treatment.

% overestimate = $7.42S - 11.4$ ($r = 0.69; p < 0.05$).

The size of various viscera generally is reduced—but less than whole-body mass, so usually there is an increase in relative visceral size (Briney and Wunder 1962; Oyama and Platt 1965b; Smith and Kelly 1963). These changes resemble the effects of fasting, except in the latter there is a decrease of even relative gastrointestinal tract size (Wilson 1954).

Size changes in chronically accelerated animals are not uniform for all organs. In growing rats, acceleration tends to increase bone size (Wunder et al. 1960); however, in growing chickens (Smith and Kelly 1963), the increase in the humerus (non-load-bearing) may be more pronounced than in the femur (load-bearing). Consequently, the acceleration-induced changes in hard tissues appear to be a whole-body phenomenon rather than a local response to the treatment. There are, however, reports of local growth responses of skeletal tissues (Thompson 1917). In skeletally mature chickens, chronic acceleration has no appreciable effect on bone size or conformation.

Muscles appear to be more selectively affected by chronic acceleration. With such treatment, extensors (antigravity muscles) tend to hypertrophy—in chickens as well as in hamsters (Briney and Wunder 1962; Canonica 1966) and mice (Bird et al. 1963). The antagonistic flexor, however, tends to become rather reduced, so that the size ratio of these paired muscles changes markedly (Burton et al. 1967) (fig. 7). These effects on muscle size are not rapid, and one year's exposure at 2 g produced only 85% of the (hyperbolically) predicted maximum change.

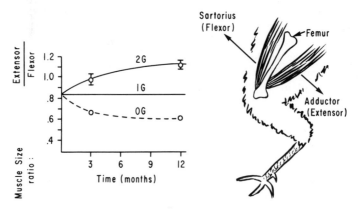

Fig. 7. Influence of the ambient accelerative force on extensor:flexor size in paired muscles. The data is derived from the adductor muscle (extensor) and its antagonistic sartorius muscle (flexor), both of which are located in the chicken's hip. The "0-g" data is the intercept from comparisons at 1, 1.5, and 2 g, after 3 mo and 1 yr exposure. The maximum effect on the "$E:F$" ratio (the asymptote) is rectilinearly related to field strength, approximately:

$$E:F = 0.6 + 0.3\,g$$

This dissimilarity in paired muscles poses questions regarding their functional properties in environments with dynamic characteristics different from those to which the muscles are adjusted. In the Gemini series, substantial difficulties were encountered in performing productive work during unrestrained weightlessness (the "extravehicular activity"). Consequently, chickens, after long residence at 1.75-2.5 g, were transferred to normal gravity and their exercise capacity (time to exhaustion) was measured on a treadmill. Initially, the previously centrifuged birds had a poorer performance, but at later times (the animals remaining on-centrifuge between exercise tests), the high-g birds exhibited a much greater exercise capacity (fig. 8). Consequently, for this kind of work performance appears to depend upon extensor muscle mass and perhaps some "learning." It should be recalled that Matthews (1953) found a "marked extensor tonus" in his rats raised at 3 g.

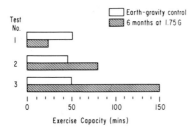

Fig. 8

Exercise capability (at normal gravity) of chronically accelerated and control male chickens. The accelerated birds had been under treatment for 6 mo, and this continued, except for test periods. The treadmill operated at 130 ft/min, which is equivalent to 9 mph on a human scale. With longer testing the performance of the controls increased, as they became physically trained.

Body Composition

Detailed chemical analyses of animals that have been chronically exposed to increased accelerative forces are rather scarce. However, in experiments involving animal dissection a marked reduction in depot fat has been reported in mice (Bird et al. 1963), rats (Casey et al. 1967; Oyama and Platt 1965b), hamsters (Briney and Wunder 1962; Canonica 1966) and chickens

(Smith and Kelly 1963). The carcass fat in male chickens, after a year's exposure, is closely and rectilinearly related to field intensity:

$$\text{carcass fat } (\%) = 11.6 - 3.8\,g$$

$$(r = 0.95; p < 0.01).$$

Associated with the diminished fat content is an increased hydration—an inverse relationship, generally found at normal gravity among animals which vary in fatness.

The carcass fat content of chronically accelerated male chickens correlates well with blood fat (NEFA) and with the concentration of some tissue enzymes involved in fat metabolism—particularly fat synthesis (Evans 1968). An increased rate of lipid synthesis, but less lipid content, is noted in the livers of chronically accelerated rats (Feller et al. 1965). Consequently, the diminished body fat content of chronically accelerated animals appears to result from increased utilization rather than from decreased formation; however no increase in tissue oxidative metabolism of fatty acid (labeled acetate) was observed (Feller et al. 1965).

When centrifuged animals are returned to earth gravity, they recover a normal fat content, as well as a quasi-normal body mass, in 1-2 mo. Nutritional characteristics indicate a very rapid onset of development of adipose tissue after reduction in the acceleration field.

Metabolic Requirements

When animals have become physiologically adapted to increased acceleration, their feed intake is noticeably increased—as reported for mice (Wunder 1961), rats (Oyama and Platt 1965b; Steel 1962), hamsters (Canonica 1966), and chickens. In mature White Leghorn chickens, and presumably generally, the maintenance feed intake (Kleiber 1962) increases with field strength. For moderate acceleration fields (1-2.5 g) this influence upon metabolizable (retained) feed appears to be rectilinear:

$$F_g = F_M + kg = 17.3 + 6.5\,g \qquad [r = 0.85; p < 0.01]$$

Where:

F_g is the maintenance feed consumption (g feed/kg body mass/day) in an acceleration field of g-strength;

F_M is the feed consumption which is independent of the accelerative force—i.e., it is "mass determined"; and

k is the proportionality constant.

Consequently, a 1-g increase in the ambient acceleration increases feed intake of adult chickens 9.7 g/kg mass/day, indicating that approximately 25% of the feed intake under natural conditions is "gravity determined." This increase in the acceleration field is rather modest, and substantially below relative feed capacity (Kleiber 1962). Feed intake of chickens can be doubled or trebled at low environmental temperatures.

A particularly interesting nutritional aspect is the influence of the acceleration field (g) on the Feed:Body Substance ratio ($F{:}BS$)—the slope of the regression of metabolizable feed intake upon changes in body mass—which decreases rectilinearly with field strength (1-3 g):

$$F{:}BS = 3.00 - 0.53\,g \qquad [r = 0.49; p < 0.01]$$

However, this appears to be only indirectly related to the acceleration, and is primarily a function of the chemical energy content of tissues being formed or used. For energies of soft tissue (T_E) between 1.3 and 1.8 kcal/g (the range from control chickens to those exposed to 2 g for a year) the relationship is:

$$F{:}BS = 4.75\,(T_E - 0.69) \qquad [r = 0.89; p < 0.05]$$

Assuming that the feed contains about 3.2 kcal per g of available energy, it is possible to calculate the partial energetic efficiency (Kleiber 1962) for tissue synthesis. A regression (through the origin) of tissue energy on its equivalent feed energy (F_E) indicates the relationship:

$$T_E = 0.11\,F_E \qquad [r = 0.95; p < 0.05]$$

Thus the partial energetic efficiency of tissue synthesis is not affected significantly by the acceleration field. A similar lack of effect of acceleration on the energetics of specific processes is indicated by observations of Bjurstedt, Rosenhamer, and Weigertz (1968), that the partial efficiency for muscular work is not affected by acute changes in the ambient accelerative force.

At present, no information is available concerning the influence of the ambient accelerative force upon requirements for specific nutrients. However, some changes might be anticipated if the basis of response to this treatment is considered to be metabolic.

Environmental and Therapeutic Modification of Responses to Chronic Acceleration

A potentially important, but so far rather neglected, enquiry into the nature of chronic acceleration stress and adaptation is the interaction between other environmental factors, or drugs, and exposure to hyperdynamic environments. Since the mode of physiological action of many drugs and other treatments is known, the nature of their relationships with an altered acceleration environment should become very informative.

Wunder, Herrin, and Crawford (1959) have demonstrated a marked interaction between temperature and acceleration in the growth of fruit-fly larvae. And Casey et al. (1967) have shown that the effects of chronic acceleration and whole-body X-irradiation are more than additive, with an enhanced radiotoxicity in hyperdynamic environments.

Among the therapeutic agents, the only marked effect observed so far with chickens is the enhanced lethality with reserpine. Adrenal corticoids (both gluco- and mineral) and B-complex vitamins (because the leg paralysis encountered in chronic acceleration sickness resembled a polyneuritis) were without effect on the survival of centrifuging chickens. A particularly interesting lack of effect was observed with iodocasein and thiouracil treatment. At normal gravity, these agents repress mature size—iodocasein by increasing energy metabolism and thiouracil by repressing it. However, in chickens exposed to 3 g, no growth response was evident from equivalent dosages of either of these agents.

Effects of Return to Normal Gravity

Inevitably, animals exposed to chronic acceleration are returned to earth-gravity situations, and this treatment is well tolerated. Over periods of weeks or months at normal gravity, the acceleration-induced anatomical and physiological changes tend to disappear. So the physiology accompanying reduction in accelerative force (at least back to the earthly situation) is not a mirror image of chronic acceleration. Similarly, orbital flights indicate that humans tolerate a reduction in acceleration from earth gravity to weightlessness, at least up to 14 days duration (and certainly much better than an increase in the ambient acceleration from 1 to 2 g). Of course, there are contrary observations, such as the rather drastic responses of humans to weightlessness as simulated by water immersion.

However, in some cases reduction in the acceleration field leads to rather bizarre postural abnormalities. Such an occurrence was noted by Matthews (1953) in rats removed from the centrifuge. In domestic birds, the usual response to removal from the centrifuge is a minor change to a ducklike posture (Smith, Winget, and Kelly 1959b) (fig. 9). In New Hampshire chicks, about 25% of the previously centrifuged birds (about a week at 1.5 g) would place their heads between their feet and somersault, repeatedly, for hours (fig. 10). This bizarre behavior would disappear after 12 hr at normal gravity or immediately upon return to the centrifuge. The incidence of these and other postural difficulties is variable. Early in the development of our acceleration-selected strain of White Leghorns, the incidence of disorientation was about 15%. A single selection for this trait (by breeding those that exhibited it) doubled its incidence to 30% (Burton and Smith 1965). In one commercial strain of Leghorns the incidence of postcentrifugation postural difficulty was 100%. Currently, this phenomenon does not appear in our acceleration-tolerant line of chickens, presumably having been genetically eliminated. This particular abnormality can be explained on the basis of a bipolar otolith that leads to a sense of inversion, rather than merely one of lightness-to-heaviness. However, we have not pursued the phenomenon with any anatomical or physiological investigations. A similar explanation has been offered by Graybiel and Kellogg (1967) for the inversion illusion of some individuals during brief periods of weightlessness.

Fig. 9. Postcentrifugation posture. The turkey poults to the right have just been removed from a field of 1.5 g—those to the left are hatch-mate controls. This "ducklike" posture is a rather typical reaction, and persists for several hours.

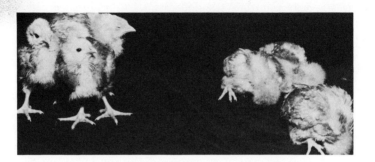

Fig. 10. Postcentrifugation postural abnormalities. The New Hampshire chicks to the right have just been removed from a field of 1.5 g—those to the left are hatch-mate controls. About 25% of the group responded by placing their heads between their feet and somersaulting repeatedly. After approximately 12 hr at normal gravity, or immediately upon return to centrifuge, they regained normal posture and locomotion.

There also appears to be some effect of chronic acceleration upon the labyrinth. Ordinarily, repeated rotatory stimulation above some threshold (60°/sec for humans) leads to a progressive diminution in response—a "habituation" (Winget and Smith 1962). However, in chronically centrifuged chickens, this decrease in response is not encountered (Winget, Smith, and Kelly 1962), similar results being obtained from repeated rotatory stimulations of 192°/sec at normal gravity. In birds exhibiting postcentrifugation postural difficulties, the labyrinthine mechanism appears to be "blocked" or "saturated," since rotatory stimulation elicits no nystagmic response.

Summary and Interpretation

Perhaps the most important outcome of chronic acceleration research was Matthew's (1953) initial observation—simply demonstrating the ability of terrestrial animals to tolerate a new stressor, one not encountered before by them or their antecedents over the previous 30 million years. However, as pointed out by Claude Bernard in 1865, what the animal adapts to is not the external phenomenon, but the internal changes that it induces. For example, in exposure to environmental thermal extremes, it is only the body-heat load, or deficit, that is effective. The influence of the ambient air temperature can be avoided by maintaining an appropriate body-heat content by means of radiation. Similarly, the effects of exposure to an increased accelerative force can be largely compensated for with a "g-suit."

Consequently, studies of the immediate responses to acceleration exposure, "acute acceleration stress," are of considerable importance to understanding the subsequent physiological adaptation to chronic exposure. Metabolic responses to accelerative forces develop rapidly, becoming maximal with exposures on the order of 1 hr. In rats and mice, such treatment increases liver glycogen and glycogen synthetase (Oyama and Daligcon 1967; Oyama, Medina, and Platt 1966) and increases blood glucose, free fatty acids, and corticosterone (Oyama and Platt 1964, 1965a). Since these changes may be abolished by starvation, adrenalectomy, hypophysectomy, or alloxan diabetes (Oyama and Platt 1964, 1965a), these acceleration-induced changes indicate an endocrine involvement. The rapid acceleration-induction of these metabolic changes is of particular interest regarding a metabolic basis for physiological adaptation with chronic exposure, as well as a metabolic basis for chronic acceleration sickness and death.

Since the organ systems of terrestrial animals are reasonably few in number and can react in a limited number of ways, it is likely that such organisms have developed regulatory mechanisms for all injurious systemic changes. Thus, for a "new" environmental variable these animals may merely have to put together a new combination of adaptations previously developed, or evolved, for resistance to systemic changes induced by other stressors. However, lacking experience, the centrifuging animal may make inappropriate combinations, and there is evidence in the results of chronic acceleration experiments for "diseases of adaptation."

In dealing with the results of chronic acceleration experiments, there is a temptation to interpret them broadly. This attitude apparently was shared by Thompson (1917), who, from the influence of body size upon structure and function, inferred the biological effect of gravity. On that basis, he predicted the effects of subgravity on such phenomena. Similarly, we have extended the analysis of our results to predict (in the purely mathematical sense) the biological effects of weightlessness. We are aware that this enthusiasm is not universal, and so, in the spirit of debate, we offer the following defense of these theses.

1. Over the range of observations, changes induced by chronic acceleration are covariant with, and appear to be dependent upon, the intensity of the ambient accelerative force. Earth gravity, physically, is merely one point on an accelerative force scale and does not represent a critical point; so a continuity of the dependent biological effects could be anticipated on the basis of the continuity of the independent physical phenomenon.

2. There is relatively little pertinent data for hypo- and hyperdynamic environments. The only available series of measurements from weightlessness to $3\,g$ were made by Roman et al. (1962), on hemodynamic phenomena in humans during exposures of less than 1 min. In these, no discontinuities were evident for weightlessness (although the relationships appeared to be approximately exponential, rather than rectilinear). If, as was discussed in the foregoing, the chronic acceleration effects represent a physiolgical adaptation to the biological effects of acute exposure, the chronic effects as well as the acute effects could be expected to have a subgravity continuity.

3. If the relationships developed at normal gravity and in greater, tolerable fields do not apply in the region from 0 to $1\,g$, there must be different and discontinuous (nonlinear) regulatory processes for accelerative forces above and below earth gravity. This would be a rather unusual situation. Of course, there may well be thresholds separating regions of "quantitative effects" and "unpredictability." For example, it has been found that a field of $0.05\,g$ is required to orient hen's egg yolks (Sluka, Smith, and Besch 1966), which is about the same order as the threshold stimulus for the otolith.

4. Whether or not the chronic acceleration "predictions" of weightlessness effects describe the real situation, they are useful. They provide a logical basis for designing satellite experiments with organisms, which reflexively, would substantiate or disprove the continuity of hypo- and hyperdynamic effects. Also, the hypergravity results will become very useful "third points" for the

rationalization of orbital experiments. If all our information is restricted to two points—weightlessness and earth gravity—few generalities will result.

Acknowledgement

The research reported for the Davis program was supported originally by the Office of Naval Research (NR 102-448) and more recently by the National Aeronautics and Space Administration (NGR 05-004-008).

References

Besch, E. L.; Smith A. H.; and Goren, S. 1965. *J. Appl. Physiol.* 20: 1232.

Besch, E. L.; Smith, A. H.; and Walker, M. W. 1965. *J. Appl. Physiol.* 20: 1241.

Bird, J. W. C.; Wunder, C. C.; Sandler, N.; and Dodge, C. H. 1963. *Amer. J. Physiol.* 204: 523.

Bjurstedt, H.; Rosenhamer, G.; and Weigertz, O. 1968. *J. Appl. Physiol.* 25: 713.

Briney, S. R., and Wunder, C. C. 1962. *Amer. J. Physiol.* 202: 461.

Britton, S. W.; Corey, E. L.; and Stewart, G. A. 1946. *Amer. J. Physiol.* 146: 33.

Burton, R. R., and Smith, A. H. 1965. *Aerospace Med.* 36: 39.

———. 1967a. *Physiologist* 10: 137.

———. 1967b. *Proc. XVI Int. Cong. Aviation and Space Med. (Lisbon).*

Burton, R. R.; Besch, E. L.; Sluka, S. J.; and Smith, A. H. 1967. *J. Appl. Physiol.* 23: 80.

Canonica, P. 1966. Masters thesis, University of South Carolina.

Casey, H. W.; Cordy, D. R.; Goldman, M.; and Smith, A. H. 1967. *Aerospace Med.* 38: 451.

Davson, H. 1966. *Textbook of general physiology*. 3d ed. Boston: Little, Brown.

Dodge, C. H., and Wunder, C. C. 1963. *Nature* 197: 922.

Evans, J. W. 1968. Doctoral thesis, University of California, Davis.

Feller, D. D., and Neville, E. D. 1965. *Amer. J. Physiol.* 208: 892.

Feller, D. D.; Neville, E. D.; Oyama, J.; and Averkin, E. G. 1965. *Proc. Soc. Exp. Biol. Med.* 19: 522.

Graybiel, A., and Kellogg, R. S. 1967. *Aerospace Med.* 38: 1099.

Kleiber, M. 1962. *The fire of life* New York: Wiley.

Matthews, B. H. C. 1953. *J. Physiol.* 122: 31P.

Menninger, B. S.; Murray, R. H.; and Robinson, F. R. 1967. *Aerospace Med.* 38: 377.

Murray, R. H.; Prine, J.; and Menninger, R. P. 1965. *Aerospace Med.* 36: 972.

Oyama, J., and Daligcon, B. C. 1967. *Endocrinology* 80: 707.

Oyama, J.; Medina, R.; and Platt, W. T. 1966. *Endocrinology* 78: 566.

Oyama, J., and Platt, W. T. 1964. *Amer. J. Physiol.* 207: 411.

———. 1965a. *Endocrinology* 76: 203.

———. 1965b. *Amer. J. Physiol.* 209: 611.

Roman, J. A.; Ware, R. W.; Adams, R. M.; Warren, B. H.; and Kahn, A. R. 1962. *Aerospace Med.* 33: 412.

Selye, H. 1950. *Stress*, Montreal: Acta Inc.

Sluka, S. J.; Smith, A. H.; and Besch, E. L. 1966. *Biophys. J.* 6: 175.

Smith, A. H., and Burton, R. R. 1965. *Physiologist* 8: 273.

———. 1967. *Growth* 31: 317.

Smith, A. H., and Kelly, C. F. 1961. *Physiologist* 4: 111.

———. 1963. *Ann. N.Y. Acad. Sci.* 110: 410.

Smith, A. H.; Winget, C. M.; and Kelly, C. F. 1959a. *Growth* 23: 97.

———. 1959b. *Nav. Res. Rev.* 12: 1.

Steel, F. L. D. 1962. *Nature* 193: 583.

Thompson, D'A. W. 1917. *On growth and form*, Rev. ed., ed. J. T. Bonner, Cambridge: Cambridge University Press, 1961.

Wagner, R. P., and Mitchell, H. K. 1955. *Genetics and metabolism*. New York: Wiley.

Wilson, P. N. 1954. *J. Agric. Sci.* 44: 67.

Winget, C. M., and Smith, A. H. 1962. *J. Appl. Physiol.* 17: 712.

Winget, C. M.; Smith, A. H.; and Kelly, C. F. 1962. *J. Appl. Physiol.* 17: 709.

Wunder, C. C. 1961. *Iowa Acad. Sci.* 68: 616.

———. 1962. *Aerospace Med.* 33: 866.

Wunder, C. C.; Briney, S. R.; Kral, M.; and Saugstad, C. 1960. *Nature* 188: 151.

Wunder, C. C.; Herrin, W. F.; and Cogswell, S. 1959. *Proc. First Biophys. Conf.*, p 639.

Wunder, C. C.; Herrin, W. F.; and Crawford, C. K. 1959. *Growth* 23: 349.

Wunder, C. C.; Milojevic, B.; and Eberly, L. 1966. *Nature* 210: 177.

Discussion

LOWENSTEIN: Are there changes in the endocrines, either in size or in appearance, that indicate a condition of acceleration-induced stress?

SMITH: Burton and co-workers have found evidence of endocrine participation in the acceleration response (*Aerospace Med.* 28: 1240, 1967). Generally there is a lymphopenia that is correlated with other functional characteristics, such as exercise capacity, sexual development, and survival (*Proc. Soc. Exp. Biol. Med.* 128: 608, 1968). With time, the lymphopenia may be lost, and such animals appear to have become physiologically adapted.

GUALTIEROTTI: What about the ability of the chickens to respond as a function of their different positions in space?

SMITH: There is no obvious behavior change in birds physiologically adapted to the treatment. They do not choose any particular orientation or position in the cage. Young chicks are very active—even playful.

WEIS-FOGH: If I understood you correctly, the difference, within the tolerable range, is an increased metabolic rate owing to extra work. Have you tried to compare your chickens with chickens exercised on the ground?

SMITH: Some experiments with chickens exercised for long periods on treadmills have been carried out by John Morse, a graduate student in our department, and the results will be available in his thesis. It appears that chronic acceleration and chronic exercise are similar in some aspects, but dissimilar in others. It also would be of interest in this regard to compare chronically accelerated animals with others exposed to some equivalent degree of hypothermia.

WUNDER: In preliminary experiments, we have found that if you put packs on the back of mice essentially to double the gravitational loading, one obtains growth patterns identical to those grown at $2\,g$.

EDWARDS: That femurs enlarge in response to chronic acceleration is understandable. But how do you account for size increased in non-load-bearing bones?

SMITH: Increase in bone size, unlike muscle hypertrophy, is not merely a local response. It is mediated by some whole-animal process, which affects all parts of the skeleton.

33
The Effects of Chronic Acceleration of Animals: A Commentary

C. C. Wunder, *University of Iowa*

The Nature of Chronic Acceleration

In speaking of chronic acceleration, we are referring to a condition in which organisms are subjected to an altered gravitational intensity, generally by means of centrifugation, over an essentially continuous period of exposure that may involve days, weeks, or even years and constitute a major fraction of the time required for an animal's development or life-span. In contrast, the previous sections of this symposium have dealt not with the long-term effects of a gravitational environment but merely with the detection of gravity or acceleration. Those sections were thus concerned with phenomena which, in terms of their effect on the animals under investigation, may involve only a fraction of a second. In the course of the discussion from the floor, it became apparent that many participants in this symposium were not familiar with the other effects of acceleration upon living organisms. Limitations of time did not permit either Dr. Smith or myself the opportunity to include in our oral presentations an introduction to some of the basic considerations of acceleration. It therefore seems appropriate to include some discussion of this aspect of the subject as an introduction to the written version of the present paper.

General Effects of Acceleration

In figure 1, a double logarithmic scale has been employed to illustrate the fields of acceleration, in g, along the ordinate that various biological preparations have survived and the time of known survival along the abscissa. The figure is intended to serve three purposes. First, it emphasizes that ability to survive is enhanced as field intensity decreases and as animal height decreases, and that survival time increases when an affected physiological process can tolerate longer suppression. Second, it indicates some of the animals for which chronic exposure has been studied. Finally, it contrasts potential differences between effects anticipated for chickens on the one hand and for animals studied at Iowa on the other.

The period of survival can be of the order of only a fraction of a second under conditions of exposure to intensities great enough to cause direct mechanical breakage. Exposures tolerable for only seconds are associated with attenuation of the heart's ability to lift blood to the brain. Death or unconsciousness can occur within minutes when acceleration imposes severe restriction upon breathing and upon aeration of the blood. In contrast to chronic exposure, these relatively short periods of acceleration might be termed acute acceleration. In recent years, the influence of acute acceleration upon the human body has become of increasing practical importance to aviation and to rocket travel. For a discussion of the effects of this type of acceleration, the reader is referred

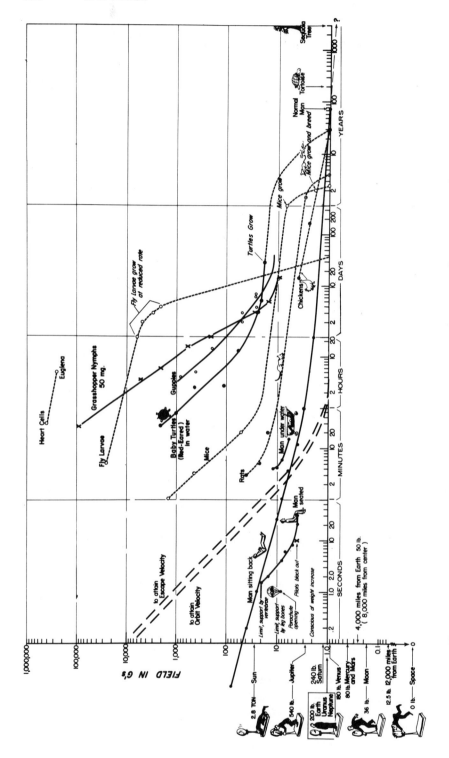

Fig. 1. Survival at various intensities of gravity (modified from Wunder 1963b; Wunder, Lutherer, and Dodge 1963).

to reviews by Lindberg and Wood (1963), Wood (1967), Wunder (1966, pp. 144-156), and Wunder, Duling, and Bengele (1968a).

Acceleration at an intensity sufficiently moderate to permit prolonged survival would have as its primary effect the immobilization of the experimental animal. Under conditions of immobilization sufficient to prevent intake of food or water, survival may be limited to hours or days.

Effects of Chronic Acceleration

When immobilization or the other previously mentioned effects of acceleration reveal themselves to be transitory in the face of persistent exposure, tolerance to chronic acceleration is possible. If the exposure is moderate enough to permit truly continuous or chronic acceleration, it is more appropriate to speak of an altered life-span extending over weeks or even years rather than of actual mortality. Chronic acceleration is accompanied by some recovery of mobility, drinking, and feeding, and by the resumption of development and the continuation of other physiological adjustments, such as the finding that mice can breed even at $2g$ (Briney and Wunder 1960; Wunder 1965a. The ability to carry on reproduction at fields up to $3g$ has been confirmed at other laboratories in studies with rats and mice (Matthews, personal communication, 1963; Lange and Broderson 1965; Vrăbiesco, personal communication, 1966; Oyama and Platt 1967). Some of the adjustments reflect something more than a mere recovery of function. We have noted what amounts to conditioning by functions of support (Wunder et al. 1960; Bird et al. 1963; Wunder and Bird 1967), of circulation (Duling 1967a, b), and of respiration (Wunder, Crawford, and Herrin 1960; Wunder 1963a) to a level different from that for normal $1 g$ animals. The preceding paper (Smith and Burton, this volume) described similar examples of conditioning with chickens.

Achieving Simulated High Gravity by Chronic Centrifugation

Equivalence of Acceleration to Gravitational Fields. On the basis of existing physical theory, inertial reaction to mechanical acceleration is indistinguishable in its effect upon mass from true gravitational fields of comparable magnitude. There should therefore be no question that the effects which have been noted in carefully controlled acceleration experiments are qualitatively the same as the effects of Newtonian gravity. The exact mechanism whereby these effects are mediated is yet to be elucidated. The effects of chronic exposure can be determined after animals immobilized by acceleration have recovered their mobility. Moreover, since the effects of intense acceleration noted in experiments with large animals can be shown from considerations of hydrostatics and similitude to be insignificant with small animals, there is no reason to suppose that the well-established effects of acute acceleration on man or large animals are necessarily significant in chronic centrifugation of smaller animals.

Precise descriptions of the equipment and techniques (Wunder et al. 1960; Wunder et al. 1959; Wunder 1965; Kelly et al. 1960; Kelly and Smith 1964) appropriate for growth and maintenance of animals during centrifugation are available elsewhere. A brief interruption of exposure, for approximately 15 min per day, is usually necessary for the feeding of animals together with other care and maintenance procedures. This schedule of exposure is diagrammed for fruit-fly larvae in figure 2, which presents a schedule that is representative for most experiments of this type.

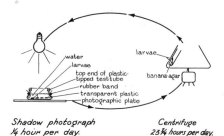

Fig. 2
Schematic schedule for exposure and observation of centrifuged fly larvae (Moressi, Herrin, and Wunder 1961).

Interruption of Exposure. The sensory detection by the animal of this brief interruption of exposure might conceivably function as a cue that stimulates nongravitational effects. Other than this, however, the interruption would be expected to have only the slightest effect upon the total physical and biological results of centrifugation. Although Cooke and Bancroft (1965) have reported that 2 wk of uninterrupted centrifugation of mice produced effects slightly different from those reported by us, the strain of the mice employed as well as other conditions of the experiments were too different to permit a valid comparison. We have noted a consistent increase in most effects of centrifugation, with sufficiently great increases in field intensity. When experimental conditions were otherwise identical, however, we have been unable to detect any dependency of the effects of centrifugation upon frequency of brief interruption of exposure. As a general practice, we have alternated the stoppage of the centrifuge for maintenance and measurement from early morning to late afternoon from one day to the next. The purpose of this procedure has been to eliminate any effects of biological rhythms that could be cued by regular and consistent interruption of the centrifuge. A large automated centrifuge has recently been completed at Moffett Field for continous, uninterrupted exposure of animals over a period of years. Comparison of the results of exposure in this centrifuge with those from other less automated equipment should permit the detection of any of the unlikely effects of these brief interruptions.

Vibration. The careful balancing of centrifuges to assure long uninterrupted service of the machinery should likewise minimize any effects of vibration. Undistorted views of animals undergoing exposure have been possible with relatively crude optical assemblies, which would not function properly in the presence of slight vibration.

Temperature and Air Currents. In our laboratory, cages of control animals face the centrifuge and are placed about the periphery of this machine in a manner that establishes identical environments relative to lighting, air currents, and the temperature as measured from either wet or dry bulb thermometers. Although some noise can be generated by the centrifuge drive, the proximity of control to experimental cages should eliminate any major artifacts arising from detectable sound.

Rotation. We have been reminded that centrifuged subjects undergo not only acceleration but also turning (Smith and Burton, chap. 32, this volume). In other words, one effect of gravity as simulated by centrifugation that does not occur under conditions of planetary gravity is a lack of uniformity of field intensity. The field intensity differs at various radial locations or during animal movement in a centrifuge. The results of such a lack of uniformity may be referred to as "rotatory artifacts." Mammals and most other vertebrates are able to detect this effect only by means of structures in the inner ear. In order to test this effect, we have compared the effect of centrifugation on hamsters exposed to a field intensity of 5 or 6 g for 1 mo. with animals that had undergone surgical labyrinthectomy and with sham-operated hamsters (Wunder, Milojevic, and Eberly 1966). Figure 3 displays a photograph, taken after 100 days of centrifugation, that permits a comparison of a typical experimental, labyrinthectomized hamster (*right, center*) and a sham-operated, centrifuged hamster of identical size (*extreme right*) with the control animals, which were either labyrinthectomized (*extreme left*) or sham-operated (*left, center*). Although centrifugation at 100 rpm caused a temporary decrease in food consumption as well as a slower rate of growth of the hamsters, the effects upon the animals which were able to detect rotation (sham-operated) were indistinguishable from the effects on those that could not (labyrinthectomized). More recent studies from our laboratory with labyrinthectomized turtles (Rice, Wunder, and Diecke 1970) also indicate that the effects of chronic acceleration, with rotation rates as high as 200 rpm, are due entirely to the high "gravity."

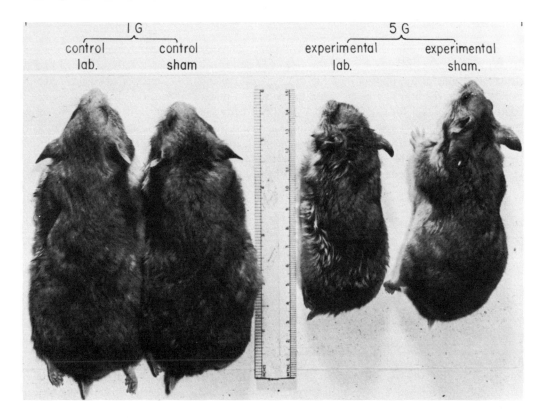

Fig. 3. Typical littermate hamsters after control conditions and exposure to 5 g for 100 days following labyrinthectomy and sham operation (Wunder, Milojevic, and Eberly 1966).

Enhancement of Effects of Acceleration with Increased Size

Physical Considerations. One of the potential differences between the effect of chronic acceleration on vertebrates in our experiments and its effect on vertebrates in the work with chickens at Davis or in studies with man being initiated elsewhere involves variation in size. That the effect should increase with height is predicted both from considerations of Pascal's Third Law of Hydrostatics and, as has already been explained (Smith and Burton, chap. 32, this volume), from Galileo's Principle of Similitude.

Contrast of Effect on Large and Small Animals. In the brief discussion of acute acceleration, reference was made to the influence of gravitational field intensity upon columns of fluid such as blood. The direct gravitational effect upon blood pressure increases in a linear manner, with the product gh of field intensity g and height h of fluid columns. At any given field intensity, therefore, the absolute magnitude of the effect upon blood would be greater with larger animals. With the exception of the giraffe (Gauer 1961), which requires unusually high arterial pressure to raise blood up its long neck from the heart to the head, most warm-blooded animals, even though they may vary greatly in size, are reported to have almost the same arterial blood pressure (Altman and Dittmer 1963, pp. 238-41). The relative effect of acceleration upon arterial pressure at any given field intensity would thus be greater with taller animals.

The effect of linear dimensions on the gravitational demands for support described by Galileo (1638) is illustrated in figure 4. Since the weight of an animal increases directly with mass m and therefore volume or the third power of length, whereas the ability to support this weight increases merely with cross-sectional area or length to the second power, the larger animal would require a relatively thicker skeleton, as illustrated by bone structure in figure 4. In that figure, the larger animal with a threefold longer bone requires a ninefold thicker bone to support its weight. The shaded figure of a bone for an animal at $3g$ has been added to indicate that under this condition the smaller animal would require the same relative thickness or stoutness of supporting structures as would otherwise be demanded by an animal of three times greater height.

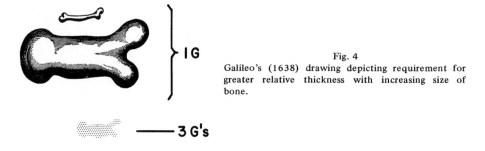

Fig. 4
Galileo's (1638) drawing depicting requirement for greater relative thickness with increasing size of bone.

The general ability of smaller organisms or structures to tolerate a given gravitational intensity is illustrated by the survival data shown in figure 1. A man leaning backward in that figure is effectively shorter than a man sitting upright and therefore can better tolerate the longer exposures shown there. In contrast, as demonstrated over thirty years ago by Beams and King (1937), structures as small as a single cell can survive $500,000\,g$ for as long as 20 min.

Because chickens exhibit high columns of blood relative to arterial pressure, the effect on them of an increase by only a few multiples of the Earth's gravity is likely to exceed that which gravity normally has upon man. Studies of chickens exposed to acceleration should reflect not only the need of muscle and bone for greater weight support (as encountered in our experiments with mice, rats, and hamsters) but also, in contrast to rodents, possible adverse hydrostatic effects on the circulatory system. Although Duling (1967a, b) was able to demonstrate cardiovascular conditioning by rats after exposure to centrifugation for one month at $3g$ the result he described could be interpreted as a response to the need for greater perfusion of blood to the muscle rather than to hydrostatic adversity. In preliminary observations of acute accelerations, mice had to be subjected to a field as intense as 25 g before they exhibited any alteration in heart rate (Shipton, Reinhardt, and Wunder 1960).

Attenuation of Size Effects with Adjustment and Selection. Under conditions of 1 g the natural challenge of gravity to small animals would be less than that normally encountered by man. In studies performed in our laboratory, the challenge encountered by experimental animals (rats, hamsters, and mice) during centrifugation theoretically approaches, but does not necessarily exceed, the challenge normally encountered by man in the 1-g environment. Although the level of this challenge is undoubtedly raised beyond its natural level for mice at 4 g, ideally there should be less of an effect than with centrifuged chickens.

Comparison of Potential Adversity on Chickens and Rodents. Perhaps the picture of physiologic stress derived from differential blood cell counts with chickens is to a large extent a reflection of hydrostatic adversity. Additional carefully planned and controlled studies might suggest a somewhat similar pattern. In initially establishing our experimental procedure, however, we were unable to observe any differential hematological effects. Vrăbiesco and his associates (Vrăbiesco, Cimpeanu, and Domilesco 1964; Vrăbiesco, Costiniu, and Enachesco 1964; Vrăbiesco and Enachesco 1964; Vrăbiesco and Domilesco 1965) in Bucharest, who were the first investigators to report any such hematological studies, detected very little effect on rats up to 5 g for exposures as long as a year or more.

The tolerance of chickens to a given centrifugal field (fig. 1) is less than that of smaller animals. Unlike chickens, the experimental rodents in our studies were able to experience immediate rather than gradual introduction to fields of from 2 to 7 g, and after weeks of exposure to centrifugation displayed only negligible mortality.

The actual effect of size upon the organisms at any given field intensity would, in practice, not necessarily be as great as that suggested by simple physical theory. If one wished to carry this reasoning to an absurd conclusion, he could extrapolate survival data from mice and conclude that man could survive only a small fraction of the earth's normal gravity. Because of the effect of size, a larger animal normally undergoes a greater gravitational challenge. Thus, through generations of natural selection as well as through constant adjustment to the environment, larger species of animals become better able to tolerate a given magnitude of the product gh. If the relatively lower gravitational load encountered by mice causes them theoretically to have less tolerance than man to the same magnitude of mg, a timely question arises: Would prolonged or chronic exposure to a low- or zero-gravity environment of the kind which might be associated with extended space travel cause man's tolerance and conditioning to gravity to decrease so drastically that upon return to earth after long voyages he could not even tolerate normal terrestrial gravity?

Laboratories Studying Chronic Animal Acceleration

Early Attempts to Study Chronic Acceleration

The approach to the study of gravitational biology by means of the chronic acceleration of different animal forms is a relatively new area of investigation. Hertwig (1899), in reporting the results of work with frog embryos, was the first to observe that exposure to chronic centrifugation would permit survival of animal material but caused modifications in its development. Before the initiation of the program at Iowa, there apparently were only two modern attempts to study the growth and survival of chronically accelerated animals. Like all later programs, they involved continuous centrifugation. The first attempt of which I am aware (Gray and Webb 1950) reported the finding that although tadpoles could survive exposure to $10\,g$ for 6 days, upon return to $1\,g$, they developed abnormal postural orientation similar to that later found with chickens (Winget and Smith 1962) and with guppies (Schmickley, Wunder, and Loomis 1964).

Developmental and Physiological Studies at Iowa

When our program was initiated in 1954, the primary goal was to exaggerate gravity's natural effect and, on the basis of that exaggeration, to investigate the normal role of gravity in controlling and guiding growth. Chronic centrifugation of motile, unrestricted organisms was established as a feasible experimental procedure for elucidating principles of gravitational biology in animal studies (Wunder 1955). This initial study, which utilized larvae of the common fruit fly, *Drosophila melanogaster*, yielded measurements of slower growth with increasing centrifugal field, demonstrating that artificial gravity can influence development. Although our laboratory has maintained its interest in the developmental adjustment an organism makes to gravity, we have extended our research interests to related adjustments accomplished by such physiological processes as respiration, circulation, and fluid balance.

New Programs

With a few exceptions (Steel 1962; Canonica 1966; Cook and Bancroft 1965; Redden 1970), the present work in this area of gravitational biology has constituted either a continuation of our original program at The University of Iowa or related work by programs established at five other laboratories. These laboratories are located at the University of California at Davis, as described by Smith and Burton in chap. 32, and at Bucharest, Moffett Field, the Soviet Academy of Sciences in Moscow, and the University of Kentucky. Although the program at Davis and ours are in some ways complementary, the Iowa laboratory is concerned primarily with studies of acceleration under conditions which merely exaggerate the natural effects of gravity, whereas the Davis laboratory has concentrated more on an investigation of the effects of less tolerable acceleratory conditions. The most productive of the foreign programs to date has been the one in Bucharest, which was established in about 1960 (Vrăbiesco, Cimpeanu, and Domilesco 1964; Vrăbiesco, Costiniu, and Enachesco 1964; Vrăbiesco and Enachesco 1964; Vrăbiesco and Domilesco 1965) and which has been concerned largely with the influence of gravity upon aging. The program at Moffett Field has been concerned with biochemical events, particularly those related to physiological stress, which reflect the responses of organisms undergoing acceleration (Oyama and Platt 1964*a*, *b*; 1965*a*, *b*; 1967; Oyama, Medina, and Platt 1966; Feller et al. 1965). The program in Moscow was established in the belief that observations made during space travel at zero gravity could be intelligently interpreted only if analyzed in the light of studies performed throughout all ranges of the gravitational spectrum. The work they have reported to date has been concerned primarily with metabolism (Gazenko and Gurjian 1964; Gyurdzhian et al. 1964; Gyurdzhian 1966). The program at the University of Kentucky has concentrated on the effects of acceleration on animal behavior and is particularly concerned with the identification of the gravitational field intensity for which various animals demonstrate a preference (Lange and Broderson 1965; Martin, Richardson, and Martin 1966).

Developmental Effects

Fly Larvae

We did not feel justified in extending our work to such animals as the mouse, hamster, rat, aquarium fish, and turtles until the techniques and developmental effects had been well established with animals such as fly larvae, which can be more readily grown and for which the results of exposure can be more simply analyzed. Before pursuing work with these other animals, we wished to establish the effect of altered gravitational intensities on fly larvae for four primary reasons: (1) Modest financial considerations were posed by the larvae's small size, simple care requirements, and rapid development; (2) By working with a terrestrial animal, the effects of gravity on growth and survival would not be clouded by buoyancy or environmental-pressure artifacts; (3) With the larvae's open circulatory system and tracheal respiratory system, the developmental analysis would not be obscured by potential difficulties in pumping air or blood; and (4) The mathematical nature of this animal's growth is very easy to analyze.

Growth for these organisms, as is indicated by the left-hand side of figure 2, was recorded by means of daily shadow photographs. The volume computed for an ellipsoid of revolution from the length and width of these shadows is represented in figure 5 in cubic millimeters by a logarithmic scale along the ordinate as a function of time in hours along the abscissa. During physical growth, the logarithm of the animal's size is a simple linear function of time. The slope of that curve, when presented in terms of natural logarithms per period of time, is equivalent to the growth constant. The ratio of the slopes for the experimental curves (C and E) to that for the solid control curve (A), when expressed in terms of percent and subtracted from one hundred, yields values of the type displayed for the curve farthest to the left in figure 7. Sufficiently great fields can cause a progressive decrease both in rate of growth and in final attained size. The time at which complete deceleration of growth occurs varies with the experimental condition, but it does take place earlier than that of the control animals. The greater slope and final size for curves C and D demonstrate that the smaller the animals are at the onset of exposure, the better they are able to adjust (Wunder et al. 1959a). The greater than normal slope of curve B for animals upon removal from centrifugation, coupled with a subnormal demand for oxygen (Wunder, Crawford, and Herrin 1960), is considered evidence of an increased efficiency of oxygen utilization for growth, which is made to partially compensate for the extra work and resulting metabolic demand imposed by the increased gravitational intensity. Most of the larvae which continued growth at normal gravity after removal from the centrifuge were able to complete sufficient development to arrive at the pupal stage. An abnormally large fraction of the larvae, however, were unable to emerge as adults.

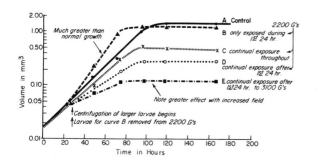

Fig. 5
Growth of fly larvae as influenced by various conditions of centrifugation (Wunder and Lutherer 1964).

In commenting upon our later work with mice, Gray and Edwards (chap. 31, this volume) noted that we had been able to confirm their finding of greater cross-section for the coleoptile of the wheat seedling and had reported comparable findings for the femur (Wunder, Crawford, and Herrin 1960). Before proceeding to a discussion of our work with other animals, one more comparison between the responses of plants and those of fly larvae seems appropriate. Figure 6 indicates that the dependence of larval growth rate upon temperature appears to decrease with increasing field intensity (Wunder, Herrin, and Crawford 1959). These results are quite similar to the recent ones shown by Gray and Edwards in figure 3 of their presentation in this volume. It is possible that the results of these temperature experiments indicate interference with some active process. In discussions during an earlier session of this symposium, Dr. Pollard urged that an Arrhenius plot be utilized in the hope of demonstrating the influence of temperature upon the gravitational phenomena affecting a living organism. But even if sufficient data yield a good Arrhenius plot, other interpretations might be possible. Various passive phenomena, such as concentration gradients and membrane permeability, might be influenced by the combined effect of temperature and gravity.

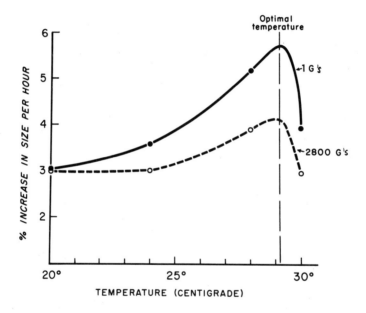

Fig. 6. Influence of temperature and gravity upon growth of fruit-fly larvae (Wunder 1966).

Other Organisms

A comparison of the influence of acceleration upon the growth and development of various organisms in a review by Wunder and Lutherer (1964) utilized the curves shown in figure 7 as a partial summarization. The figure presents a comparison of the influence of field intensity (in g on an arithmetic scale along the abscissa) upon the relative increase or decrease in growth rate (in percent along the ordinate). The curve for wheat is based upon the data of Gray and Edwards

(1955) for coleoptiles maintained in a centrifuge at 28°C during the first 24 hr of growth. Other curves are based upon the measurements from our laboratory for various animals during the first 24 hr of exposure. The turtles were the Red-Eared species (*Pseudemys scripta elegans*) grown in one inch of water at 28°C during the winter months (Dodge and Wunder 1963). The apparent stimulation of turtle development by exposure to a field intensity of 5 g is less evident during seasons other than winter (Wunder, Dodge, and Duttweiler 1962; Wunder et al. 1965), though this seasonal effect is not consistent (Rice, Wunder, and Diecke 1970). The mice employed were the NLW strain grown at 22°C (Wunder, Lutherer, and Dodge 1963).

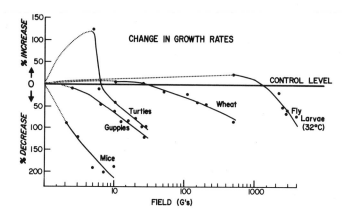

Fig. 7. Immediate (first 24 hr) change in growth rate upon exposure to centrifugation (Wunder and Lutherer 1964).

Two effects represented by figure 7 should be noted. Increases in g-intensity, when sufficient, progressively reduce growth rates. The application of moderate g under appropriate conditions, however, appears to have a net stimulating effect on the growth of wheat, of fly larvae, and particularly of turtles of the species indicated. Since this data is merely for the first 24 hr of exposure, it indicates a misleadingly great effect on the growth rate of mice.

It is tempting to interpret the loss of body mass in the tortoise (Testudo horsfieldi Gray) in the Soviet translunar flight (Gazenko et al. 1969) with a zero-g extrapolation from our 5-g data for the turtle. However, there are experimental and taxonomic differences involved, and the Soviet workers acknowledge that factors other than free fall or space flight would have caused the loss in mass (Gaydamakin et al. 1969).

Unlike the more constant effect of gravity on most cold-blooded animals, its effect on mice and other warm-blooded animals during the first day of exposure, at a given physical g-intensity, is greater than the effect which develops in the course of longer exposure. This polyphasic response of mice is well illustrated by the 4-g curve in figure 8, in which the precentage of change in mass of developing mice during exposure is plotted along the ordinate. Time in days of continued exposure is indicated on the abscissa. The results shown there are qualitatively similar to the growth curves for all chronically accelerated warm-blooded animals. Dr. Smith has assured me that the Davis laboratory could reproduce all portions of this figure from similar measurements taken from experiments performed with centrifuged chickens.

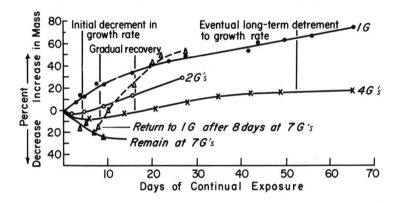

Fig. 8. Growth of white mice exposed to various gravitational conditions (Wunder and Lutherer 1964).

Adjustments to Chronic Centrifugation

From the dip in the 4-g curve of figure 8, it is evident that a period of time is necessary before these animals become adjusted to a gravitational load that exceeds any field intensity to which they have been previously exposed. High *g* has its greatest effect during the first days of exposure, during which there occurs a reversible depression of many physiological processes. At first these animals behave almost as if they had been immobilized. Rats (Bengele 1969) and mice (Wunder, Meyer, and Mason 1970) newly introduced into centrifugation exhibit a more reduced water consumption than is noted with pair-fed control animals. They also exhibited a polyuria. Russian reports suggest that the initial period of exposure is characterized by reduced oxygen intake and decreased lung ventilation (Gyurdzhian 1966), together with a urinary loss of muscle nitrogen (Gazenko and Gurjian 1964; Gyurdzhian et al. 1964). Oyama, Platt, and Holland (1968) have found a decreased core temperature for rats, indicating a depressed metabolism during the early stages of exposure. This has been confirmed by Duling (1967a). It should be noted with reference to figure 9 that the weight loss and slow initial growth can be reproduced by a decreased dietary intake (Wunder 1961). A number of these changes may reflect adjustments made by the animal so that it will have less body mass to support under conditions of increased gravitational intensity.

Although there is some recovery of food consumption after a few days, at one time we considered the prolonged period of subnormal growth to be the result of an incomplete return to normal food intake. On the other hand, we noted that aquatic turtles developed ravenous appetites and often underwent rapid growth as a result of their increased food consumption (Dodge and Wunder 1962). Moreover, when the Davis laboratory made available measurements from studies with chickens which revealed an eventual excessive food consumption, and when these findings were later confirmed by Oyama and Platt (1965a) in work with rats, as well as by our own unpublished results from experiments with mice, we were compelled to acknowledge that growth was not limited solely by food intake but also by the diversion for work against gravity of food materials otherwise available for growth.

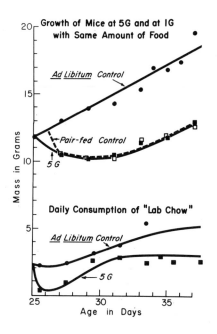

Fig. 9
Growth and food consumption of mice at $5\,g$ and under conditions of paired-feeding at $5\,g$ (Wunder 1961).

In contrast to the growth response of centrifuged plants in the studies described by Gray and Edwards (chap. 31), the growth of animals maintained at increased field intensity, as shown in figures 5 and 8, tends to level off rather more quickly at high g. There are at least three possible ways of interpreting this pattern of growth response. Endocrinologists who advocate Seyle's concept of a "Multiple Adaptions Syndrome" (Selye 1950; Applezweig 1961) might consider the final phase of growth response to be identical with the so-called "exhaustion phase of physiologic stress-response." The second possible explanation is that which accounts for the growth pattern by reference to precocious aging. It does appear that centrifuged animals age rapidly. Vrăbiesco, Cimpeanu, and Domilesco (1964) in Bucharest have reported a large number of observations which indicate that the exaggeration of "gravitational physiologic stress" beyond its natural level in experiments with rats has accelerated the aging process and thus decelerated development at an earlier chronological age. Extending the results of this research, Vrăbiesco found that animals evolved from the surviving members of several generations of rats maintained at $5\,g$, when returned to normal g, demonstrated abnormally slow aging and long life-spans (personal communication 1966). Something more reversible than merely precocious aging is suggested by the observation we made some time ago that fly larvae (curve B of fig. 5) and mice (broken curve of fig. 8) demonstrate accelerated growth upon removal from centrifugation. The observation made by Oyama and Platt (1964b) in work with rats and by the Davis laboratory in work with chickens that this reversibility persists after growth has leveled off at high gravity is a persuasive argument for a third interpretation. This may be described, in the words of Smith and Burton (chap. 32), as "that of equilibrium between factors favoring growth and the gravitational factors opposing it." The observations we made early in our program concerning fly larvae, however, indicate certain changes that are less reversible than altered equilibrium. Comparable observations were made by the Emory laboratory in describing the results of studies with coleoptiles of wheat (Edwards and Gray 1956). After removal from the centrifuge, in which their growth had been delayed, the fly larvae demonstrated such an accelerated growth that they temporarily surpassed in size the developing control animals (Wunder 1960). But although the experimental larvae were growing faster, their oxygen demands were less than control levels (Wunder, Crawford, and Herrin 1960). What this suggests is that increased metabolic efficiency partially compensates for the extra work exerted in opposing gravity. The efficiency for temporary physical growth appears to have been achieved at the eventual expense of morphological processes. The experimental larvae returned to $1\,g$ after one or more days of exposure to high g grew to a slightly smaller final size. As noted earlier even though most of these animals develop sufficiently to enter the pupal stage, a significant number of them are unable to emerge as adults.

In addition to eventual recovery of food intake, a number of other adjustments made during continued exposure have been noted with respect to mice and other rodents. In studies undertaken in our laboratory to determine the influence of gravity on fluid balance, Bengele (1969) noted that within one or two days of exposure to $3\,g$ rats were again able to drink water at a normal or even at an enhanced pace. Although the observation we reported of a slight decrease in hematocrit has been interpreted by Vrăbiesco, Cimpeanu, and Domilesco (1964) as reflecting a decreased red-cell production, it might, on the other hand, reflect a dilution of blood by increased plasma volume so as to avoid the possibility of circulatory shock. Duling (1967a, b) has noted several indications of cardiovascular conditioning in rats after exposure to $3\,g$ for one month. The total peripheral resistance appeared to rise in a way that would prevent venous pooling and maintain arterial pressure for circulation to the skeletal muscle. There is some evidence, based on measurements performed on the hind-limbs of these animals, that skeletal muscle decreases relative circulatory resistance and therefore permits better perfusion of blood to muscle tissue. The most dramatic observation he reported was a threefold rise in vascular response of the arterial pressure reflexes.

There is preliminary evidence which suggests that mice removed from centrifugation (Wunder 1963), like wheat coleoptiles and fly larvae, can develop an increased efficiency of oxygen metabolism. Vrăbiesco, Cimpeanu, and Domilesco (1964) have reported that rats undergoing continuous exposure to centrifugation finally demonstrate an increased core temperature, a finding that is consistent with results showing an increased dietary intake and increased oxygen intake.

In my discussion of figure 4, I referred briefly to Galileo's (1638) prediction that an increasing gravitational load demands greater stoutness of supporting structures. In chapter 32, Smith and Burton have cited evidence by Thompson which demonstrates that larger animals do indeed possess relatively larger skeletons. All this conforms to the enhanced cross-section noted for both wheat coleoptiles (Gray and Edwards 1955) and the femurs of developing mice (Wunder et al. 1960) during centrifugation. As Smith and Burton (chap. 32, this volume) have already pointed out, this observation has been confirmed by and extended in their studies with chickens. Although skeletal growth may be accelerated in young animals, the eventual size that the femurs can attain at the end of development is not necessarily greater than or even as great as that of the femurs of the control animals. On the other hand, the results shown in figure 10 do, in terms of radiographic density, indicate a greater mineral content in the bones of centrifuged animals. In agreement with results from studies with chickens described by Smith and Burton (chap. 32), the stimulation of skeletal development is not limited merely to the weight-bearing bones; other portions of the skeleton also appear more dense.

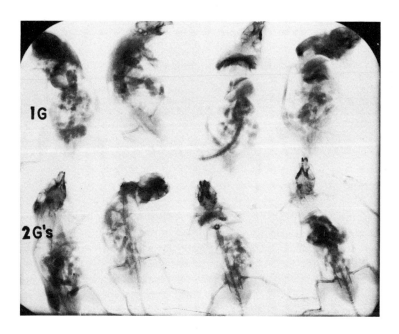

Fig. 10. X-ray photograph of mice grown under control conditions (*top*) and in a centrifuge at 2 g (*bottom*). (Courtesy of Dr. Duane Graveline, Wright-Patterson AFB.)

Within 1 wk after the onset of exposure, the gastrocnemius muscle of immature mice shows a greater relative size (Bird et al. 1963; Wunder and Bird 1969c). Confirmation for this finding with more dramatic data for another extensor muscle has been furnished by Burton et al. (1967) in studies which utilized mature chickens. Where this effect has been noted with mice it has been with particular reference to the concentration of noncollagen nitrogen [NCN]. In normal gravity the [NCN] of animals ranging in size from young mice to mature rats increased with (body mass)$^{1/3}$, suggesting a compliance with the demands of similitude (Wunder 1969a, b). For centrifuged animals the [NCN] relative to the concentration in control animals of the same age and size is shown in figure 11. The relative concentration in the muscle is plotted on a logarithmic scale along the ordinate as a function of the gravitational field in g on another logarithmic scale along the abscissa. All mice were 5 wk old at the beginning of the exposure. After 1 to 8 wk of exposure to fields of from 1.5 to 7 g, there was a consistent increase in the relative concentration beyond the control level of unity. Although the experimental levels approached the level of satisfaction of the demands of similitude and gravitational adversity shown by the broken line, other considerations prevented complete satisfaction of this demand. In the figure, NCN refers to absolute concentration of noncollagen nitrogen, B to body mass, X to experimental values, and C

to control values. The coefficient n' relates the logarithm of noncollagen nitrogen concentration in the control animals to their body mass. The variable G refers to the intensity of the inertial field. The variables X and C enclosed by brackets also refer to the concentration of noncollagen nitrogen in the experimental and control animals respectively.

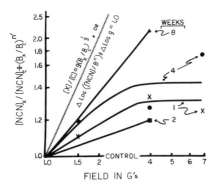

Fig. 11
Relative noncollagen nitrogen concentrations (NCN) as corrected for body size in gastrocnemius muscle of mice after centrifugation (*solid lines*) at 1.5, 4, or 7 g for periods of from 1 to 8 wk in comparison to the relative corrected concentration from theoretical considerations of similitude. See text for detailed interpretation. (From results of Bird and Wunder 1969c.)

Extrapolations to Zero G

It is possible that the recent biosatellite experiment will permit Gray and Edwards to extend the curve for wheat in figure 7 into the low-g side of the gravity spectrum. Owing to the contrasting effects of growth retardation at sufficiently high fields and growth enhancement at moderate fields, however, I am not prepared at this time to extrapolate any of the curves from figure 7 to zero g. Several years ago, our laboratory prepared for the Air Force a comparative survey of the various methods of weightlessness simulation in biomedical research (Wunder 1964, 1965b, 1966, pp. 157-169; Wunder, Duling, and Bengele 1968b). The advantages and disadvantages of chronic centrifugation as outlined at that time are still applicable.

The use of chronic centrifugation for investigating the potential effects of exposures to low-g conditions divides itself into three, or actually, four approaches. Figure 6 from the preceding paper (chap. 32) may be taken as a recent example illustrating the first approach, which proceeds by extrapolating to the opposite conditions from results obtained under conditions of exposure to increased field intensities. The same paper also alluded to the second approach: the reintroduction to a lower gravitational state (i.e., a 1-g state) of animals exposed to high g. When we eventually develop theoretical equations that will enable us to extrapolate all data to the same zero-g predictions, I will be more satisfied with both approaches. There has not yet been enough work in the area of chronic centrifugation to guarantee such an agreement of predictions. If we want actual experimental measurements from the low-g side of the gravity spectrum, we must resort to a third approach involving the utilization of a 1-g-satellite-based-centrifuge, since it is absolutely necessary that control animals be maintained in a 1-g environment before any meaningful net effect of weightlessness can be discerned from the gross effects of space flight. Some investigators may attribute to weightlessness many of the results of manned space flight; but unfortunately these results cannot be positively distinguished from other possible effects of space flight arising from such factors as immobilization and confinement. Last, as a fourth approach, we must remember that any satellite system is, after all, merely a fabulously expensive centrifuge system and one which has not actually removed material from planetary gravity. As Dr. Frier reminded us in the first session of this symposium (chap. 1), a satellite merely negates the gravitational effect by the vector addition of counterbalancing angular acceleration.

Depending upon how one chooses to extrapolate our data, the following factors, in addition to altered growth, might conceivably be involved in the effects of weightlessness: degeneration of muscle, particularly of the respiratory diaphragm; weakening of the ligaments; kidney and fluid imbalance; inefficient metabolism; obesity; postural disorientation; demineralization of bone, together with kidney-stone formation; altered life-span; cardiovascular deconditioning, with insufficient plasma volume; unsatisfactory peripheral resistance; abnormal vascular reflexes.

At this point let me emphasize that, in spite of my profound conviction that all zero-g effects suggested by centrifugation cannot be overlooked in the planning of manned space flights, extrapolation of curves or equations, particularly empirical ones, is unsound. As some of the centrifuge data in figure 7 is neither linear nor progressive with field strength, extrapolations or even interpolations could be highly misleading.

Our present understanding of gravitational biology is such that the results of chronic centrifugation merely warn of the conceivable effects of weightlessness without indicating the probability of their occurrence. Should even qualitatively accurate predictions prove to be quantitatively inaccurate, a false security concerning the effects of weightlessness may have tragic consequences.

Summary

Chronic centrifugation is a new area of investigation, confined to a relatively few laboratories working with several different species of animals. The studies at The University of Iowa have established that such g fields:

1. Influence the growth rate of animals, progressively decreasing the rate at sufficiently high fields but sometimes stimulating growth at moderate intensities in a manner dependent upon temperature and size but largely independent of the nongravitational (rotational) artifacts of centrifugation;

2. Can sometimes provoke a temporary reduction in food intake, which is a partial cause of the initial decrement in growth observed with mammals;

3. Results in eventual adjustments, which can include enhanced caloric intake, increased efficiency of oxygen metabolism, relatively larger supporting bones and muscles, enhanced circulatory pressure reflexes, together with altered circulatory resistance and altered fluid balance;

4. Influence life-expectancy in a manner dependent upon field strength.

Acknowledgments

Initial studies on gravitational biology in this laboratory were supported in part by the Central Scientific Fund, College of Medicine, The University of Iowa. The continued studies have been supported by the Iowa and National Offices of the American Cancer Society, the National Institutes of Health, the State University of Iowa Foundation, and the National Aeronautics and Space Administration. The author wishes to express his appreciation to his students, co-workers, and associates at The University of Iowa who have contributed to this work.

References

Altman, P. L., and Dittmer, D. S. 1963. *Biology data book*. Washington, D.C.: Fed. Amer. Soc. Exp. Biol.

Applezweig, M. H. 1961. In *Psycho-physiological aspects of space flight*, ed. B. E. Flaherty, pp. 139-57. New York: Columbia University Press.

Beams, H. W., and King, R. L. 1937. *Biol. Bull.* 73: 99-111.

Bengele, H. H. 1969. *Amer. J. Physiol.* 216: 659-65.

Bengele, H. H.; Moore, W. W.; and Wunder, C. C. 1969. *Aerospace Med.* 40: 518-20.

Bengele, H. H., and Wunder, C. C. 1969. *Proc. Soc. Exp. Biol. Med.* 130: 219-23.

Bird, J. W. C.; Wunder, C. C.; Sandler, N.; and Dodge, C. H. 1963. *Amer. J. Physiol.* 204: 523-26.

Briney, S. R. and Wunder, C. C. 1960. *Proc. Iowa Acad. Sci.* 67: 495-500.

Burton, R. R.; Besch, E. L.; Sluka, S. J.; and Smith, A. H. 1967. *J. Appl. Physiol.* 23: 80-84.

Canonica, P. G. 1966. Effect of prolonged hypergravity stress on myogenic properties of the gastrocnemius muscle. Masters thesis, University of South Carolina.

Cooke, J. P., and Bancroft, R. W. 1965. *Aerospace Med.* 36: 843-50.

Dodge, C. H., and Wunder, C. C. 1962. *Proc. Iowa Acad. Sci.* 69: 594-99.

———. 1963. *Nature* 197: 922-23.

Duling, B. 1967*a*. *Amer. J. Physiol.* 213: 467-72.

———. 1967*b*. The effects of four weeks of centrifugation on cardiovascular function in the albino rat. Ph.D. diss., The University of Iowa.

Edwards, B. F., and Gray, S. W. 1956. *J. Cell. Comp. Physiol.* 48: 405-20.

Feller, D. D.; Neville, E. D.; Oyama, J.; and Averkin, E. G. 1965. *Proc. Soc. Exp. Biol. Med.* 19: 522-25.

Galilei, Galileo. 1638. *Discorsi e Dimostrazioni Matematiche intorna â due nuove Scienze*, pp. 158-72 (as translated by H. Crew and A. de Salvio, 1914. New York: Macmillan Co., pp. 118-34).

Gauer, O. H. 1961. In *Gravitational stress in aerospace medicine*, ed. O. H. Gauer and G. D. Zuidema, pp. 43-45. Boston: Little, Brown.

Gaydamakin, N. A., et al. 1969. Translated by D. Koolbeck, TDBRS-3, FTD-HT-23-618-69, Foreign Technol. Div., Wright-Patterson Air Force Base, Ohio, 1970.

Gazenko, O. G.; Antipov, V. V.; Parfenov, G. P.; and Saksonov, P. P. 1969. *Aerospace Med.* 40: 1244-47.

Gazenko, O. G., and Gurjian, A. A. 1964. *On the biological role of gravity* (a Soviet translation into English of a paper presented at COSPAR symposium, Florence, May 1964). Moscow: Zak, 2144. USSR Acad. Sci.

Gray, S. W., and Edwards, B. F. 1955. *J. Cell. Comp. Physiol.* 46: 97-123.

Gray, S. W., and Webb, R. G. 1950. *Anat. Rec.*, vol. 108, no. 3.

Gyurdzhian, A. A. 1966. Combined action of acceleration and hypoxia. Presented at 17th Congress of the International Astronautics Federation, Madrid, Spain, October 1966 (abstracted in *ATD Press* 4(98): 9-11, 1966. Washington, D.C.: Library of Congress).

Gyurdzhian, A. A.; Apanasenko, Z. I.; Baranov, V. I.; Kuznetsova, M. A.; and Radkevich, L. A. 1964. *Izdvo Nauka, Moscow* pp. 48-59.

Hertwig, O. 1899. *Arch. Mikrosk. Anat.* 53: 415-44.

Kelly, C. F., and Smith, A. H. 1964. Chronic acceleration research unit (final report for NSF Grant G-2251). University of Calif. (Davis).

Kelly, C. F.; Smith, A. H.; and Winget, C. M. 1960. *J. Appl. Physiol.* 15: 753-57.

Lange, K. O., and Broderson, A. B. 1965. In *Institute of Environ. Sci. Ann. Tech. Meeting Proc.*, pp. 497-509.

Lindberg, E. F., and Wood, E. H. 1963. In *Physiology of man in space*, ed. J. H. U. Brown, pp. 61-111. New York: Academic Press.

Martin, R. C.; Richardson, W. K.; and Martin, W. L. 1966. *J. Eng. Psychol.* 5: 22-24.

Matthews, B. H. C. 1953. *J. Physiol. Lond.* 122: 31P (abstr.).

Montgomery, P. O'B.; Van Orden, F.; and Rosenblum, E. 1963. *Aerospace Med.* 34: 352-54.

Moressi, W. J.; Herrin, W. F.; and Wunder, C. C. 1961. *Proc. Iowa Acad. Sci.* 68: 603-615.

Oyama, J.; Medina, R.; and Platt, W. T. 1966. *Endocrinology* 78: 556-60.

Oyama, J., and Platt, W. T. 1964a. *Amer. J. Physiol.* 207: 411-14.

———. 1964b. *Nature* 203: 766-67.

———. 1965a. *Endocrinology* 76: 203-9.

———. 1965b. *Amer. J. Physiol.* 209: 611.

———. 1967. *Amer. J. Physiol.* 212: 164-66.

Oyama, J.; Platt, W. T.; and Holland, V. B. 1968. *Fed. Proc.* 27: 634.

Redden, D. R. 1970. *Am. J. Physiol.* 218: 310-313.

Rice, J. O.; Wunder, C. C.; and Diecke, F. P. J. 1970. *Proc. Iowa Acad. Sci.* 76: (In Press).

Schmickley, D. L.; Wunder, C. C.; and Loomis, A. 1964. *Abstracts of 8th Annual Biophys. Soc. Meeting, paper T. E. 9.*

Selye, H. 1950. *The physiology and pathology of exposure to stress.* Montreal: Acta.

Shipton, H. W.; Reinhardt, W. E.; and Wunder, C. C. 1960. *Central Association of EEG Proceedings* 7: 755 (abstr.).

Steel, F. L. D. 1962. *Nature* 193: 583.

Vrăbiesco, A.; Cimpeanu, L.; and Domilesco, C. 1964. *Extrait Rev. Francaise Gerontol.*, pp. 254-59.

Vrăbiesco, A.; Costiniu, M.; and Enachesco, G. 1964. *Extrait Actes XV Cong. Int. d'Astronautique, Warsaw,* pp. 33-48.

Vrăbiesco, A., and Domilesco, C. 1965. In *International Conference on Gerontology*, pp. 737-62. Budapest: Hungarian Academy of Sciences.

Vrăbiesco, A., and Enachesco, G. 1964. *Fiziologia Normala Si Patalogica* 10: 271-76.

Winget, C. M., and Smith, A. H. 1962. *J. Appl. Physiol.* 17: 712-18.

Wood, E. H. 1967. *Aerospace Med.* 38: 226-33.

Wunder, C. C. 1955. *Proc. Soc. Exp. Biol. Med.* 89: 544-46.

———. 1960. *Proc. Iowa Acad. Sci.* 67: 488-94.

———. 1961. *Proc. Iowa Acad. Sci.* 68: 616-24.

———. 1962. *Aerospace Med.* 33: 866-70.

———. 1963a. *Abstracts of 7th Annual Meeting of the Biophysics Society*, paper W.E. 6.

———. 1963b. In *McGraw-Hill Yearbook of Science and Technology*, pp. 292-94.

———. 1964. *Survey of chronic weightlessness simulation in biological research.* U.S. Air Force: HQARSC-TDR-64-1.

———. 1965a. In *Methods of animal experimentation*, ed. W. I. Gay, 2: 371-449. New York: Academic Press.

———. 1965b. *Proc. Inst. Environ. Sci.* (Annual Tech. Meeting), pp. 593-602.

———. 1966. *Life into space: An introduction to space biology.* Philadelphia: F. A. Davis.

Wunder, C. C., and Bird, J. W. C. 1969a. *Life Sci.* 8: 707-12.

———. 1969b. *Proc. Iowa Acad. Sci.* 76: 368-75.

———. 1969c. *Proc. Iowa Acad. Sci.* 76: 376-83.

Wunder, C. C.; Briney, S. R.; Kral, M.; and Skaugstad, C. 1960. *Nature* 188: 151-52.

Wunder, C. C.; Crawford, C. R.; and Herrin, W. F. 1960. *Proc. Soc. Exp. Biol. Med.* 104: 749-51.

Wunder, C. C.; Dodge, C. H.; and Duttweiler, C. G. 1962. *Amer. Zool.*, vol. 2, paper 120.

Wunder, C. C.; Dodge, C. H.; Eberly, L.; and Cogswell, S., Jr. 1965. *Abstracts of 9th Annual Biophys. Soc. Meeting*, p. 23.

Wunder, C. C.; Duling, B.; and Bengele, H. 1968a. In *Hypodynamics and hypogravics: The physiology of inactivity and weightlessness*, ed. Michael McCally, pp. 1-69. New York: Academic Press.

———. 1968b. In *Hypodynamics and hypogravics: The physiology of inactivity and weightlessness*, ed. Michael McCalley, pp. 71-108. New York: Academic Press.

Wunder, C. C.; Herrin, W. F.; and Cogswell, S. 1959. *Proc. Natl. Biophys. Conf.*, Columbus, Ohio, 1957, pp. 639-46.

Wunder, C. C.; Herrin, W. F.; and Crawford, C. R. 1959. *Growth* 23: 349-57.

Wunder, C. C., and Lutherer, L. O. 1964. *Int. Rev. Gen. Exp. Zool.* 1: 333-416.

Wunder, C. C.; Lutherer, L. O.; and Dodge, C. H. 1963. *Aerospace Med.* 34: 5-11.

Wunder, C. C.; Meyer, F. N.; and Mason, M. E. 1970. *Physiologist*. In press.

Wunder, C. C.; Milojevic, B.; and Eberly, L. 1966. *Nature* 210: 177-79.

Discussion

SPARROW: What do you do about noise and vibration? Isn't there quite a bit of both, particularly at the higher accelerations?

WUNDER: Not if the centrifuge is carefully balanced. If there were a major effect of vibration it would be detectable in the inner ear. Experimental differences are not affected by destruction of the inner ear (Wunder, Milojevic, and Eberly 1966). The growth curves of the labyrinthectomized animals are essentially identical to those of the normal ones.

BROWN: What about the experimenter's ear? Can't he hear a sonic vibration?

WUNDER: Oh, there is noise from the motor and drive apparatus, but the control animals are in the same room. We put the control cages as close to the centrifuge as we can, equalizing for illumination, evaporation, and temperature.

WEIS-FOGH: You may occasionally be subjecting your preparation to other effects owing to the bulk elasticity of water. You are actually doing an experiment which has some of the effects of high pressure. Have you calculated what these pressures amount to?

WUNDER: Roughly, I think it would correspond to a few feet of water. And to get the pressure effects you are talking about you would have to go up to about 100 to 1,000 atmospheres (See Wunder 1966, pp. 103-4). The pressure effects might be important when very high accelerations are used (Montgomery, Van Orden, and Rosenblum 1963).

JOHNSON: Are you familiar with the work of Megel? He studied the effects of vibration and acceleration in the rat and found that in combination their effects were additive. There was an increase in core temperature and the rat would die. Have you made any such measurements?

WUNDER: Oyama, Platt and Holland (1968) have made some measurements of core temperatures of rats, as have we (Duling 1967*b*). During the first few days of exposure there is a drop in core temperature. Now Vrăbiesco, Cimpeanu, and Domilesco (1964) and Vrăbiesco, Costiniu, and Enachesco (1964) have done this after a year or so of exposure; they claim there is a general increase in metabolism, and that the core temperature rises slightly, but it doesn't happen right away.

WILKINS: The animals are being treated for 6 or 8 mo. Does this machine ever stop?

WUNDER: We stop for 15 min a day to service the machine and animals.

WILKINS: Could I just make one point about stopping the centrifuge for 15 min every day? I realize why this was necessary in these experiments, but I think one must be very careful how one interprets experiments in which you are, in fact, giving a 24-hr cyclical signal of centrifugal force. It is possible that a pulse of 1 g against a background of 2 g has effects on the circadian oscillating systems in living organisms, and that this might give rise to significant effects in your experimental animals.

WUNDER: To avoid some of this difficulty we varied the time of day that we stopped the centrifuge. With the fly larvae exposed to the order of 3,000 g we could detect no difference in the growth constant whether the centrifuge was stopped after 24 or after 48 hr.

WILKINS: What I would like to see done is to stop the centrifuge once a day and see if the animals grow any differently from the ones which have been centrifuged constantly. I think this is rather important.

WESTING: Dr. Edwards, you have reported that strong centrifugal forces, low temperatures, or exposure to X-rays depresses the growth of wheat seedlings. It is curious to me that when these diverse insults were applied in combination their effects seemed to be nonadditive, suggesting that they all act, to some extent, upon the same system. The low temperature experiment of Dr. Wunder's with larvae also, I believe, indicates a similar phenomenon.

EDWARDS: Yes, there is proportionately less reduction with two factors than with either factor alone. This lack of additivity does suggest an effect on a common limiting reaction, a concept that it would be interesting to expand upon experimentally.

SCHÖNE: The papers this morning discussed effects of chronic acceleration on metabolism and growth. I would like to draw attention to the effect of long-lasting acceleration on orientation as well. The mechanisms of orientation have been developed during evolution under the continuous influence of the earth's acceleration of 1 g. The information about the tilt position of the body delivered by the gravity sense organs is based on the calibration of the receptor response in units of tilt. This calibration is valid only for the situation of 1 g. This can be shown in experiments on space orientation in man. If, for instance, the subject is tilted 30° to the right he turns a luminous line and adjusts it at 28° to his left in order to bring it into a position which appears vertical to him. This shows that the information from the gravity receptors about the gravity direction is in fairly good accord with the real gravity direction. If this experiment is performed in a centrifuge under an increased magnitude of gravity, for instance at 2 g, the result is quite different. At 30° of body tilt to the right, the subject tilts the line to 50° to his left. This has not only the consequence of divergence of subjective and true vertical but also implies a loss of space constancy. When tilted from 0° to 30° right, a line standing and fixed in a true vertical position, that is, 30° left from the subject, appears to shift its position. This is because his subjective vertical is now at 50° left from him, and he therefore perceives the line to change its tilt position through 20° from (subjective) toward the right. Thus, if he moves his head from side to side, space-fixed objects seem to sway in the direction of movement, showing that space constancy is lost. Preliminary experiments of long duration (1-2 hr) under increased g, in which optical references for space orientation were offered, gave no indication of a decrease of the difference between subjective and true vertical.

VII "Space" Oriented Studies

34
Simulated Weightlessness Studies by Compensation

S. A. Gordon and J. Shen-Miller, *Argonne National Laboratory*

The discussion after Dr. Freier's presentation made us conscious of the ambiguity of the term *weightlessness* in our title. We had in mind the nonaccelerating vehicle and the absence of differential stress of gravitational origin—conditions under which the accelerative forces are insufficient to produce a detectable georesponse in the organism.

Geomorphism and geotropism in free fall cannot be studied under terrestrial conditions, as the drop fall and the parabolic trajectories of aircraft yield only seconds or perhaps a minute of useful experimental time. These durations are less than the stimulation times required even at 1 g by most plant organs (Audus 1962), aside from the confounding effect of the accelerations preceding and following the terrestrial free fall. Hence, any terrestrial environment that would simulate an extended absence of geostimulus could provide potentially useful bases for extrapolation to the free-fall state, and also contribute to our understanding of the mechanics of gravity perception.

As Dr. Brown pointed out in his Introduction, the botanist has long had a device for such simulation—the clinostat. It is based on the stimulation-time requirement of plants to which we have just alluded: though plant organs can perceive minute changes in their orientation with respect to earth, they require relatively long stimulation or "presentation" times for detectable reaction to result. Thus, geotropic response is a consequence of not only the direction but also of the duration of the acceleration. One can, therefore, grow the organism in an environment where the direction of the gravitational field constantly changes *before* the organism is directionally stimulated. The clinostat provides such continuous reorientation (Larsen 1962). In the simplest form of the device the plant is turned about a single horizontal rotational axis (fig. 1). By such reorientation the polarizing effect of the directional component of the field can be nullified or "compensated"—the field is omnipresent but becomes, biologically, multidirectional. Several restrictions must be imposed on this pattern of motion, as Dr. Brown indicated. The radii of the plant from the center of rotation and the rotational rate must be such that the centrifugal forces generated are not in themselves great enough to be perceived by the plant as a geostimulus. The plant recognizes the principle of physical equivalence and cannot distinguish an acceleration by terrestrial gravitation from an acceleration of the same magnitude by centrifugation (cf. Westing 1964; Shen-Miller, Hinchman, and Gordon 1968). On the other hand, if the rotational rate is very low, geosensitive organs will respond to a continuously effective but shifting transverse stimulation by gravity.

For the past several years we have been examining the properties of plants subjected to continuous reorientation of field direction by clinostat. In this paper we shall discuss briefly two of the areas investigated: the effects of gravity compensation on tropism and the forces required for geotropic response.

Fig. 1. A multiple-unit "single-axis" clinostat.

Gravity Compensation and Tropic Response

Does nullifying the unidirectional effect of gravity alter the reactivity of the seedling to gravity and light when these stimuli are imposed unilaterally? Oat seedlings were germinated and grown in the dark for 3 days on a clinostat, basically similar to that shown in figure 1. The seeds were arranged so that the direction of growth of the organs, particularly the shoot, was along the rotational axis. The rotation rate was 2 rpm. Toward the end of this 3-day growth period the shoot is growing by cell elongation; at this stage of development the seedling consists of a somewhat elliptical shoot, or coleoptile, and several roots. At this point the seedlings were unilaterally stimulated by gravity: rotation was halted for various lengths of time and then resumed. In a similar manner, other seedlings were stimulated phototropically by unilateral irradiation with blue light. The tropic curvatures of the shoots elicited by these unilateral stimuli were then measured, and compared with the curvature of similarly stimulated shoots that had been rotated with their longitudinal axis in normal vertical orientation. As illustrated in table 1, the response to both gravity and light was enhanced when the seedlings were gravity-compensated before tropic stimulation.

Table 1. Effect of Gravity Compensation on the Tropic
Response of Etiolated Oat Coleoptiles Exposed
Unilaterally to Gravity and Light

	Degrees Curvature		
Stimulus	Horizontally Rotated	Vertically Rotated	p
Gravity[1] ($1\,g$, 30 min)	15.4	7.9	$<.01$
Light[2] (first positive)	57.8	51.0	$<.01$

Sources: Dedolph, Naqvi, and Gordon 1965 (gravity); Shen-Miller and Gordon 1967 (light).
[1] Response 60 min after stimulation.
[2] Broad-band blue glass (6.4 K · ergs · cm^{-2}; means of 6 hourly responses).

Tropism is a consequence of differences in growth rate—inequalities of cell expansion—of laterally adjacent tissues along the longitudinal axis of a plant organ. These cellular expansions, though metabolically based, are limited primarily by growth hormone, by the amount of auxin synthesized in the tip of the organ and transported in a morphologically downward direction. A lateral difference in concentration of the hormone occurs upon tropic stimulation, a difference that is classically accepted as the basis of the lateral inequalities of growth rate. Thus, enhancement of tropic response by compensation might be a result of a greater lateral inequality in the amounts of auxin transported basipolarly from the tip, of an enhanced lateral inequality in reactivity to the hormone in the subapical tissues, or of both.

There are experimental evidences that both metabolic and hormonal economies of the plant are altered by gravity compensation. Measurements of the CO_2 evolved from oat seedlings growing in the dark under horizontal and vertical axes of rotation (Dedolph et al. 1966) showed consistently higher specific CO_2 outputs with horizontal rotation, suggesting that respiration is enhanced by compensation. That this rise is coupled to phosphorylation is indicated by the increases in phosphate uptake and esterification in those seedlings (ibid.). Furthermore, coleoptile segments from compensated seedlings show an enhanced absorption of apically-applied auxin (Dedolph, Naqvi, and Gordon 1966), an absorption that is curtailed by anoxia (Naqvi, Dedolph, and Gordon 1965). Collectively, then, these are indices of a heightened metabolic activity accompanying gravity compensation.

Consistent with these metabolic responses is a greater reactivity of compensated tissues to applied auxin. Decapitated coleoptiles to which auxin is applied unilaterally curve more if the seedling has been compensated before application of the hormone (Dedolph, Naqvi, and Gordon 1966). An analogous effect of compensation in enhancing the curvature response of sunflower hypocotyls to unilaterally applied auxin was described by Diehl et al. (1939), and by Brain (1942) for curvatures of lupine hypocotyls induced by a single asymmetric cotyledon and plumule. In these experiments with hypocotyls, however, the magnitude of the effect of compensation is obscured by negative geotropism of the organ, since the vertical, erect controls were not compensated for the confounding effect of gravity after the beginning of the reaction to growth substance. A more direct demonstration of the effect of compensation on reactivity is shown by measurements of the growth rates of intact coleoptiles to whose tips auxin had been symetrically applied (Shen-Miller and Gordon 1967). Both with illuminated and nonilluminated coleoptiles, the growth response to the exogenous hormone was greater in compensated organs than in those vertically rotated. Relatedly, Brauner (1966) has observed that the influence of auxin on the elongation of Helianthus hypocotyl sections is also increased by previous horizontal rotation. As compensation appeared not to affect materially either the amount of auxin supplied by the organ tip or the capacity of subapical tissues to transport auxin, it was concluded that the enhancement of tropic response by compensation was caused primarily by increases in metabolism, and not by modification of hormone availability (Dedolph, Naqvi, and Gordon 1966).

Later, however, a more extensive study demonstrated that compensation does decrease significantly the amount of native auxin translocated from the isolated shoot tip, and decreases slightly the rate at which ^{14}C-labeled auxin is transported basipolarly from the tip in the *intact* organ (Shen-Miller and Gordon 1967). The decrease is compatible with the compensation-induced reductions in growth rate observed for both illuminated and nonilluminated intact coleoptiles at the same stage of ontogeny (ibid.). The lengths of Torenia and tomato stems, as well as the fresh and dry weights of their shoots, are also curtailed by compensation (Lyon 1965). It is compatible also with the enhanced root growths in compensated Vicia (Veen 1964) and Avena (Dedolph, Naqvi, and Gordon 1965), if we assume concentrations of auxin supraphysiological for root growth in the erect plant (Leopold 1955; cf. chap. 13, this volume).

A decrease in auxin level would be consistent also with attributing the enhancement of tropism by compensation to a reduction in rate of recovery or straightening of the curved organ; de Witt (1957) has shown that the rate of reattainment of linear form is positively correlated with its auxin content. A fall in auxin titer would also raise the *ratio* of hormonal concentrations in the tissues proximal to those distal to stimulus direction if the lateral difference, in moles of auxin, remains the same. Analogously, the drop in titer should increase the ratio of growth produced by the lateral gradient of the hormone, since elongation of intact shoots is proportional to the log of auxin concentration (see Shen-Miller and Gordon 1967). In this connection, a comparative study is needed of the amounts of auxin actually "diffusing" from the proximal and distal halves of geotropically stimulated apices from previously compensated organs.

How does compensation reduce the amount of auxin translocated from the tip? It does not change significantly the rates at which auxin can be transported down *subapical* tissues (Dedolph, Naqvi, and Gordon 1966). Thus, there must be either a diminished biosynthesis or an enhanced depletion (inactivation, immobilization, competitive diversion) of the hormone. It is clear that compensated tissues respond more sensitively to applied auxin. It does not follow, however, that this response reflects a greater withdrawal of auxin from the endogenous basipolar flow by the growth process. Though there is evidence that the hormone is inactivated or immobilized in its function as a growth substance (Went and Thimann 1937; Gordon 1957), compensation causes a reduction in the growth rate of the mature coleoptile and the dicot shoot. It could well be, however, that one component of accelerated metabolism in the compensated plant is a correlated catabolism of the hormone, competitive with the growth function. As far as we are aware, there is no information at hand on the direct influence of compensation or, indeed, of any gravitational treatment on the endogenous rate of hormone biosynthesis per se.

In short, we would interpret the enhancements of tropic response by gravity compensation as mediated both by a reduction in auxin level and by an enhanced responsivity to the hormone. Presumably the greater reactivity to auxin of elongating tissues under compensation derives from their higher metabolic rates.

The cause of the enhancements in metabolism by gravity compensation is not certain. Do they derive from an auxin-induced rise in respiration, a continuous activation of the geosensor, or an interaction between the two? On the basis of correlations between the uniformity of starch grain distribution with the extent of compensation and metabolic enhancement, Dedolph et al. (1967) in our laboratory have elaborated upon the concept that the gravitational sensor consists of a group of metabolically active particulates, and that their more uniform distribution within the cell by compensation leads to a steeper solute concentration gradient between the particles and the surrounding cytoplasm. This would reduce potentially limiting diffusion gradients of either substrates or products of particle function. It does not necessarily follow that the starch grains *are* the gravity sensors (see Pickard and Thimann 1966). The starch particle distributions may be looked at simply as a model of the functional system, which could well involve entities of smaller dimension (Audus 1962; Gordon 1963; Pollard 1965 and this volume, chap. 2).

Though the hypothesis is provocative, it rests on the assumption that tropism is limited by such solute gradients. Support of this assumption is required. We should also consider the possibility that statolith motion itself, rather than spatial distribution, may be the determinant not only of the metabolic effects but also of geotropic perception, that is, it may be that the polarization derives from asymmetry in the intracellular distribution of sensor motion rather than position. The influence of compensation on starch grain location in the cell, however, does bear on the enhancement of tropic response, regardless of how the geosensor is activated physiologically. If the uniform distribution of the amyloplasts in compensated cells is an index to the distribution of the geosensor, a unilateral geostimulus will cause stratification in a greater volume of the cell than would be the case in noncompensated organs. Thus an enhanced lateral polarization after a geostimulus would be anticipated in the compensated organ.

We have indicated that compensation-induced enhancements of metabolism per se appear not to be channeled into growth of the shoot: growth either is not materially affected or is diminished. In these various clinostat experiments, intact plants are usually grown with the roots anchored in a solid or semisolid medium and the shoots unsupported. When tilted to the horizontal, the shoot sags, subjecting the stem to reoriented shear, compression and stretch stresses, and to a continuous reorientation of these stresses, as well as torque, as the horizontal axis is rotated under compensation. It would be odd indeed if metabolic changes were not induced by such physical distortions of an organ whose anatomy and physiology are reflections, ontogenetically and phylogenetically, of dissimilar force vectors. The phenomenon of compression wood is in itself evidence of a diverted metabolic flow. And so the respiratory enhancements accompanying compensation are probably strain phenomena to some as yet undefined extent, divorced from the nullification of the tropic stimulus. For this reason we suggest that it would be desirable to examine the influence of full support of aerial organs on parameters of growth and metabolism in gravity-compensated plants.

Finally, the increased response of the compensated shoot to a geotropic or phototropic stimulus does not in itself imply that the sensitivities of the sensors have been altered. Whether they are or not could be readily determined experimentally by examining the intercepts as well as the slopes of dose-response curves. We do suggest, however, that the tropic responses of preformed organs to acceleration and light in the free-fall or "weightless" state would be greater not only because of the absence of a countergeostimulus but also because of the changes anticipated in their levels of auxin and metabolism. We leave unresolved the question of where in the hormone-metabolic interaction the activated geosensor enters the physiology of the geotropic process.

Thresholds for Geoperception

Experiments were and are being designed for extraterrestrial vehicles to determine the direct effects of the free-fall state, and to test the possibility of interaction of that state with other environmental stresses. If the absence of differential stress of gravitational origin is to be tested, then the patterns of motion of the vehicle must be constrained—the accelerations of vehicular motion cannot impose forces of sufficient magnitude and duration that they affect the organism. The motional constraints of the recently flown Biosatellite II were such that the accelerations the test organisms experienced during flight did not exceed about 10^{-5} g. This acceleration limit was based on several studies on the relative forces required to produce a geotropism in normally oriented plants ($\sim 10^{-5}$ to $10^{-2} g$: Chance and Smith 1946; Lyon 1961; Gordon 1963). These studies have been reviewed (Gordon and Shen-Miller 1966) and their inferences questioned largely on the bases of applicability or of apparatus design.

As a consequence, we have reinvestigated (Shen-Miller, Hinchman, and Gordon 1968) the magnitudes of acceleration required to produce observable geogrowth responses in the gravity-compensated oat seedling. Using an apparatus that enables the imposition of a centrifugal field on seedlings that were simultaneously compensated by rotation on a horizontal axis, the effects on growth of forces parallel and transverse to the longitudinal axes of the seedling organs were examined. We will describe briefly only the experiments involving transverse geostimulation.

A sketch showing the principle of the apparatus employed is shown in figure 2. The seeds were planted at the surface of a bed of agar, with their embryos oriented so that the initial direction of

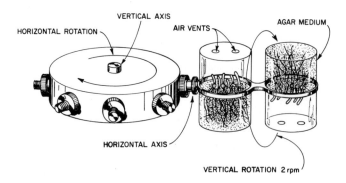

Fig. 2. Clinostat assemblies and paths of motion to impose various centrifugal forces on gravity-compensated seedlings.

the germinating shoot was perpendicular to the horizontal axis of rotation of the growth chamber. During a 70-hr growth period, the seedlings were simultaneously "centrifuged" so that transverse relative centrifugal forces ranging from 10^{-7} to $3\,g$ could be applied simultaneously with compensation. Figure 3 shows the geotropism of compensated shoots elicited by various transverse accelerations. The curvatures of the shoots merge into noise levels of response at relative forces between 10^{-3} and $10^{-2}\,g$. From a probit/log acceleration regression fit of the percentages of the plants that responded (by reorientation in a centrifugal or centripetal direction), the estimated minimal effective acceleration is $1.4 \times 10^{-3}\,g$. It is of interest that the seedlings exposed to "0" g, that is, those compensated without horizontal angular acceleration, are indistinguishable in orientation from those seedlings exposed to relative centrifugal forces less than about $5 \times 10^{-3}\,g$.

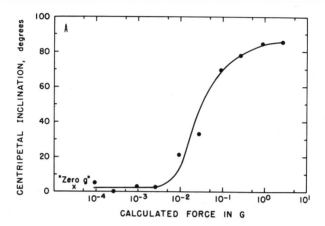

Fig. 3. Curvature responses of shoots of gravity-compensated oat seedlings exposed to various accelerations by centrifugation.

Roots were also examined for the directional distribution of their orientation with reference to the direction of the centrifugal field. They proved to be about an order of magnitude more sensitive than the shoot. Figure 4 shows the percentages of the populations of roots that were inclined in centrifugal and centripetal directions at the termination of the exposures. Between 10^{-4} and $10^{-3}\,g$

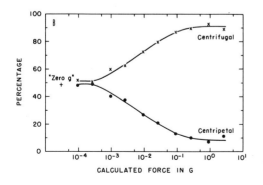

Fig. 4
Distribution of root orientations of gravity-compensated oat seedlings exposed to various accelerations by centrifugation.

was required for significant georesponse. The probit/log acceleration transform yielded an estimated minimal acceleration of 1.3×10^{-4} g. Here again the effect of relative centrifugal forces less than the calculated intercept were indistinguishable from the response to "0" g.

We do not know, as yet, where in the ontogeny of the seedling geosensitivity begins. Nor is there definition of the force-time-response relations of organs compensated throughout their development. However, if we assume that detectable response of the root begins after a transverse acceleration of $\sim 10^{-4}$ g is applied for 3 days, as in the present experiments, the force-time product, or the *stimulus*, is about 25 $g \cdot$sec. Threshold stimuli of similar magnitude may be derived from the observations of Lyon (1961), where a force vector of vibrational origin was applied across Zea roots rotating on a horizontally oriented wire, and from the presentation time data of Larsen (1957), where the roots of Artemesia were challenged by 1 g.

We have used the term threshold in a statistical sense, that is, an acceleration magnitude that can be associated with the appearance, in the populations employed, of growth responses that are significantly different from the unstimulated controls. This association does not mean that activation of the geosensor is a discrete, all-or-none, phenomenon. In this context, the plotting of response versus log relative centrifugal force in figures 3 and 4 may be to some degree distortive since, by nonlinear constriction of the force range, it introduces an apparent x-axis intercept in a function that could move linearly or asymptotically to the origin. Experimentally, however, as the stimulus becomes small, one faces the problems of noise and variability in measuring low-order responses. The present approach does reveal an ordered response, definable by a parametric constant, which approximates a limit of acceleration and duration for the tropism of compensated organs. In this sense, our results indicate that under continuous exposure of compensated oat seedlings for 3 days, accelerations greater than 10^{-4} g for the root and 10^{-3} g for the shoot are required to produce geotropism. Incidentally, we would consider geotropism to be the most sensitive index known of plant response to acceleration.

That the plant can perceive accelerations on the order of one ten-thousandth that of gravity means that any postulate as to the identity of the sensor must be compatible with sensitivities of this order. Furthermore, we may infer that the displacements induced in the geosensor by unidirectional continuous accelerations of such minute magnitude are not nullified by thermal motion or by protoplasmic movements such as streaming or localized sol-gel transitions. Consideration of these inferences tends subjectively to support the idea that the primary geosensing apparatus involves a statolith relatively massive with respect to the medium. As yet, however, we cannot discount the possibility of statolith function of small inclusions or organelles, whose redistribution is physically consistent with the kinetics of georesponse in a 1 g field, and yet whose asymmetries of distribution are sufficiently stable that they can accrete in minute force fields if such fields are applied unidirectionally for extended periods. What this implies is a statistical stability of function and of location within the cell of the geosensor, despite the heaves, surges, and limpid flows of the plasm.

Acknowledgement

We are indebted to the United States National Aeronautics and Space Administration and the Atomic Energy Commission for their support of the work reported above, and to Mr. Ray Hinchman for the photomicrographs in figure 5.

References

Audus, L. 1962. *Symp. Soc. Exptl. Biol.* 16: 197-226.

Brain, E. D. 1942. *New Phytologist* 41: 81-90.

Brauner, L. 1966. *Planta* 69: 299-318.

Chance, H. L., and Smith, J. M. 1946. *Plant Physiol.* 21: 452-58.

Czapek, F. 1895. *Jahrb. Wiss. Botan.* 27: 243-339.

Dedolph, R. R; Naqvi, S. M.; and Gordon, S. A. 1965. *Plant Physiol.* 40: 961-65.

Dedolph, R. R; Wilson, B. R.; Chorney, W.; and Breen, J. J. 1966. *Plant Physiol.* 41: 1520-4.

Dedolph, R. R.; Naqvi, S. M.; and Gordon, S. A. 1966. *Plant Physiol.* 41: 897-902.

Dedolph, R. R.; Oemick, D. A; Wilson, B. R.; and Smith, G. R. 1967. *Plant Physiol.* 42: 1373-83.

Diehl, J. M.; Gorter, C. T.; Iterson, G. van; and Kleinhoonte, A. 1939. *Rec. Trav. Botan. Neerl.* 36: 711-28.

Gordon, S. A. 1957. *Quart. Rev. Biol.* 32: 3-14.

———. 1963. In *Space Biology,* ed. F. A. Gilfillan. Corvallis: Ore. State Univ. Press.

Gordon, S. A., and Shen-Miller, J. 1966. In *Life Science and Space Research IV*, ed. A. H. Brown and M. Florkin, pp. 22-34. Washington, D.C.: Spartan Books.

Larsen, P. 1957. *Physiol. Plantarum* 10: 127-63.

———. 1962. In *Handbook of Plant Physiology*, vol. 17(2), ed. W. Ruhland and E. Bünning, pp. 34-73. Berlin: Springer-Verlag.

Leopold, A. C. 1955. *Auxins and Plant Growth*. Berkeley: Univ. Calif. Press.

Lyon, C. J. 1961. *Science* 133: 194-95.

———. 1965. *Plant Physiol.* 40: 953-61.

Naqvi, S. M; Dedolph, R. R; and Gordon, S. A. 1965. *Plant Physiol.* 40: 966-68.

Pickard, B. G., and Thimann, K. V. 1966. *J. Gen. Physiol.* 49: 1065-86.

Pollard, E. C. 1965. *J. Theoret. Biol.* 8: 113-23.

Shen-Miller, J. 1970. *Planta* 92: 152-63.

Shen-Miller, J., and Gordon, S. A. 1967. *Plant Physiol.* 42: 352-60.

Shen-Miller, J.; Hinchman, R.; and Gordon, S. A. 1968. *Plant Physiol.* 43: 338-44.

Veen, B. W. 1964. *Acta Botan. Neerl.* 13: 91-96.

Went, F., and Thimann, K. V. 1937. *Phytohormones.* New York: Macmillan Co.

Westing, A. H. 1964. *Science* 144: 1342-44.

Witt, J. L. de. 1957. *Acta Botan. Neerl.* 6: 1-45.

Discussion

WESTING: I wonder whether the very low threshold value of geoperception of centrifuged plants that are simultaneously being clinostated is not a possible argument against a statolith hypothesis of geoperception. I should think that if statoliths were involved, they would be tumbling around continuously and not able to initiate the chain of events leading to geotropic response without appreciably higher inertial forces than you demonstrate to be necessary.

GORDON: Implicit in your question is the idea that geoperception requires a contact of the statolith with the cell periphery. But geosensing might simply be a consequence of the spatial distribution of either the *position* or *motion* of the statolith. As to whether the forces from 10^{-4} g are great enough to redistribute suspended particles, remember that 10^{-4} g for 2 days represents a stimulus of 17 g·sec, which is, roughly, the minimum stimulus Larsen found for 1 g with roots of Artemesia.

AUDUS: Is there any relation between the tropic sensitivity and the orientation of the bundles in relation to centrifugal force?

GORDON: I suspect that there is, but don't know so experimentally. In this work the direction of the vector parelleled the major diameter of the elliptical coleoptile, that is, it was parallel to the plane of, and initially perpendicular to, the bundles.

WILKINS: Have you pictures of what exactly is happening to the starch grains? If you put some shot in a jar of treacle and begin rotating it, you will find that the shot tends to keep on sinking to the bottom all the time, but as you go a little bit faster and a little bit faster there eventually will come a speed where they will all be uniformly distributed. I think it would be very valuable to know in this connection what the actual distribution of the starch grains was in the cells of your tissue. If you rotate very fast of course the shot stays located in a lump.

GORDON: We do have a picture of the starch grain distribution in clinostated material. Figure 5 shows longitudinal sections of coleoptiles taken from Avena seedlings in clinostats with horizontal or vertical axes of rotation, and fixed within 15 sec. Notice that compensation produces a relatively uniform distribution of the amyloplast, whereas the coleoptile of shoots in normal orientation shows a packing of the grains in the lower portion of the cell. I believe Dr. Edwards has photomicrographs which show a starch distribution similar to that of the cells of the compensated material for shoots fixed in the free-fall state of the biosatellite.

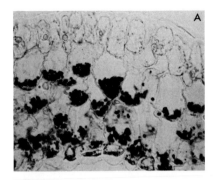
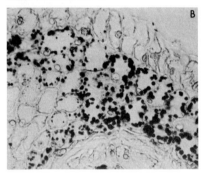

Fig. 5
Longitudinal sections of the tips of Avena coleoptiles, fixed within 15 sec after removal of the tissues from clinostats with horizontal (*A*) or vertical (*B*) axes of rotation. Approx. 150X.

JOHNSSON: Assume that we can apply the so-called reciprocity rule to the experiments just described:

$$f \times t_p = \text{constant},$$

where

$$f = \text{transversely applied force}$$

and

$$t_p = \text{presentation time}.$$

The constant is of the magnitude of 4 $g \cdot$min for Avena coleoptiles (room temperature). The hyperbole

$$f = \frac{4\,g \cdot \text{min}}{t_p}$$

also implies that you can calculate the smallest detectable f during an experiment lasting for t_p min. If we have an exposure time of about 70 hr, as in your experiments, this would mean that the critical force is about $4\,g/70 \times 60$, which is something like $10^{-3}\,g$, exactly what you have gotten experimentally. This also points out that the time during which the force is applied is of great pertinence when we are talking of this "threshold force."

GORDON: I agree. Dr. Shen-Miller is further investigating these force-time interactions to determine if reciprocity holds for a seedling compensated throughout its ontogeny. If it does hold, presumably one should be able to accelerate a satellite up to the range of normal g for brief durations and still retain the plant in the free-fall state with respect to geotropism. [Note: The above mentioned investigation with gravity-compensated seedlings has demonstrated that the force-presentation time product is a constant (Shen-Miller 1970)].

KALDEWEY: Concerning the enhancement of geotropism by a reduction in auxin content, which would lead to a higher ratio between the two sides of the organ—it is of interest that in *Fritillaria meleagris* the negative geotropic curvature occurs in a region of low auxin content, *below* the main zone of elongation.

AUDUS: One point that might be germane to the situation concerns the effects you mention of the compensation upon respiration. Some work I did a very long while ago shows that if you distort plant cells (I was working particularly with leaves and you don't have to distort them very much), you can raise the respiration 2 or 3-fold. This respiratory increase will persist for up to 24 hr or more. Now I think it's a possibility that in your centrifugal field the mechanical stresses on your coleoptile might in fact account for all this change in respiration.

GORDON: I don't know what the effect of centrifugal force is on respiration. And does the distortion of an orthogeotropic organ by reorienting it to the horizontal raise the rate? However, compensation alone, without any additionally imposed centrifugal stress, raises the rate of respiration. So your point is certainly germane to the clinostat. You touch on a question that is relevant to space biology. How far can you push inferences from clinostat rotation to the free-fall state? In the clinostat one nullifies the tropic effect of the directional component of gravity, but the force continues as a distributed vector.

HOWLAND: I am concerned about the effect on the plant of gyroscopic torques generated by your apparatus. You will agree that when you rotate something about two nonparallel axes you may generate gyroscopic torques.

GORDON: I wonder what the magnitudes of these torques are at the rotational rates used. Conceivably they might be sufficient to affect the plant; empirically they are not great enough with respect to their directional vectors to elicit geotropism.

BROWN: Concerning identification of the g threshold, it seems to me that a good definition of this, since you have both intensity and time to worry about, is finding a region on this intensity-time relationship where nonlinearity is encountered for the first time. Now this quick calculation, which was made a short time ago, indicated that you have not yet found such a thing. Is this the reason you put vocal quotation marks around "threshold"? Would you suggest that the term may be inappropriate?

GORDON: Such a rate discontinuity might indeed be a useful parameter of minimal stimuli. The reason I was hesitant is semantic. The term threshold has the connotation of a discrete energy level in the yes-no sense. Because of this we would prefer that another word be used. [Note: A rate discontinuity has been observed in stimulus-response curves for compensated Avena coleoptiles. Using this discontinuity as an index of presentation time, reciprocity still holds (Shen-Miller 1970).]

35
Growth Responses of Plants to Gravity

Charles J. Lyon, *Dartmouth College*

My work with plant growth and geotropism during recent years has been directed toward problems of auxin distribution and transport in plants. My use of the clinostat in terrestrial experiments has been comparable to that of the Argonne group, but I have never felt that I was simulating a condition of weightlessness. I have used the principle of rotating plants on a horizontal clinostat only to eliminate the transport effect of gravity on the redistribution of auxin as the hormone appears in a plant organ. Other possible effects of gravity on growth processes cannot be known until we have obtained the results of growth experiments in spacecraft.

The auxin transport effect of gravity is responsible for the erect axis of a typical terrestrial plant. The same effect operates in lateral organs in conjunction with a special transport process of leaf and branch epinasty to produce the typical plagiotropic organs of a leafy, branched plant. These results are a major feature of the gravimorphism of an erect plant but they also reveal two phases in the mechanism of evolution in the origin of tall, widespread units of aerial vegetation. Without an auxin transport by gravity and an opposing transport by an unknown process, our civilization would be limited to the use of a much smaller amount of solar energy, intercepted by a shallow layer of green plants near the surface of the earth.

Commentary

It is unfortunate that Gordon and Shen-Miller used Avena seedlings as test plants in their ingenious apparatus for estimating the sensitivity of plant organs to acceleration forces smaller than 1 unit of terrestrial gravity. There is also some question about the sensitivity of their method for determining if a force of the order of 10^{-3} or 10^{-4} g produces a measurable growth response in roots immersed in agar. As reported in a study of the orientation of wheat seedling organs in relation to gravity (Lyon and Yokoyama 1966), the roots of Avena seedlings are rather insensitive to gravity.

The argument that certain data for a lower threshold value in the geotropic sensitivity of corn roots to acceleration forces (Lyon 1961) were subject to error through "sag" of the vibrating wires fails to take into account the fundamental requirement for the use of the horizontal clinostat. Only the axis of rotation of the instrument need be horizontal (Lyon 1967) unless the alleged sags in the vibrating wires were so great as to prevent constant rotation of the corn seedlings about the axis of the clinostat. The illustration in the 1961 report also showed that the wires did not sag. The requirement for a maximum acceleration of 10^{-5} g in the NASA Biosatellite was set well below the level of sensitivity reported for the corn seedlings and was easily met during flight. The conservative level removed all doubt of the elimination of a significant acceleration force during the period of free fall. It is relevant that the results from the orbited experiments with wheat seedlings and pepper plants in the NASA Biosatellite II were essentially identical to the results from plants rotated on clinostats at 6 rph as simultaneously run controls. This confirms the reliability of the clinostat as a simulator of free-fall conditions in relation to the growth responses of such plants.

I would concur that the reported increase in respiration and other aspects of metabolism in the Avena seedlings rotated on the clinostat (cf. chap. 34, this volume) may not be the result of rotation in a horizontal position. That increases in these processes were also found in plants rotated about a vertical axis at the same rate of 2 rpm suggests an effect of the relatively high rotational speed. A combination of vibrational and shearing stresses, some of which have long been known to increase respiration, would help to account for the pronounced effect on phototropic curvature response after rapid rotation about either a vertical or a horizontal axis.

The observation of a reduction in basipetal transport of auxin while the Avena coleoptiles were rotating in the horizontal position is in agreement with my results of growing tomato and Torenia plants on clinostats for weeks, starting at a small seedling stage (Lyon 1965). The clinostat plants were much shorter than erect controls with the same number of leaf-bearing nodes. The internodes were shorter by reduction in length of cells.

This evidence in favor of an auxin transport factor, acting by reduction in rate of basipetal movement in the absence of a gravity vector, was further supported by axial curvatures (cf. Lyon 1965) in the plants that grew on clinostats. This characteristic of plants grown from seedlings without an effect of gravity on basipetal transport has since been confirmed with Coleus seedlings which developed strong axial curvatures with the clinostat rotating at the relatively slow rate of 1 rph.

Action of Gravity in Maintaining Erect Axes

It has long been known that a coleoptile or seedling stem makes a 90° growth curvature by negative geotropism if the axis of the young plant is placed in a horizontal position for a suitable length of time. The auxin transport basis for the necessary imbalance in growth, first reported by Dolk (1930), has since been thoroughly confirmed. Unless modified by phototropic curvatures or the effects of such mechanical forces as a prevailing wind, the axis of a growing plant remains erect as though the force of gravity continued to act on the processes which control its orientation.

When I began some years ago to attach potted plants to horizontal clinostats, with the stem more or less horizontal, I found that axial curvatures appeared in the growth zone of the stem within hours (Lyon 1962). This uneven growth of stems had never been reported for clinostat experiments. The absence of a gravity vector along the axis of the immature internodes resulted in an imbalance of growth comparable to that known to occur in petioles and branches during clinostat experiments. The three examples of growth curvatures that develop rapidly in such genera as Torenia, Coleus, and Dahlia are shown here in figure 1.

The direction of an axial curvature can be predicted only if leaves are removed from one side of the growth zone. The more rapid growth on the foliage side suggests a supply of some essential material that is distributed so evenly in an erect stem that curvature does not develop. The same effect can be obtained, however, by removing all leaves and replacing the terminal bud with a cap of 1% indoleacetic acid (IAA) in lanolin. Uneven growth in the immature internodes demonstrates an imbalance in the supply of the auxin to the two halves of the stem, with curvatures as in figure 2. The stems of control plants for such tests remain straight because they receive an even supply of the auxin through the action of gravity in equalizing the basipetal transport that is otherwise unequal in opposing sides of the curved stems.

Fig. 1. Axial curvature and epinasty of leaves and branches in Torenia after 24 hr on clinostat. Erect control plant at left.

Fig. 2. Variable degrees of axial curvatures in defoliated Coleus axes after 24 hr on clinostat. Auxin supply in terminal caps of IAA lanolin paste. (From Lyon 1965.)

Quantitative evidence for this effect of gravity is shown in table 1. The data for the radioactivity found in the convex and concave halves of stem curvatures as illustrated in figure 2 were obtained by counts of radiocarbon introduced with the IAA in the terminal caps of auxin paste. Since each molecule of radioactive auxin (IAA-2-C^{14}) carried a radioactive atom, the distribution of these atoms within extracts from the curved halves of the stems after the curvatures had developed on clinostats (rotated at 10 rph) described the uneven supply of the auxin to the stem tissues. The degree of curvature has been found to vary with the ratio of auxin transported to the convex and concave sides from a uniform supply at the distal end of the stem. The cause of the uneven transport must be internal, but it has not yet been identified.

TABLE 1. Radioactivity from C^{14} in Bisected Axial Curvatures of Coleus

Expt. No.	Hrs. of Expt.	Cpm in 0.02 ml Each of 0.50 ml Extracts[a]		Ratio Convex/Concave
		Convex Side	Concave Side	
1	46	436.2	310.8	1.403
2	41	173.8	58.0	2.997
3	42	80.0	32.6	2.454
4	47	165.4	82.3	2.010
5	47	130.2	64.0	2.034
6	47	91.4	70.6	1.295
7	47	326.4	106.3	3.069
8	47	145.4	98.9	1.470
9	47	43.2	33.7	1.282
Mean				2.002

Source: Lyon 1965.
[a]Separate extracts in chloroform, acid-water, and ethanol.

The action of gravity in equalizing the distribution of auxin within an erect stem is shown even more clearly by experiments in which the radioactive auxin was provided from a patch of 1% IAA-2-C^{14} in lanolin applied to one side of the youngest internode of a defoliated and disbudded Coleus stem. As is illustrated in figure 3, the stem curves away from the side with the auxin paste if the growth takes place on a clinostat but remains straight when the stem is left erect to gravity. Table 2 shows the corresponding unequal distribution of the IAA-2-C^{14} to the opposing sides of the axial curvatures and the equalized transport of the uneven auxin supply to the two sides of the erect, straight stems.

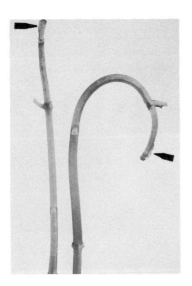

Fig. 3
Lateral supply of IAA in lanolin produces growth curvature on clinostat but not in stem erect to gravity. (From Lyon 1965.)

TABLE 2. Transport of a Lateral Supply of IAA-2-C^{14}

Expt. No.	Hrs. of Growth	Cpm in 0.05 ml Each of 0.50 ml Extracts			
		On Clinostat		Erect to Gravity	
		IAA Side	Opposite Side	IAA Side	Opposite Side
1	18.0	100.2	71.0	107.4	110.0
2	24.0	132.7	96.5	110.2	109.7
3	23.5	106.1	74.3	218.0	208.0
4	24.0	414.1	285.9	387.9	384.2
Mean		188.3	131.9	205.9	203.0

Source: Lyon 1965.

The same equalizing effect of gravity appears when the experiment with a terminal cap of radioactive auxin is modified by applying a patch of 1% triiodobenzoic acid (TIBA) in lanolin to one side of the defoliated Coleus stem, just below the capped end. The result is growth curvature toward the side with the TIBA, a known inhibitor of auxin transport, only when the experimental plants are rotated on a horizontal clinostat. The distribution of the radiocarbon from the transported auxin (see table 3) was found to be very uneven in favor of the unblocked (concave) side of a stem growing on a clinostat but very closely balanced in the erect, straight stem below the level of the TIBA block.

TABLE 3. Transport of IAA-2-C^{14} Past Lateral Block of TIBA

Expt. No.	Hrs. of Growth	Cpm in 0.050 ml Each of 0.50 ml Extracts			
		On Clinostat		Erect to Gravity	
		Free Side	TIBA Side	Free Side	TIBA Side
1	23.0	45.4	33.4	84.9	87.8
2	26.0	17.9	11.5	29.0	28.1
3	23.5	19.8	14.7	61.0	58.7
4	23.0	65.2	42.5	62.8	65.8
Mean		36.4	25.0	61.7	62.4

Source: Lyon 1965.

This action of gravity in keeping the axis essentially straight in a terrestrial plant seems to be the result of the negative geotropism effect working throughout the life of the plant. If an unequal supply of auxin is transported to one side of a stem, either from unequal sources above the growth zone or by differential basipetal transport, the lateral transport of auxin by gravity corrects the effect of slightly more rapid growth on the side with an excess of auxin. This lateral transport can hardly be separated from a basipetal transport effect of gravity, a similarity previously noted by Hertel and Leopold (1963).

Branch Epinasty

Elimination of the effect of gravity on auxin transport when a branched plant is grown on a horizontal clinostat produces an epinastic growth curvature in the immature section of each active branch. A typical form is illustrated in the Torenia plant of figure 1. Until recently the tendency to epinasty could only be described as an inherent tendency that opposed negative geotropism and aided in the orientation of plagiotropic organs.

If the source of endogenous auxin is taken away by removing the leaves and apical bud from a branch, the same epinastic curvature can be provided for by applying IAA in lanolin paste to the end or lower side of the youngest internode. When the branch develops from a bud over a period of a week or more on a clinostat, as illustrated for a Fuchsia plant in figure 4, the curvature is formed only when the leaves are at least one-third of their full size but before they show epinasty in their petioles.

Fig. 4
Leaf and branch epinasty in Fuchsia after branches had developed from buds on a horizontal clinostat.

The physiological basis for the more rapid growth of the upper side of a branch in the absence of a unidirectional gravity vector has been found (Lyon 1963a) to be a lateral transport into the upper side, regardless of the source of the auxin. The results of a series of tests with lanolin pastes of IAA and TIBA is summarized in table 4 and illustrated in part in figure 5. The curvatures in defoliated branches were measured 24 hr after growth on a clinostat in darkness, following the application of the auxin paste to the end of the branch or a side of the youngest internode.

The data of table 4 show how uneven growth produces similar epinastic curvatures with the auxin coming from the lower side (from IAA in lanolin or from a leaf) or from a terminal cap of 1% IAA in lanolin. Much greater curvatures are produced if the auxin is supplied to the upper side, but the transport from this source of auxin is effectively blocked by a patch of TIBA paste on the same side. A ring of TIBA between the auxin supply and the growth zone is quite effective in preventing the IAA from reaching the upper side of the elongating internodes (cf. fig. 5e).

More specific information, however, about the path of movement of the auxin is provided by the data (see also fig. 5f) for the results of applying IAA to the lower side of the branch, with a small patch of TIBA paste also applied to the upper side and slightly nearer the growth zone. The effectiveness of this transport block shows that the auxin is moved almost directly across the branch, with the normal basipetal transport also acting to give a vector direction of about 45° toward the upper side in this test.

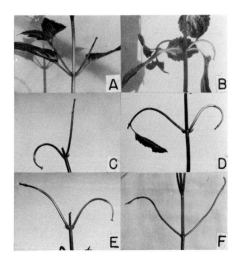

Fig. 5
Epinastic responses of Coleus branches to exogenous auxin after 24 hr on clinostats. (From Lyon 1963a.) a. Branches and leaves on erect stem. b. Leafless branch with cap of 1% IAA in lanolin has same curvature as leafy branch. c. Control branch without IAA and curved branch with cap of 1% IAA paste. d. Curvatures with one leaf vs. terminal cap of 1% IAA paste. e. Two branches with IAA caps. Ring of TIBA (*left*) stops most epinastic curvature. f. Two branches with lower-side patch of IAA. Proximal patch of TIBA on upper side (*right*) blocks epinastic curvature.

TABLE 4. Epinastic Responses of Leafless Branches

Treatment	No. of Tests	Degrees of Curvature after 24 Hr
Control without auxin	20	1.9 ± 0.1
1% IAA in terminal cap	21	68.1 ± 7.5
IAA in patch on lower side	36	90.1 ± 6.8
1 leaf left on upper side	17	66.7 ± 5.3
IAA cap + 2% TIBA ring	19	16.6 ± 4.0
Lower side IAA + proximal TIBA patch on upper side	24	15.8 ± 4.8
Lower side IAA + opposite patch of TIBA	5	33.4 ± 13.7
IAA patch on upper side	10	113.6 ± 4.6
Upper side IAA + upper TIBA	10	6.3 ± 2.4

Source: Lyon 1963a.

The use of radioactive IAA in the lanolin emulsion made it possible to measure the distribution of the auxin molecules within the curved sections of branches after epinasty had controlled growth on a clinostat. The curved portion of each branch was bisected into upper and lower halves and extracted with three solvents, and the content of radiocarbon in each half was computed from the radioactivity as measured in a scintillation counter. The data of table 5 show that the IAA delivered to the upper and lower halves of the immature branch internodes was in the ratio of about 9:5.

This ratio is particularly interesting and is to be expected on theoretical grounds because it provides for the necessary extra supply of auxin to the upper side required to offset the continuous loss to lateral transport by gravity when the plant stands erect to gravity. The epinastic ratio of 9:5 maintains a balanced growth in the branch at its characteristic angle if the distribution ratio for the effect of gravity on auxin is accepted at about 35% upper and 65% lower side. The same epinastic ratio of about 9:5 has also been measured for the distribution of extractable auxin from the branch curvatures (Lyon 1963a). It agrees well with the 5:9 ratio for the auxin distribution in geotropic curvatures.

TABLE 5. Radioactivity in C^{14} Extracts from Bisected Curvatures

Test No.	Tissue Extracted in	Cpm in 0.01 ml Samples of Concentrated Extracts	
		Upper Half	Lower Half
1	Chloroform	1258	590
	Acid-water (not saved)		
	Ethanol	818	497
	Total	2076	1067
2	Chloroform	724	280
	Acid-water	1028	748
	Ethanol	601	419
	Total	2353	1447
3	Chloroform	637	130
	Acid-water	771	454
	Ethanol	325	274
	Total	1733	858
4[a]	Chloroform	943	835
	Acid-water	642	545
	Ethanol	281	211
	Total	1866	1591

Source: Lyon 1963a.
[a]With older branches that showed weak epinasty.

Leaf Epinasty

Epinastic curvatures in leaves are well known to anyone who has grown a leafy plant on a horizontal clinostat. The folded-back positions of the older leaves in figures 1 and 4 are commonly observed after rotation for less than 24 hr. The cause seems to be the delivery to the upper side of the petiole of an excess of auxin comparable to that found in branches (Lyon 1963b).

Epinastic responses of debladed petioles of Coleus to applications of IAA and TIBA in lanolin were tested and measured by the procedures used for Coleus branches. The growth curvature behavior of the petioles after auxin applications on the end or lower side near the end, with blocking effects from TIBA in a ring or on the upper side, was almost identical with that of defoliated branches. Auxin supplied to the lower sides of petioles produced curvatures as rapidly as auxin applied in terminal caps of lanolin paste. The path of movement from the lower epidermis seems to be a diagonal at about 45° with the axis, as in a branch. The nature of this transport process in petioles and the interaction with TIBA blocks is partially illustrated in figure 6.

The pattern of results of auxin transport from leaf blades into the petioles was determined with the aid of radioactive IAA. IAA-2-C^{14} was used in the lanolin emulsion of tests with branches and debladed petioles but was applied to intact leaf blades in a thin layer over part of the lower epidermis, between the principal veins. This isotope work with epinastic curvatures on clinostats rotating at either 1 rph or 6 rph was carried out with fifteen multileaved plants of Coleus, tomato, or poinsettia for each set of measurements.

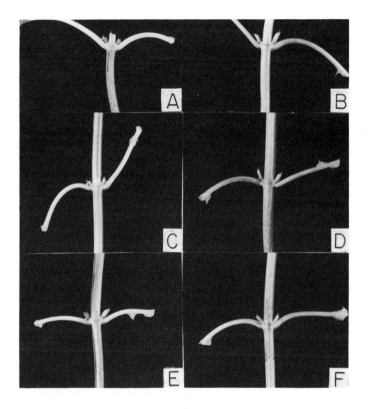

Fig. 6. Epinastic responses of Coleus petioles to 1% IAA in lanolin, with and without blocking effect of 2% TIBA in lanolin. (From Lyon 1963b.) a. Intact leaf vs. cap of IAA paste. b. Control without IAA (left) vs. epinasty with IAA patch on lower side. c. Curvature with cap of IAA (left) is checked by ring of TIBA paste. d. Curvature with lower-side patch of IAA is checked (right) by proximal TIBA on upper side. e. Curvature with lower-side patch of IAA is scarcely affected (right) by proximal TIBA on same side. f. Curvature with lower-side patch of IAA is partially checked (right) by opposite patch of TIBA.

The distribution of the radioactivity as an index of IAA transport was determined from the relative C^{14} content in triple extracts of the upper and lower tissues of the curved petioles that in most cases had received the radioactive auxin for over 40 hr. The data for one set of measurements each for the three species of plants are shown in table 6. The ratio of IAA-C^{14} supplied to the (upper) and concave sides of the petioles varied from 1.22 for poinsettia to 2.20 for nonvascular Coleus slices, but in all cases the ratio was large enough to account for the imbalance of growth that produced the epinastic curvatures. The principle is the same for both leaf and branch epinasty, but the basis for the upward, lateral transport of auxin is unknown for both organs.

TABLE 6. Radioactivity of Petiole Extracts after Epinasty[a]

Petiole Tissue	Extracted in	Cpm from 0.010 ml of Conc. Extract		
		Upper Half	Ratio U/L	Lower Half
Tomato (halves)	Chloroform	136.2		109.9
	Acid-water	147.1		115.7
	Ethanol	70.4		37.3
	Total	353.7	1.35	262.9
Poinsettia (halves)	Chloroform	305.9		173.9
	Acid-water	659.1		480.8
	Ethanol	655.8		613.6
	Total	1620.8	1.28	1268.3
Coleus (slices)	Chloroform	54.7		19.5
	Acid-water	641.8		366.2
	Ethanol	206.9		63.0
	Total	903.4	2.01	448.7

Source: Lyon 1963b.
[a] Representative tests for each species used.

Summary

The horizontal clinostat was used in studies of auxin transport and distribution in relation to the evolutionary mechanisms by which the typical terrestrial plant came to grow tall, with its foliage spread for efficient capture of solar radiation. Clinostat speed was kept slow enough to avoid the introduction of stress and vibrational factors. The reduction of basipetal transport of auxin reported by Shen-Miller and Gordon for coleoptiles is in agreement with the stunting effect on internodal length that appears when plants are made to grow for weeks without the additive effect of gravity on such transport.

Gravity has also been found to equalize the downward movement of auxin in the stems of plants which are held erect by negative geotropism that depends on a lateral, downward transport of auxin. This mechanism, however, would have allowed only oligotropic branches and petioles if an opposing lateral, upward transport of auxin had not evolved in these organs. It became the basis for a growth tendency which opposes negative geotropism and produces epinastic curvatures in the absence of an effective gravitational force.

Acknowledgment

Most of the experimental studies for my contributions to this field were supported entirely by the National Aeronautics and Space Administration.

References

Dolk, H. E. 1930. Geotropie en Groeistof. Diss., Utrecht. Eng. transl. 1936. *Rec. trav. bot. neérland.* 33: 509-85.

Hertel, R., and Leopold, A. C. 1963. *Naturwissenschaften* 50: 695-96.

Lyon, C. J. 1961. *Science* 133: 194-95.

———. 1962. *Science* 137: 432-33.

———. 1963a. *Plant Physiol.* 38: 145-52.

———. 1963b. *Plant Physiol.* 38: 567-74.

———. 1965. *Plant Physiol.* 40: 953-61.

———. 1967. *Plant Physiol.* 42: 875-80.

Lyon, C. J., and Yokoyama, K. 1966. *Plant Physiol.* 41: 1065-73.

Discussion

GALSTON: I believe that in one table you showed that in the erect plant TIBA did not block the downward flow of auxin.

LYON: Yes, because the TIBA was applied to a Coleus stem as a lateral block near the disbudded tip. The IAA was supplied from an apical cap. The tissues were taken for analysis from the older internodes below both applications. With the effect of gravity within an erect stem, the IAA moved down and then in part over to the side of the stem below the TIBA block. The radioactivity in the extracts from the two sides showed that equal amounts of IAA were supplied to the tissues below the level of the TIBA block. This is what I call the equalizing effect of gravity on auxin transport in an erect stem.

GALSTON: So there could actually have been a block which could be circumvented by lateral redistribution below the block.

LYON: I believe so. I don't believe the IAA goes through the block. I think the table of radioactivity count would show a restriction of movement into the region immediately under the block. When the TIBA has been applied on one side of an erect stem, less IAA comes down through that stem than through a control stem and even less comes down through when it is on the clinostat.

LARSEN: Growth responses to gravity are not always eliminated when the plants are rotated parallel to a horizontal clinostat axis. Admittedly, the geotropic stimulations above and below the axis would be expected to balance each other, but if a root has previously received a unilateral stimulation and produces a curvature while being rotated, it is no longer parallel to the axis. If it has curved 45°, it will be sticking up at 135° when exactly above the axis, a position that will subsequently give rise to more curvature than the stimulation induced below the axis (at 45°). As a result we get a straightening out of the root during the rotation, if the rate of rotation is slow enough. If we rotate somewhat faster, this straightening does not develop. The organs do not remain long enough in one position to receive a stimulus, and either a curvature will remain as it is, or if there was a previous geotropic induction, the curvature will become larger.

LYON: I don't deal with previous stimulations. That's where we differ in the use of a clinostat.

LARSEN: But if an orthotropic plant organ is bent at an angle, this will be corrected (straightened out) on a slow clinostat. The meaning of "slow" probably depends on the organ. For roots of Lepidium and Artemesia 2 rph is slow, and 2 rpm is fast enough to prevent a countermovement; so induced curvatures will develop freely. We do not know the critical rotation rate for all organs, but it should be worthwhile to determine the effects of various clinostat speeds for different plants.

36
Effect of Net Zero Gravity on the Circadian Leaf Movements of Pinto Beans

Takashi Hoshizaki, *University of California at Los Angeles*

The mitotic process, as is well known, has a diurnal variation in frequency, and this variation has been associated with the biological rhythm. The mitotic changes reported for horizontal rotated and vertical rotated *Vicia faba* seedlings by Gordon and Buess (1968) indicate that a phase shift occurred when the seedlings were subjected to a simulated weightless environment. I will later present evidence supporting this hypothesis, evidence that strengthens the possibility that a phase shift may occur in the circadian rhythms of an astronaut in orbit.

It was this possibility that focused my interest on the interaction of gravity and biorhythms. I have been studying in particular the effect of simulated weightlessness on the circadian rhythm of plant leaf movements. In Pinto beans, the leaves are horizontal during the day and move down to a vertical position at night. In continuous light and constant temperature, these circadian movements persist for many days. For these reasons the Pinto bean is a useful test organism in biorhythm studies (Hoshizaki and Hamner 1964).

In the course of experiments on the circadian leaf movements of Pinto beans, and in the corollary experiment on the effect of gravitational changes on the leaf movements, I have found that the time at which the rotational treatment was started in relation to the phase of the circadian rhythm has a profound effect on leaf movements. Initiation of the simulated weightless environment during one phase of the rhythm has little effect on the leaf movements of the Pinto beans. If simulated weightlessness is initiated in another phase, the leaf movements cease; they do not resume until two or three days have passed. I believe these results must be taken into consideration in the performance and analysis of any biological experiments that will be performed in space, whether they concern plants, animals, or men. Depending upon the time of the day in which the organism is treated or projected into orbit, the response of the organism can vary greatly. Hence, interpretation of results may be quite misleading if one does not consider the rhythmic component.

The experiments were performed in the following manner. Pinto bean plants were grown in the greenhouse under a long-day photoperiod until the first primary leaves were near complete expansion. The plants were then brought into a growth chamber having continuous light with an intensity of around 750 f.c. and a constant temperature of $80°F \pm 0.5°F$. The apparatus used to simulate a weightless condition for plants has been described previously (Hoshizaki et al. 1966). Briefly, it consists of a rectangular frame which is rotated about a horizontal axis (major axis) at a rate of one revolution per four min. Thus the plants are tumbled end over end once in four min. On this frame are also placed turntables whose axes rotate perpendicular to the major axis. These turntables rotate at the rate of one revolution per min. The experimental plants are placed on these turntables and are rotated about their longitudinal axis. Plants placed on the apparatus will experience two motions—rotation about their longitudinal axis while simultaneously being tumbled end over end. Calculations of the gravity vectors on one leaf have shown that when the plant has made one major revolution, that is, when four min have passed, the net vector forces of gravity are equal to zero. Thus we have called our machine the net zero gravity apparatus or, in short, the Nogravitron.

The control plants were placed on the apparatus so that the control and the experimental plants would be exposed to similar levels of vibration, illumination and temperature. Four plants were used as controls and four as the experimental plants. Only the primary leaves were studied in these experiments; since there are two primary leaves to each plant, eight leaves were used for each set.

Time-lapse photography was used to obtain leaf movement data. When the plants reach a vertical position, a cam on the rectangular frame tripped a switch and a single-frame exposure was made by a 16 mm movie camera. Thus the data on the film was obtained only when the experimental plants were in a vertical position. The data was then read off the developed film and plotted.

The plants, after being placed on the Nogravitron, were permitted to acclimatize to the growth chamber environment for a period of three days before the simulated weightlessness treatments was started. Time-lapse photographic data was also taken during this acclimatization period. A kymograph was attached to one of the control plants to monitor the circadian leaf movements and to determine the exact time when the Nogravitron treatments were to be started. In the latter half of the experiments the plants were given light-dark cycles of 12 hr of light and 12 hr of dark just prior to the Nogravitron treatment. This facilitated the determination of the exact time when the Nogravitron treatment was to be started in relation to the rhythm phase. In the latter half of the experiments the Nogravitron treatment was started at two-hour intervals from the beginning of the last light period. The results of these experiments were similar to the results of the first group of experiments.

Figures 1 and 2 summarize many experiments conducted over a period of three years. In these figures the night position of the leaf is indicated when the curve is up and the day position is indicated when the curve is down. In figure 1, the effect of the initiation of simulated weightlessness during the night phase is presented. Each curve represents the average value of eight leaves. The solid line indicates the experimental plants and the broken line indicates the control plants. As the time of initiation progressively passes through the night phase, a shift, or possibly an interaction of two rhythms, is discernible. A distinct shoulder is seen in the leaf movements of the rotated plants when rotation is started very late in the night phase. On the other hand, when simulated weightlessness is started during the early morning phase, there is a loss of leaf movement for a period of about three days (fig. 2). However, upon the recovery, the movements in both the experimental plants and the stationary control plants are in phase with each other. Initiation of the simulated weightless condition during the noon phase of the rhythm shows a distinct change from the previous treatment: there is no loss of the rhythm but the phasing shifts immediately, so that the phase relationship between the experimental plants and the control plants are 180° out of phase. Later initiation of simulated weightlessness brings about a similar 180° shift between the rotated plants and the stationary control plants.

Effect of Net Zero Gravity on the Circadian Leaf Movements of Pinto Beans 441

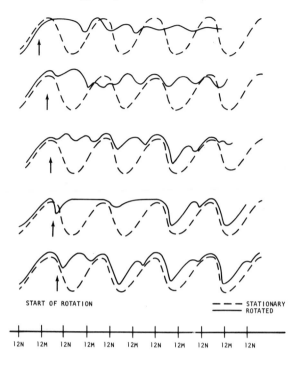

Fig. 1
Leaf movement of Pinto bean as affected by the onset of rotation during the night phase of the circadian rhythm. The curves are the average of eight primary leaves from four plants. The leaf movements of the stationary control plants are depicted by the broken lines. The leaf movements of the rotated plants are depicted by the solid lines. The arrows indicate the onset of simulated weightlessness by rotation treatment. Local time is indicated on the abscissa.

Fig. 2
Leaf movement of Pinto bean as affected by the onset of rotation during the day phase of the circadian rhythm. See figure 1 for details.

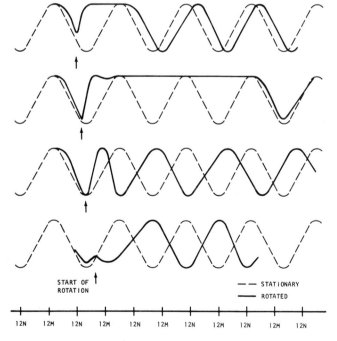

The data presented here indicate that the phase relationships of circadian rhythms, at least in Pinto bean leaf movement, play a critical role in determining the response of the leaf movement to an altered gravity environment. I suggest that analogous responses may occur in organisms that are placed in biosatellites. If data from space experiments are to be critically examined and evaluated, one should do so in the light of these shifts in rhythm at organ, as well as intracellular (Gorden and Buess 1968), levels of organization.

References

Gordon, S. A., and Buess, E. 1968. In *Life Sciences and Space Research*. W. Vishniac and F. Favorite, eds. North Holland Publishing Co., Amsterdam.

Hoshizaki, T., and Hamner, K. C. 1964. *Science* 144: 1240-41.

Hoshizaki, T.; Adey, W. R. and Hamner, K. C. 1966. *Planta* 69: 218-29.

Discussion

BROWN: How were the chambers illuminated? I ask because I recall vaguely some work of Bünnings, with a similar phenomenon, indicating phase alterations that depended very much on the color of illumination. This might be another variable that could lead to some interesting experiments.

HOSHIZAKI: The light came from standard cool-white fluorescent tubes, not supplemented by incandescent bulbs. However, in the present work the light conditions remained constant from the time the plant was placed in the chamber. Bünning did show a wave-length-dependent lengthening and shortening of the rhythm.

WILKINS: The figures were 28 hr in the darkness; in red light they went down to 26 and in the far-red they went up. In the present experiments is it possible that there is a feedback from the endogenous rhythm into a shading effect on the pulvinus?

HOSHIZAKI: None that I know of.

GALSTON: One thing puzzles me. You spoke about effects on mitosis; yet the pulvinar movement that you are observing has to do, I believe, only with turgor changes in preexisting cells, with no mitosis involved. Could you clarify what you meant?

HOSHIZAKI: When I ran into this gravitational effect of a shift in the rhythm of leaf orientation, I wondered if there are other processes within the plant, or even in other organisms, that may be affected similarly. When I heard Dr. Gordon describe their observations on the mitotic shift induced by the horizontal clinostat, it indicated to me that at least two separate growth phenomena, apparently operating independently, responded to gravity compensation by phase shifts.

37
The Experiments of Biosatellite II

J. F. Saunders, O. E. Reynolds, *National Aeronautics and Space Administration*
and F. J. deSerres, *Oak Ridge National Laboratory*

Prior to Biosatellite II, the concept of space bioscience in the United States tended to be identified with the technical support of man in space and other man-related problems. The design of the biosatellite and its subsequent development provided biologists with a unique instrument for the study of a fundamental and challenging biological problem—the relationship of gravity to life processes.

In Biosatellite II, various living systems were removed from the normal influence of the earth's gravitational field. Two remarkable features of these experiments were that the scientific exploration of the space environment was inseparable from the experiments carried out on earth, and that the scientific effort was both interdisciplinary and concerted (Saunders, Reynolds, and Smith 1971).

Launched on 7 September 1967 from Cape Kennedy and recovered on 9 September by a U.S. Air Force plane over the Pacific Ocean near Hawaii, Biosatellite II had as its objectives the study of the effects on living systems of the unique factors of the space environment (fig. 1). More specifically, there were 13 experiments designed to study the influence of "weightlessness" on various biological processes, and the interaction of weightlessness and radiation provided by an on-board 85-strontium gamma ray source (Hewitt 1968; 1971). The seven radiation-weightlessness experiments in the forward section (figs. 1 and 2), were carried out with the blue wild flower *Tradescantia* sp (clone 02), the parasitic wasp *Harobracon junglandis*, the mold *Neurospora crassa*, the flour beetle *Tribolium confusum*, larvae and adults of the fruit fly, *Drosophila melanogaster*, and the lysogenic bacteria *Escherichia coli* and *Salmonella typhimurium*. The bacteria packages were located in front of the *Drosophila* containers (fig. 2). The six weightlessness experiments in the aft compartment (figs. 1 and 3) were performed on pepper plants *Capsicum Annuum*, wheat seedlings, *Triticum vulgare*, fertilized eggs of *Rana pipiens*, and multinucleated amoebae, *Pelomyxa carolinensis*. In addition, there were nonirradiated, weightless controls for each experiment in the forward section.

The effects of exposure to all of the other recognized space flight parameters (with the exception of weightlessness) were tested prior to the mission to give background data for each experiment on responses to vibration, linear acceleration, temperature excursion, and other factors. At the time of the actual flight, several sets of earth control experiments were run simultaneously with the orbital period.

Acceleration and vibration levels (fig. 4) and temperature and relative humidity in the spacecraft were maintained under satisfactory control and within limits used in preflight testing (Look 1968).

Unfortunately, it was necessary to terminate the flight about one day early, owing to impending bad weather in the Pacific recovery area and a communications problem with the spacecraft. Therefore, experiments which were designed for a 72-hour flight were in orbit for only 45 hrs. In some experiments, the shortened flight duration decreased the expression of changes brought on by weightlessness.

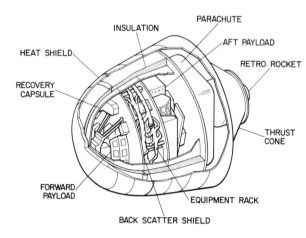

Fig. 1
The Biosatellite II reentry vehicle, showing location of the experiment payloads.

Fig. 2. Experiments of the forward section exposed to 85-strontium gamma radiation during weightlessness.

Fig. 3. Weightlessness experiments of the aft compartment, shielded from the on-board radiation source.

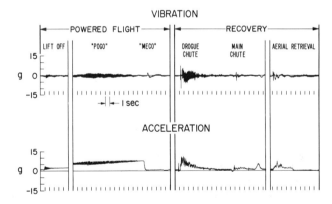

Fig. 4
Biosatellite II vibration and acceleration profiles measured along the longitudinal (thrust) axis during powered flight and recovery.

The experimental results showed that there is interaction between radiation and one or more factors encountered during space flight. This interaction varied between increasing the effects of radiation severalfold, to decreasing the effect slightly but significantly. Whereas several of the experiments had a background of control studies, assuring that vibration alone could not have been the only interacting factor, others did not have sufficient preflight control data. The six experiments for the study of the effects of weightlessness alone showed a close correspondence to results obtained by exposure to rotation on a clinostat (Reynolds and Saunders 1968; Saunders 1968; Thimann 1968).

To ascertain that the effects observed were due actually to weightlessness and not some other condition of space flight, a series of postflight experiments were conducted at the Ames Research Center under conditions simulating, as closely as possible, the environments encountered during flight. The experiments were divided into three phases (Dyer 1969). The objective of the first phase was a comparison of the radiation-scattering characteristics of the satellite capsule and the capsule used as the earth control. Radiation doses were measured within the experimental packages, at the positions occupied by the test organism, with lithium fluoride (LiF) dosimeters. The doses within the packages were compared with those measured at the locations of the LiF dosimeter tubes mounted on each package.

In the second phase, the test organisms were placed in the experiment packages recovered from the flight capsule and in the experiment packages used in the earth control capsule. Simultaneously, the two vehicles and their experiments were subjected to the radiation exposures and the levels of temperature and relative humidity actually observed during flight. The intent of the experiment was to determine whether any of the effects attributed to weightlessness were actually due to a difference between the flight and earth control capsule environment.

The third-phase experiment also included test organisms in both capsules. Both received flight-level doses of radiation. While the recovered biosatellite capsule was exposed to flight acceleration and vibration profiles, the earth control capsule served as a static control, as it did at Cape Kennedy. The intent of this experiment was to determine whether any of the effects attributed to weightlessness might actually derive from vibration or from the brief exposure to high g forces during launch and reentry.

As a result of the three postflight experiments, it was possible to draw the following conclusions. In the larvae of *Drosophila melanogaster* ^{85}Sr irradiation during space flight produced an increase in mortality and sex-linked lethal recessive mutations, as well as chromosome breakage, translocation, and disjunction (Oster 1971). In the adults of *D. melanogaster*, significantly fewer radiation-induced chromosome translocations were found in the pre- and postflight earth control experiments than were found in the flight experiment. An increase in the frequencies of flies with a missing wing and deformed thorax was observed in those flies hatched after irradiation in the satellite. However, there was a lower frequency of loss of genetic markers from the Y chromosome in the irradiated flight specimens than in all earth control specimens (Browning 1971).

The experiment with *Habrobracon junglandis* showed that the fertilizing capacity of sperm was enhanced and the fecundity and hatchability of both transitional and primitive oogonia were increased after irradiation during flight. Space flight alone as well as irradiation during space flight increased the lifespan of females. Insofar as mating behavior was concerned, males from both compartments of the biosatellite were disoriented up to 48 hrs after flight. Hatchability of oocytes in metaphase I and early prophase I was reduced in both unirradiated and irradiated flight specimens (Grosch 1970; von Borstel et al. 1971).

A higher percentage of radiation-induced wing abnormalities was found in *Tribolium confusum* flight specimens than in those irradiated in any of the earth control experiments (Buckhold and Slater 1969). This effect is not due to vibration, temperature excursions, or altered rate of development in space. There is a suggestion that more dominant lethal mutations and a lowering of fertility result from radiation combined with the factor(s) of space flight (Tobias, Buckhold, and Slater 1971).

In the *Tradescantia* experiment weightlessness alone interfered with the mitotic spindle mechanism(s) in microspores, megaspores, and root tip cells. Enhanced interactions between radiation and space flight factors resulted in higher levels of pollen abortion, micronuclei in pollen, and stamen hair stunting, suggesting injury during the more sensitive stages of meiosis and mitosis. Clearly, differences exist between flight and nonflight samples, both with and without irradiation (Marimuthu, Sparrow, and Schairer 1970; Marimuthu, Schairer, and Sparrow 1970; and Sparrow, Schairer, and Marimuthu 1971).

In *Neurospora crassa*, there was no significant effect of weightlessness alone or of weightlessness combined with radiation on the inactivation of spores or the induction of recessive lethal mutations at specific loci in spores supported on millipore filters. This differs considerably from effects observed in Gemini XI wherein spores were flown in suspension (deSerres and Webber 1970; deSerres and Smith 1970; and deSerres 1971).

Growth rates of *Salmonella typhimurium* appeared to be more rapid during space flight than those in earth control experiments. In *S. typhimurium*, free phage induction was consistently lower in the flight populations than in the earth control. The greater bacterial densities are believed to be a function of random cell distribution in the liquid medium under reduced gravity conditions. The *Escherichia coli* experiment was incomplete since the growth phase did not reach maximum owing to the early call down of the satellite (Mattoni et al. 1971).

Exposure of wheat seedlings to free-fall appeared to have no effect on basic growth processes and biochemical reactions that control the rates of meristematic activity (Lyon 1968). However, these wheat seedlings did show an increase in peroxidase activity, an increase that can be correlated with degree of epinasty. Figure 5 illustrates the disorientation of roots and shoots of flight specimens. The diameters of the flight wheat-seedling organs are smaller than those of earth controls grown on the clinostat, with or without vibration. Flight seedlings and vibrated-clinostat seedlings had fewer gross abnormalities than all other earth controls. Fewer cells were in mitosis in flight root tips, with a most marked reduction in the frequency of cells in early prophase. The root tips, both primary and lateral, of the flight seedlings possessed more elongate cells. Their nuclei as well as those of the coleoptile were larger (Conrad 1971; Gray and Edwards 1971; Johnson 1971; and Lyon 1971).

No significant differences were observed in the development of frog eggs (Young 1968; Young et al. 1971) and in the amoeba (Abel, Haack, and Price 1971; Ekberg et al. 1971). Nor were there any significant differences in biochemical and physiological responses of the pepper plants during their exposure to the space environment and the earth control clinostat (Johnson and Tibbitts 1971).

In the Biosatellite II experiments, 65 well-defined genetic endpoints were studied. There is no doubt, however, that some of these endpoints yielded convincing evidence for an interaction between radiation and one or more factors encountered during space flight, presumably weightlessness, especially on processes of mutagenesis. Furthermore, evidence was also obtained that some space flight factor, again presumably weightlessness, induced biological damage or change that was not duplicated in any of the earth-control experiments (including the clinostat controls). It was surprising that many fundamental biochemical and developmental processes appeared to be unaffected by the space flight.

Fig. 5
Disorientation of wheat seedling organs after 45 hrs of flight in Biosatellite II.

The findings considered to be a significant expression of weightlessness alone include the following:

1. Chromosome translocations in *Drosophila* larvae at a stage of cell division where translocations are not normally seen.

2. Death in *Tradescantia* microspores, occurring late in their development and not when undifferentiated.

3. Increased frequency of abnormal nuclei (disturbed spindle) in *Tradescantia* (Marimuthu, Sparrow, and Schairer 1970).

4. A lowered mitotic rate (early prophase), increased cell length, and enlarged nuclear volume in wheat seedling root tip cells.

5. Decreased malformations (absence of specific organs) in the wheat seedlings.

It should be stressed that the Biosatellite II was intended only as a pilot experiment. The effects of weightlessness alone need to be explored more thoroughly, in a larger variety of experimental organisms and especially for longer periods. In this respect the present findings are foundations for the design of experiments for inclusion in future automated and manned spacecraft, orbital laboratories, and interplanetary missions.

Acknowledgment

The contribution of F. J. deSerres was jointly sponsored by the National Aeronautics and Space Administration (Order No. R-104-T8) and the U.S. Atomic Energy Commission under a contract with the Union Carbide Corporation.

References

Abel, J. H., Jr.; Haack, D. W.; and Price, R. W. 1971. In *The Experiments of Biosatellite II*, ed. J. F. Saunders. Washington, D.C.: NASA SP-204, in press.

Browning, L. S. 1971. In *The Experiments of Biosatellite II*, ed. J. F. Saunders. Washington, D.C.: NASA SP-204, in press.

Buckhold, B., and Slater, J. V. 1969. *Rad. Res.* 37: 567-76.

Conrad, H. M. 1971. In *The Experiments of Biosatellite II*, ed. J. F. Saunders. Washington, D.C.: NASA SP-204, in press.

deSerres, F. J., and Webber, B. B. 1970. *Rad. Res.* 43: 452-59.

deSerres, F. J., and Smith, D. B. 1970. *Rad. Res.* 42: 471-87.

deSerres, F. J. 1971. In *The Experiments of Biosatellite II*, ed. J. F. Saunders. Washington, D.C.: NASA SP-204, in press.

Dyer, J. W., ed. 1969. Biosatellite Project Historic Summary Report. Moffett Field, California: NASA Ames Research Center, pp. 191-97.

Ekberg, D. R.; Silver, E. C.; Bushay, J. L.; and Daniels, E. W. 1971. In *The Experiments of Biosatellite II*, ed. J. F. Saunders. Washington, D.C.: NASA SP-204, in press.

Gray, S. W., and Edwards, B. F. 1971. In *The Experiments of Biosatellite II*, ed. J. F. Saunders. Washington, D.C.; NASA SP-204, in press.

Grosch, D. S. 1970. *Mutation Res.* 9: 91-108.

Hewitt, J. E. 1968. *Bioscience* 18: 565-69.

———. 1971. In *The Experiments of Biosatellite II*, ed. J. F. Saunders. Washington, D.C.: NASA SP-204, in press.

Johnson, S. P. 1971. In *The Experiments of Biosatellite II*, ed. J. F. Saunders. Washington, D.C.: NASA SP-204, in press.

Johnson, S. P., and Tibbitts, T. W. 1971. In *The Experiments of Biosatellite II*, ed. J. F. Saunders. Washington D.C.: NASA SP-204, in press.

Look, B. C. 1968. *Bioscience* 18: 560-64.

Lyon, C. J. 1968. *Plant Physiol.* 43: 1002-7.

———. 1971. In *The Experiments of Biosatellite II*, ed. J. F. Saunders. Washington, D.C.: NASA SP-204, in press.

Marimuthu, K. M.; Sparrow, A. H.; and Schairer, L. A. 1970. *Rad. Res.* 42: 105-19.

Mattoni, R. H. T.; Ebersold, W. T.; Eiserling, F. A.; Keller, E. C., Jr; and Romig, W. R. 1971. In *The Experiments of Biosatellite II*, ed. J. F. Saunders. Washington, D.C.: NASA SP-204, in press.

Oster, I. I. 1971. In *The Experiments of Biosatellite II*, ed. J. F. Saunders. Washington, D.C.: NASA SP-204, in press.

Reynolds, O. E., and Saunders, J. F. 1968. *Cur. Mod. Biol.* 2: 147-47.

Saunders, J. F., ed. 1968. *Bioscience* 18: 537-661.

Saunders, J. F.; Reynolds, O. E.; and Smith, G. D. 1971. In *The Experiments of Biosatellite II*, ed. J. F. Saunders. Washington, D.C.: NASA SP-204, in press.

Sparrow, A. H.; Schairer, L. A.; and Marimuthu, K. M. 1971. In *The Experiments of Biosatellite II*, ed. J. F. Saunders. Washington, D.C.: NASA SP-204, in press.

Thimann, K. V. 1968. *Proc. Nat. Acad. Sci.* 60: 347-61.

Tobias, C. A.; Buckhold, B.; and Slater, J. V. 1971. In *The Experiments of Biosatellite II*, ed. J. F. Saunders. Washington, D.C.: NASA SP-204, in press.

von Borstel, R. C.; Smith, R. H.; Whiting, A. R.; and Grosch, D. S. 1971. In *The Experiments of Biosatellite II*, ed. J. F. Saunders. Washington, D.C.: NASA SP-204, in press.

Young, R. S. 1968. *Space Science Rev.* 8: 665-89.

Young, R. S.; Tremor, J. W.; Willoughby, R.; Corbett, R. L.; Souza, K. A.; and Sebesta, P. D. 1971. In *The Experiments of Biosatellite II*, ed. J. F. Saunders. Washington, D.C.: NASA SP-204, in press.

VIII Summations

38
Responses to Gravity in Plants: A Summary

A. W. Galston, *Yale University*

The varied responses of plant organs to the gravitational stimulus can best be understood in terms of the strategy of the higher green plant. In such an analysis, the following three considerations are paramount:

1. The green plant is essentially a mechanism for converting quanta of radiant energy into ATP and other photosynthetic products.

2. Higher plants are anchored to one location; although their parts may move relative to a fixed point and to each other, the entire organism is fixed in one position on the earth's surface.

3. Growth in higher plants, which makes many movements possible, occurs in localized regions called meristems. Meristems producing growth in length occur at ends of stems and roots; meristems producing growth in diameter occur between wood and bark.

The efficient capture of light quanta demands the fabrication of large, flat, thin structures, abundantly supplied with absorbing pigments and so positioned as to intercept maximum numbers of quanta. The plant has responded to this requirement by making diageotropic leaves, raising them above the ground on negatively geotropic stems, orienting them around the stem to insure minimum mutual shading, and endowing both stem and leaf petiole with phototropic sensitivity such that they are constantly moving, through differential growth patterns, into areas of maximum light intensity. The result of all this is the beautiful mosaic of light-absorbing leaves with which we are familiar.

The stationary habit requires a firm anchor for positioning and holding erect the large, negatively geotropic stem. The predominantly positively geotropic root serves this function as well as the auxiliary nutritional role of water and mineral salt absorption from the soil. In both of these functions, the main positively geotropic root is aided by lateral, mainly diageotropic branch roots and root hairs.

Finally, the localization of the main growth centers at apexes of roots and stems means that movements of plant parts must occur through differential growth activity on the two sides of the organ, just behind the apex. This activity may consist in differential rates of cell division or cell elongation or even, as in reaction wood, qualitatively different deposition of secondary cell-wall materials. The effect of such differential growth is the "steering" of the stem apex away from and the root apex toward the center of the earth.

Let us note that the assumption of the different responses to gravity—positive for roots, negative for stems, and diageotropic for branches and leaves—is an aspect of development which is not at all well understood. We do know that the patterns of response arise at specific ontogenetic stages. For example, the early embryonic cells developed as a result of division of the fertilized egg in the embryo sac are unresponsive to gravity; yet their progeny, the young root and shoot, are responsive. Similarly, the zygote of Fucus is at first indifferent to gravity, but after the first cell division the resulting daughter cells are either positively or negatively geotropic. Let us note further that the geotropic characteristics of any cell, tissue, or organ are alterable by natural or man-induced changes in the environment. For convenience, I have summarized some such transformations in table 1. Although the changes are numerous, varied, and long recorded in the literature, I believe it is fair to say that we know virtually nothing about the basis for the change in behavior. What we can say is that all tropistic responses are the result of modified circumnutational patterns; that is, during normal growth of cylindrical plant organs there are temporary, random inequities in the growth rates of the two sides of the organ which are averaged into overall symmetrical growth. Tropistic curvatures result from unequal self-correction of deviations from the vertical under the influence of a physical gradient in the outside world (Darwin 1876).

TABLE 1. Examples of Transformations of Geotropic Behavior

Nature of the Change	System in Which the Change Occurs	Agent Producing the Change
- to 0	Main stem → branch	Natural
- to +	Corn stem → prop root	Natural
+ to 0	Main root → branch root	Natural
+ to -	Convolvulus root → stem	Natural
0 to +	Leaf → root	Natural
0 to -	Rhizome → bud	Natural
0 to -	Pine leader excised → lateral branch responds	Surgery[a]
+ to -	Corn plants sprayed with TIBA → prop roots respond	Chemical[b]
- to 0	Pea stem treated with ethylene	Chemical[c]
0 to +	Aegopodium rhizome	Light[d]

Note: (+) = positively geotropic; (-) = negatively geotropic; 0 = diageotropic.
[a] Westing 1959.
[b] Galston 1957; see especially plate XXV.
[c] Borgström 1939.
[d] Bennet-Clark and Ball 1951.

Components of the Geotropic Response

The gravitational field must be sensed by some organelle or structure of the plant cell. The sensed stimulus must then be transduced into some physical or chemical effect which can be amplified in such a way as to regulate growth and produce the ultimate response to the stimulus. Whether transduction is separate from amplification and whether, if separate, it precedes or follows amplification can at present only be conjectured. If we accept for discussion the sequence: sensor → transducer → amplifier → *growth regulation*, then for the geotropic righting of a prostrate stem, for example, we can hypothesize that the sensor organelles are special starch grains or other statolithic bodies of such density that they fall to the bottom of the cell under a force of 1 g. Somehow this simple event is transduced first into a lateral displacement of auxin to the gravitationally lower side so that the auxin concentration below becomes about twice as great as it is above (Dolk 1936). Later, the organ also develops a transverse electrical potential of about 100 mv, the lower side being electrically positive (Bose 1907). Both the auxin displacement and the transverse electrical potential can result in amplification. Auxin probably controls growth through a process involving altered configuration of a strategically located allosteric macromolecule; since such macromolecules may themselves be catalytic, it is easy to see how the displacement of relatively few auxin molecules could have a major effect on the distribution of metabolic activity and growth. Also, the development of a transverse electrical potential might influence growth by effecting (electrophoretic) redistribution of additional auxin or other growth-regulatory materials. The asymmetrical growth patterns clearly cause the curvature which reorients the growing point vertically. Once attained, the vertical orientation is thereafter maintained, although deviated from somewhat, during normal circumnutation.

Statoliths as Sensors

The evidence that special dense falling bodies, including special starch grains, serve as gravity receptors is overwhelming (Audus 1962). Such bodies, and their movements during organ displacement, have been abundantly documented. We even have two instances discussed in this volume (chaps. 5, 13) in which the removal of the statoliths by centrifugation or surgery removes sensitivity to gravity without markedly affecting growth rate. Unfortunately, neither of these systems, dealing as they do with rhizoids and roots respectively, can be easily rationalized with the auxin regulation of growth.

The massive evidence favoring statoliths of the approximate size of starch grains (Audus 1962; chap. 13, this volume) cannot be considered vitiated either by the claim (Pickard and Thimann 1966) that geotropic sensitivity of coleoptiles remains after the complete depletion of starch grains or by the reminder (chap. 9, this volume) that several taxa of georesponsive plants lack conventional starch. Let us remember that the Chara rhizoid statoliths not only are *not* starch grains, but apparently function by upward displacement of wall-synthesizing Golgi vesicles. Thus, even if starch grains were *completely* absent in the experiments of Pickard and Thimann (a claim virtually impossible to prove) either the plastids in which they are normally formed or other bodies of similar density could act as organelles of geoperception. This question has been adequately discussed elsewhere in this volume, and no useful purpose will be served by adding to that discussion here. In my view, the statolith theory is still the most convincing one available.

Transduction and Amplification

It is clear that sometime after the statolith has fallen to the bottom of cells in certain cylindrical plant organs, a transverse electrical potential of perhaps 100 mv is set up. This used to be considered as the physical basis for the redistribution of auxin, but the work of Grahm and Hertz (1964), Wilkins and Woodcock (1965), and others proves that transverse potentials do not develop in the absence of auxin, and probably not until *after* auxin redistribution occurs. Thus, the basic problem in understanding transduction is to explain how falling statoliths in the many individual cells across an organ can cause an asymmetry in auxin distribution across the organ *as a unit*. We do not have any good leads to this problem, and again, I can do little but call attention to the summaries presented earlier in this volume. It has been claimed (Audus and Lahiri 1961) that auxin redistribution by itself is insufficient to account for the observed differences in growth rate on the two sides of the stimulated organ. The asymmetric distribution of other substances, such as growth inhibitors, has thus been invoked as a possible additional mechanism contributing to geotropic curvature. Although this certainly must be kept in mind, a recent analysis of the techniques involved in analyzing transverse gradients (Burg and Burg 1967) indicates that the concentration difference of auxin at the peripheral upper and lower layers *is* by itself large enough to account for curvature. Inclusion in older analyses of the more median tissue, in which neither auxin redistribution nor growth difference is very marked, serves only to mask the large gradients which exist across the organ.

With respect to amplification, the development of the transverse potential is a possible agent which could account for *part* of the asymmetric distribution of auxin and other molecules capable of electrophoretic displacement. The amplification may also result from the oligodyamic action of auxin, which probably exerts its regulatory effect either through control of the synthesis of proteins which are themselves catalytically active or through the activation of already existing catalytic proteins. Growth inhibitory components, possibly also oligodynamic, may also function. Once again, speculation as to the exact molecular events would be easy, but probably redundant and not too instructive. So, in making a general statement on the physiology of geotropism, we are limited to saying that (*a*) some mechanical georeceptor (statolith) is displaced during stimulation. This results (*b*) in an asymmetric distribution of auxin and possibly other substances. This is followed (*c*) by the development of a transverse electrical potential and ultimately (*d*) by asymmetrical growth on the two sides of the organ and (*e*) curvature. There is abundant opportunity and obvious need for further work in this field.

Lessons from Tendrils

Frequently, answers to problems in biology came from unexpected sources. For example, the analysis of nyctinastic movements in leaves has recently provided information which may possibly explain the mode of action of phytochrome in controlling plant growth and development (Fondéville, Borthwick, and Hendricks 1966) and the analysis of the control of cell division of tobacco pith cells in vitro has provided a plausible explanation of the senescence of leaves (Richmond and Lang 1957). So perhaps I may be excused for introducing into this discussion some results obtained in an analysis of the movement of pea tendrils in response to mechanical stimuli (Jaffe and Galston 1966*a, b*; 1967*a, b*; 1968*a*). Indeed, this apparent digression is really directly on the subject, since tendrils do furnish the plant with an economical means of insuring negative geotropic orientation of massive and cumbersome stems which otherwise could remain erect only at the expense of producing large numbers of vertically oriented, heavily lignified cells such as are found in the secondary xylem of trees.

The tendrils of the leaves of light-grown Alaska pea plants vary with the node at which the leaf is found (fig. 1). The long, unbranched tendril at node 5 is most adaptable for experimental study. Under our conditions (23°C, continuous illumination with about 3000 ft-candles daylight fluorescent tubes supplemented with incandescent bulbs, 70% relative humidity), such tendrils are completely elongated and maximally sensitive between 10 and 13 days after planting of the dry seeds (fig. 2). When stimulated with 5 gentle strokes with a glass rod on the ventral (concave) surface just below the hook, such tendrils curve rapidly, showing a visible response within 2 min, maximal response at 32 min, and significant uncoiling by 64 min after stimulation (fig. 3). Excised tendrils respond somewhat more slowly, but the ultimate magnitude of their response is just as great as tendrils in situ (fig. 4). The dorsal surface is less sensitive than the ventral surface, and if rubbed simultaneously with the ventral surface, inhibits coiling. If the dorsal surface is rubbed 10 min after the ventral surface, no inhibition of coiling is noted (fig. 5). Interruption of the vascular bundle on the dorsal side by surgery greatly diminishes coiling, whereas similar treatment of the ventral surface produces little effect (fig. 6). The temperature optimum for the coiling response of shaken excised tendrils is between 20° and 30°C, depending on the time of the reaction (fig. 7). The pH optimum for the growth of excised tendrils is about 6.5, which is precisely the minimum point for curvature (fig. 8). Auxins such as indole-3-acetic acid and 2,4-dichloro-phenoxyacetic acid facilitate curvature of shaken excised tendrils (fig. 9) and have recently been shown to substitute completely for mechanical stimulation (Reinhold 1967; fig. 10). The curvature is due initially to ventral shrinkage and is later emphasized by differential dorsiventral elongation (table 2). Certain cells occurring predominantly about 3 files below the ventral epidermis are possible thigmoreceptive loci, since their contents appear to coalesce at one end of the cell during fixation, especially after mechanical stimulation (Jaffe and Galston 1966a).

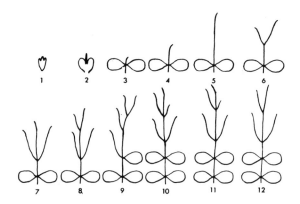

Fig. 1
Appearance of leaves and tendrils arising from the different nodes of light-grown Alaska pea plants.

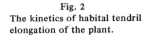

Fig. 2
The kinetics of habital tendril elongation of the plant.

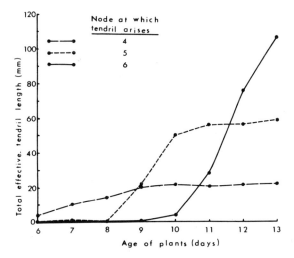

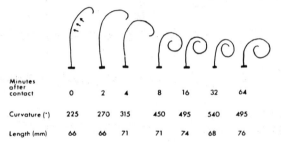

Minutes after contact	0	2	4	8	16	32	64
Curvature (°)	225	270	315	450	495	540	495
Length (mm)	66	66	71	71	74	68	76

Fig. 3
Appearance of a tendril at various times after rubbing. Arrows indicate position of stimulation by 5 strokes with a glass rod. The curvature is estimated to 45° and the length measured with a calibrated planimeter.

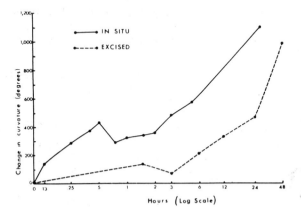

Fig. 4
Kinetics of curvature of excised and in situ tendrils. Each point is the mean of 3 replications of 10 tendrils each.

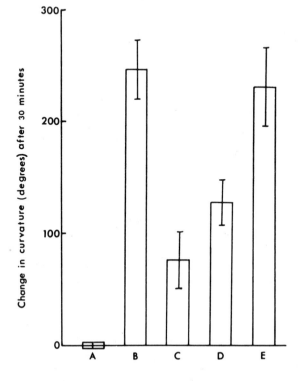

Fig. 5
Thigmotropic sensitivity of dorsal and ventral surfaces. Vertical lines indicate the magnitude of the standard error of the mean: a) unrubbed control; b) rubbed on ventral side only; c) rubbed on dorsal side only; d) dorsal and ventral sides rubbed simultaneously; e) rubbed on dorsal side 10 min after rubbing on ventral side.

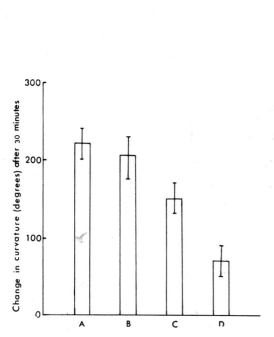

Fig. 6. The effect of dorsal or ventral notching or both on curvature during coiling of tendrils in situ: *a)* unnotched control; *b)* ventral side notched; *c)* dorsal side notched; *d)* both sides notched.

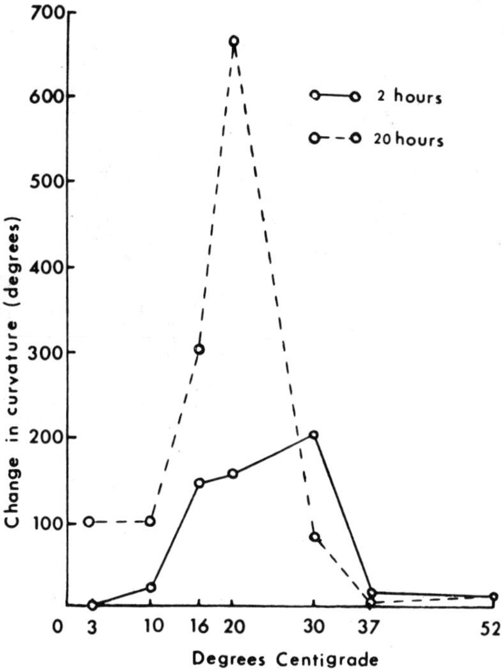

Fig. 7. The effect of temperature on curvature during coiling of excised tendrils.

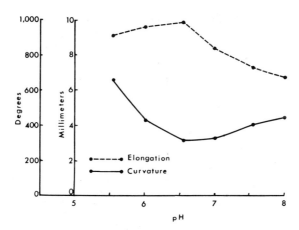

Fig. 8. The effect of pH on curvature and elongation of excised tendrils after 20 hr.

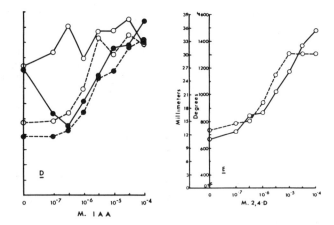

Fig. 9. Dosage responses of coiling and curvature to selected growth substances incubated for 20 hr in the light and in the dark. *Solid line* indicates curvature; *dotted line*, elongation; *open circle*, light; *solid circle*, dark.

Fig. 10. Curvature of unstimulated excised pea tendrils in response to applied IAA. *Top row, left to right:* 0, 10^{-7}, and 10^{-6}M IAA; *Bottom row, left to right:* 10^{-5}, 10^{-4}, and 3×10^{-4}M IAA.

TABLE 2. Changes in Length of Dorsal and Ventral Sides of Coiling Tendrils

Measurement	Period		
	0-1/2 hr	0-2 hr	1/2-2 hr
Change in total curvature of tendril (degrees)	176 ± 24	324 ± 42	148 ± 51
Change in length (mm) of: Ventral side	-0.12 ± 0.10	0.17 ± 0.04	0.29 ± 0.13
Dorsal side	0.18 ± 0.15	0.57 ± 0.08	0.39 ± 0.17
Difference (mm) between ventral and dorsal sides	0.30 ± 0.12	0.40 ± 0.09	0.10 ± 0.07

Note: Each datum is followed by its standard error.

If plants are kept in the darkness for 16 hr or more, their tendrils react less than if they are kept in continuous light. If the plants are returned to the light for several hours after the dark period, they curve even more than continuously illuminated tendrils (fig. 11). This probably represents a physiological adaptation to the normal alternation of light and dark periods in nature. Sucrose fed to excised darkened tendrils only partially replaces light, but high concentrations of ATP or related nucleoside triphosphates partially substitute for light (figs. 12, 13). The ATP content of tendrils decreases after stimulation (table 3), and there is a concomitant, though not stoichiometric, rise in inorganic phosphate (table 4). The ATPase activity of the tendril declines during coiling and rises again during recovery (fig. 14). A cell-free ATPase prepared from tendrils shows a decreased viscosity (i.e., it contracts) after the addition of ATP (fig. 15). Tendrils also have high concentrations of a particular flavonoid, quercetin triglucosyl p-coumarate (QGC), which declines markedly during coiling (table 5). QGC inhibits coiling when applied to excised tendrils in vitro (table 6); so its breakdown seems to be required if coiling is to occur.

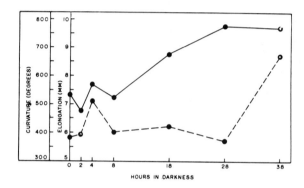

Fig. 11
Effect of duration of dark treatment on curvature (*solid line*) and elongation (*dotted line*) of tendrils transferred to the light and incubated there for 2 hr.

Fig. 12
Interactions of 0.1 M sucrose with mM adenosine (A), AMP, ADP, and ATP in coiling of tendrils incubated for 2 hr. Curvature and elongation in the ordinates are expressed as percentage change from the control. *Open bars* indicate incubation in the light; *black bars*, incubation in the dark. The mean tendril measurements at excision were 58.9 mm ± a standard error of 0.7 mm, and 74° ± a standard error of 4°. All data are given relative to the dark controls lacking sucrose. The measurements of these controls were 1.6 mm and 414°.

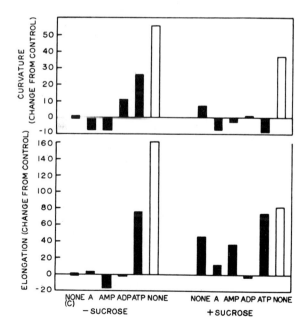

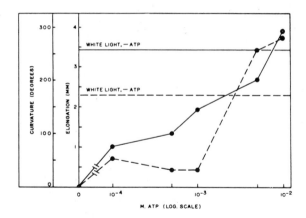

Fig. 13. Elongation (*dotted line*) and curvature (*solid line*) response of tendrils to various concentrations of ATP. Each point on the curves is given relative to the control, which contained no ATP. The net changes in these controls were 2.6 mm and 291°. Except for the light controls, incubation was for 2 hr in the dark. The mean tendril measurements at excision were 50.3 mm ± 1.4 mm and 86° ± 6°.

TABLE 3. Effect of Darkness and Coiling on ATP Levels in Tendrils

Treatment	Mean Degrees Curvature	mμmoles ATP per g fr wt
Overnight in darkness, unstimulated	69 ± 6	480 ± 134
Kept in light, unstimulated	74 ± 2	690 ± 168
Kept in light, stimulated and contracted[a]	350 ± 6	170 ± 124

Note: Each datum is followed by its standard error.
[a]Stimulated and allowed to coil in situ for 30 min.

TABLE 4. Effect of Coiling on Inorganic Phosphate (P_i) Levels in Tendrils Incubated in the Light

Treatment	Mean Degrees Curvature	mμmoles P_i per g fr wt
Unstimulated tendrils	66 ± 20	320 ± 17
Contracted for 10 min[a]	191 ± 2	360 ± 59
Contracted for 20 min[a]	267 ± 6	510 ± 59
Contracted for 30 min[a]	342 ± 8	440 ± 69

Note: Each datum is followed by its standard error.
[a]Stimulated and allowed to coil in situ.

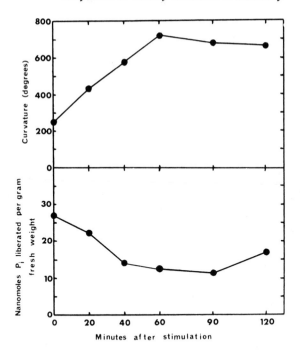

Fig. 14
Correlation of kinetics of coiling (*above*) with the kinetics of decrease in ATPase activity (*below*). For the ATPase assay, samples were extracted from coiling tendrils at various times after stimulation, the reaction mixtures incubated for 30 min, and the P_i measured at 650 nm using the phosphomolybdate chromogenic reagent. The correlation coefficient between the 2 sets of data was 0.935 and was significant at the 1% level.

Fig. 15
The effect of a final concentration of 1.18 mM ATP on the specific viscosity of the tendril extract.

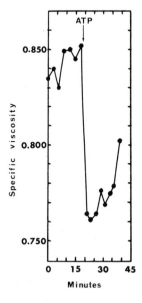

TABLE 5. Correlation of Kinetics of Coiling with Decrease in QGC in Tendrils Allowed to Coil In Situ in Continuous Light

Min after Stimulation	Curvature (degrees)	Total Flavonoids (μmoles/g fr wt)
0	248	2.32
20	429	2.00
40	573	1.43
60	709	1.41
90	677	0.94
120	663	0.65

Note: Correlation coefficient = -0.874 (significant at the 5% level).

TABLE 6. The Effect of Aqueous Extracts of Tendrils and of QGC on the Coiling of Excised Tendrils in the Light

Addendum	Net Degrees Curvature in 20 Hr
None	852 ± 156[a]
Aqueous extract of unstimulated tendrils	548 ± 57
Aqueous extract of coiled tendrils	674 ± 144
None	720 ± 60
QGC, 10 μM	488 ± 100

Note: The aqueous extracts were made from excised tendrils which were either unstimulated (ca. 75° curvature) or shaken and allowed to coil for 2 hr (ca. 522° curvature).
[a]Standard error.

The cut bases of stimulated excised tendrils excrete more electrolytes and label from previously absorbed ^{14}C-sucrose and acetate than do tendrils at rest (table 7). The major excreted cation is H^+, and its loss is potentiated by preincubation with such substances as benzoate, whose loss is also facilitated by stimulation (table 8). Benzoate pretreatment also enhances coiling in response to stimulation. Tendrils supplied with tritiated water and then stimulated lose more label than unstimulated tendrils (table 9); most of the label comes from the ventral side, which contracts during coiling (table 10). The label does not pass from ventral side to dorsal side.

TABLE 7. Release of Electrolytes, ^{14}C-Sucrose and ^{14}C-Na Acetate into the Bathing Solution from Floating Excised Tendrils

Curvature (degrees)		Measurement	Unstimulated	Coiled
Unstimulated	Coiled			
39	113	14C-Na acetate (cpm per g fr wt) n = 2,[a] 30 min incubation	917 ± 141[b]	1,187 ± 96
25	113	^{14}C-Sucrose (cpm per g fr wt) n = 5, 30 min incubation	282 ± 95	445 ± 84
75	366	Electrolytes (specific conductance, 23 micromhos) n = 6, 2 hr incubation	12.0 ± 0.0	15.4 ± 1.3

Note: The tendrils were in 7 to 10 ml of buffer or distilled water and were either unstimulated or shaken to induce coiling. Immature tendrils, incapable of coiling, showed no such enhancement of loss by shaking, indicating that the effect is not due to accelerated removal of diffusing solutes.
[a]n is the number of times the experiment was replicated.
[b]Standard error.

TABLE 8. The presence of Electrolytes, Cations, and Substances Absorbing at 226 mµ in the Effluent of Coiled Tendrils or Tendrils at Rest for 15 min

Experiment	Measurement	Tendrils unstimulated	Tendrils coiled
1	Curvature (degrees)	5 ± 7[a]	258 ± 13
	Fr wt per tendril (mg)	9.1 ± 0.7	8.6 ± 0.7
	Electrolyte efflux (μMhos)	0.6 ± 0.3	5.7 ± 1.8
	K^+ efflux (ppm)	0.67 ± 0.09	0.46 ± 0.07
	Na^+ efflux (ppm)	1.08 ± 0.11	0.89 ± 0.40
	Mg^{++} efflux (ppm)	0.73 ± 0.05	0.76 ± 0.04
2	Curvature (degrees)	50 ± 14	292 ± 14
	Change in pH of effluent during 15-min incubation period	-0.10 ± 0.06	-0.30 ± 0.08
	Change in absorbancy at 226 nm	0.06 ± 0.01	0.13 ± 0.02

Note: Three or 5 excised tendrils per vial were pretreated overnight with 100 μM benzoic acid.
[a]Standard error.

TABLE 9. The Effect of Coiling on the Release of Tritiated Water from the Cut Base of Excised Tendrils Pretreated Overnight with 100 μm Benzoic Acid

Treatment	Curvature (degrees)	Cpm per Tendril in Effluent
Unstimulated tendrils	45	3,510
Coiled tendrils	342	4,370
Least significant difference at 5%		376

Note: After absorbing tritiated water for 2 hr, the tendrils were stimulated and allowed to coil for 15 min. The experiment was replicated 14 times with single tendrils.

TABLE 10. The Effect of Coiling on Changes in the Amount of Tritium Label in the Dorsal and Ventral Halves of Tendrils

	Unstimulated Tendrils		Coiled Tendrils	
Experiment	Dorsal Side	Ventral Side	Dorsal Side	Ventral Side
1	3,839	2,775	3,433	2,626
2	3,398	2,679	3,085	2,503
3	3,384	3,384	3,985	2,853
4	829	1,070	1,115	1,180
5	1,001	1,365	846	800
6	601	933	449	333
average	2,180	2,035	2,160	1,715
Percentage change in label due to coiling			-1%	-16%

Note: After being allowed to absorb tritiated water for 2 hr, tendrils were stimulated and allowed to coil for 15 min. The tendrils were then frozen and cut down the middle and the radioactivity in each half was measured by liquid scintillation counting.

How can all these facts be put together into a plausible scheme, and how can such a scheme aid in the interpretation of geotropism? First, let us construct an overall hypothesis to explain tendril movement. Mechanical stimulation, perceived by specialized cells under the ventral epidermis, is transduced into an activation of a membrane-localized ATPase. ATP breakdown and ATPase contraction change membrane characteristics, resulting in a release of H^+ (Mitchell 1966) accompanied (and limited?) by available anions. Water then diffuses from the cells, both in response to lowered osmotic concentration inside the cell and because of a dramatic lowering of the permeability barrier normally presented by the membrane. This causes collapse of the turgor of the ventral cells, resulting in the initial observed contraction of the ventral surface. We must also hypothesize some role of high auxin levels both in initiating and facilitating curvature; perhaps this also involves an activation of ATPase. The role (if any) of the flavonoids is obscure, but it appears that an inhibition of ATPase activity by QGC may be involved; if this is true, then destruction of QGC may be the trigger releasing ATPase activity.

How much of this scheme is true, and how much of it may be referable to geotropism is anyone's guess. Let me suggest that the falling of statoliths against the membrane could activate ATPase, which in turn could result in the localized excretion of H^+, accompanied by the indoleacetate anion. Out of this could come the asymmetric distribution of IAA, which in turn could generate the transverse potential and the unequal growth requisite for tropistic curvature. Although this theory is a bit on the wild side, it does have the virtues of novelty and testability. It seems almost too much to ask that it possess verity as well; nonetheless, we intend to find out.

Acknowledgment

I wish to acknowledge support from the National Science Foundation and the Whitehall Foundation, which made this work possible, and the excellent collaboration of Dr. M. J. Jaffe and Patricia Lucas.

References

Audus, L. J. 1962. *Soc. Exp. Biol. Symp.* 16: 197-226.

Audus, L. J., and Lahiri, A. N. 1961. *J. Exp. Bot.* 12: 75-84.

Bennet-Clark, T. A., and Ball, N. G. 1951. *J. Exp. Bot.* 2: 169-203.

Borgström, G. 1939. *The transverse reactions of plants*. Lund, Sweden: C. W. K. Gleerup.

Bose, J. C. 1907. *Comparative electrophysiology*. London.

Burg, S. P., and Burg, E. A. 1967. *Pl. Physiol.* 42: 891-93.

Darwin, C. 1876. *The movements and habits of climbing plants*, 2d ed. rev. New York: D. Appleton.

Dolk, H. E. 1936. *Rec. Trav. Bot. Néerl.* 33: 509-85.

Fondéville, J. C.; Borthwick, H. A.; and Hendricks, S. B. 1966. *Planta (Berl.)* 69: 357-64.

Galston, A. W. 1957. In *The experimental control of plant growth*, ed. F. W. Went, pp. 313-17. Waltham, Mass.: Chronica Botanica.

Grahm, L., and Hertz, C. H. 1964. *Physiol. Plantarum* 17: 196-201.

Jaffe, M. J., and Galston, A. W. 1966a. *Pl. Physiol.* 41: 1014-25.

———. 1966b. *Pl. Physiol.* 41: 1152-58.

———. 1967a. *Pl. Physiol.* 42: 845-47.

———. 1967b. *Pl. Physiol.* 42: 848-50.

———. 1968a. *Pl. Physiol.* 43: 537-42.

———. 1968b. *Ann. Rev. Pl. Physiol.* 19: 417-34.

Mitchell, P. 1966. *Res. Rept.* no. 66/1. Bodmin. Cornwall: Glynn Research Ltd.

Pickard, B. G., and Thimann, K. V. 1966. *J. Gen. Physiol.* 49: 1065-86.

Reinhold, L. 1967. *Science* 158: 791-93.

Richmond, A. E., and Lang, A. 1957. *Science* 125: 650-51.

Westing, A. H. 1959. Studies on the physiology of compression wood formation in Pinus. Ph.D. diss., Yale University.

Wilkins, M. B., and Woodcock, A. E. R. 1965. *Nature* 208: 990-92.

39
Gravity and the Animal: A Summary

O. Lowenstein, *University of Birmingham*

Taking the lead of my predecessor, I might describe animals as organizations of matter whose function is to convert cosmic energy into insight. To do this the animal has to move about in its environment. The hallmark of the animal's organization, therefore, is a fantastic degree of complexity. The cortex now addressing you has a structure of ten thousand million neurons working together for the purpose of making this kind of a pronouncement. These neurons are, of course, connected with inputs. In one of the first papers it was stated almost hesitantly by a botanist that there may be more than one input in geotropism. You have heard in some of the zoological contributions that the number of inputs can be legion and that they do not all, in fact, arise from the environment; many of them arise by way of feedback in the organism itself. So it isn't surprising that some of the phenomena you heard described were less than transparent at first sight.

I might usefully start by an examination of whether prophesy has come true because, as you will remember, our dealings were opened by the pronouncement of a prophet, Dr. Brown. He told us what we had to expect. Let us see whether things have turned out according to expectation. Dr. Brown said, for example, that he would not be surprised to find that most of us wanted to see everything reduced to a molecular mechanism. He anticipated, in fact, that all of us—the audience as well as the speakers—were probably indulging in reductionism of an extreme kind. Now this is only partly true. Molecular mechanisms have so great an attraction that their investigation, and the popularization of the results of such investigations, have led to a revolution in university teaching; no one is satisfied today unless a college course begins with molecular biology. This is strange, and what makes it stranger still is that molecular mechanisms are commonly found difficult to teach. Why, then, put them right at the beginning? There is a danger, of which the educators among you will be well aware, namely, that this approach is rapidly leading to a revival of medieval modes of teaching. We teach texts, and so we have good excuses when we cannot demonstrate the experimental evidence to our students; they must go home and learn their texts. As a zoologist I am not convinced that I would be satisfied to see the animal phenomena around me reduced to molecular mechanisms. I recognize, however, a sequence of emergent planes of complexity, each of which can be an object of analysis and, quite usefully here, the end point of analysis.

This symposium has made clear that we no longer work in watertight compartments, and it is quite obvious that we are sufficiently aware of the importance of insight into molecular mechanisms. But we are not too disappointed, especially as zoologists, if our analysis stops short of the ultimate. I can think, for example, of mechanisms that might be labeled *self-preservation, information gathering, learning,* and the like. Each of these labels is sufficiently complex to provide enough interest for a lifetime. Quite wisely, Dr. Brown predicted that there would be an emphasis on behavior in the zoological papers. All plants behave, but animals do so to a greater extent, a fact connected with the animal's greater mobility. An animal uses its mobility to collect information from as wide a variety of environments as possible, in consequence of which the animal can redeploy its activity at very short notice. It could not do so, however, if it lacked another great asset, the nervous system.

Any hubris on the part of zoologists and animal physiologists has been dampened by the fact that plants also have action potentials. However slow, these are decent, honest action potentials. But what happens in the nervous system of an animal organized on even a relatively low level has no counterpart in plant growth. In those papers that dealt with flight, the multiple inputs became rather obvious. The animals' orientation toward the plumb line was guaranteed by a range of sensory mechanisms that were by no means confined to mechanoreceptors. Vision played a great part, and, in some instances, there was evidence that olfactory input might be important.

Nevertheless, it is good to be at home, and home means the primary orientation to gravity. Home means the surface of the earth. Home means a balanced roosting on a tree branch. Home also means swimming in the water in an anatomically fixed position that is also oriented toward gravity. It would have been a mistake if the papers from the zoological side had been confined to the statocyst or its functional mechanism. However, this system is relatively easy to analyze and has served in this symposium to elucidate one fact which is of use for the future. This fact is the nature of the possible first, or primary, transducer. Wherever we have statocysts, we have a strange ciliary organelle with 9 + 2 filaments and a basal body which in many instances shows directional polarization. This organelle has a knack, as we have learned, of turning up throughout the invertebrates; it seems to be a common feature in a great number of organs.

I am especially pleased about this because of the battle between possible contradictory hypotheses on the function and construction of the hair cell in the labyrinth. To mention one aspect of this uncertainty, we have been faced with a quandary in respect to the secondary sensory cell. We would very much like to assume that the bending in one or the other direction of the hair bundle, including the kinocilium, is the first step in the transduction process. We would like to identify the kinocilium chiefly as the locus of original depolarization, were it not for two facts. One I have already mentioned in my paper (chapter 24), namely, that the chemical environment of the vestibular endorgan is unsuitable for an ordinary potassium-sodium mechanism; the second is that in the cochlea of the ear the hair cells have no kinocilia. The next step is to consider the kinocilia as the centers of organization responsible for the health of the cell, and say that the bundles of stereocilia may be the locus of depolarization. As the latter are pushed about, their deformation may initiate processes which are either conducted along the cell membrane or along a type of intracellular organelle. Prior to the synaptic region there is only a dc analog change in potential in response to the stimulus. This process may be similar to what occurs in plant cells. The dc potential change then may cause the release of a chemical transmitter which depolarizes the endings of the sensory neurons to produce trains of all-or-nothing electric potentials. It is these digital action potentials which then comprise the coded message from the sensory cell to the central nervous system.

There are many reasons why this hypothesis is tempting. Nature doesn't like big jumps. Thus it may be useful to think in terms of an intracellular conduction mechanism consisting of charged organelles that transfer the charge step by step. Such gradualism may have much to commend it.

And yet this very symposium, which I have been privileged to attend, has fortified in my mind the idea that whatever the physicochemical difficulties at the top of the hair cell, the kinocilium or its basal body ought to be the locus of primary depolarization. The repeated emphasis on this type of structure in all the plant and animal systems described has encouraged me to bet on the kinocilium. This conviction is something worth taking home. And so we botanists and zoologists, will have to grapple with this possibility, a possibility that bolsters further my feeling of true companionship with my botanical cousins.

The Participants

Solon A. Gordon, Chairman of the symposium
  Division of Biological and Medical Research, Argonne National Laboratory, Argonne, Illinois 60439 U.S.A.
Melvin J. Cohen, Cochairman of the symposium
  Department of Biology, Kline Biology Tower, Yale University, New Haven, Connecticut 06520 U.S.A.
Frank G. Favorite, Coordinator of the symposium
  National Academy of Sciences-NRC, Washington, D.C. 20418 U.S.A.

W. Ross Adey, Department of Anatomy, School of Medicine, University of California, Los Angeles 90024 U.S.A.
Leslie J. Audus, Botany Department, Bedford College, University of London, U.K.
Geoffrey H. Banbury, Botany Department, University Laboratories, South Road, Durham, U.K.
Allan H. Brown, Department of Biology, University of Pennsylvania, Philadelphia 19104 U.S.A.
Russell R. Burton, Department of Animal Physiology, University of California, Davis 95616 U.S.A.
William J. Davis, Division of Natural Sciences-I, University of California, Santa Cruz 95060 U.S.A.
David S. Dennison, Department of Biological Sciences, Dartmouth College, Hanover, New Hampshire 02138 U.S.A.
Frederick J. deSerres, Biology Division, Oak Ridge National Laboratory, Oak Ridge, Tennessee 37831 U.S.A.
Betty F. Edwards, Department of Anatomy, Emory University, Atlanta, Georgia 30322 U.S.A.
Bud Etherton, Department of Botany, University of Vermont, Burlington 05401 U.S.A.
George D. Freier, Department of Physics, University of Minnesota, Minneapolis 55455 U.S.A.
Arthur W. Galston, Department of Biology, Yale University, New Haven, Connecticut 06520 U.S.A.
Lennart Grahm, Department of Electrical Measurements, The Lund Institute of Technology, Lund, Sweden
Stephen W. Gray, Department of Anatomy, Emory University, Atlanta, Georgia 30322 U.S.A.
Torquato Gualtierotti, Laboratorio di Fisiologia Umana, Università di Milano, Italy
Rainer A. Hertel, Institut f. Biologie III der Universität Freiburg, 78 Freiburg i. Br., West Germany
C. Hellmuth Hertz, Department of Electrical Measurements, The Lund Institute of Technology, Lund, Sweden
G. Adrian Horridge, Department of Behavioral Biology, Research School of Biological Sciences, Canberra City, A.C.T., 2601, Australia
Takashi Hoshizaki, Brain Research Institute, University of California, Los Angeles 90024 U.S.A.
Howard C. Howland, Division of Biological Sciences, Cornell University, Ithaca, New York 14850 U.S.A.
Leszek S. Jankiewicz, Laboratory of Fruit Tree Physiology, Research Institute of Pomology, Skierniewice, Poland
Dale Jenkins, Ecology Program, The Smithsonian Institution, Washington, D.C. 20546 U.S.A.
*Samuel P. Johnson, North American Aviation, Downey, California 90241 U.S.A.

Anders Johnsson, Department of Electrical Measurements, The Lund Institute of Technology, Lund, Sweden
Harald Kaldewey, Botanisches Institut, Universität des Saarlandes, 6600 Saarbrücken 15, West Germany
Poul Larsen, Plantefysiologisk Institut, Aarhus Universitet, Aarhus C., Denmark
Michael S. Laverack, Gatty Marine Laboratory, St. Andrews University, Fife, Scotland, U.K.
A. Carl Leopold, Department of Horticulture, Purdue University, West Lafayette, Indiana 47907 U.S.A.
O. E. Lowenstein, Department of Zoology and Comparative Physiology, University of Birmingham, Edgbaston, Birmingham 15, U.K.
Charles J. Lyon, Department of Biological Sciences, Dartmouth College, Hanover, New Hampshire 03755 U.S.A.
Hubert Markl, Zoologisches Institut der TH, 6100 Darmstadt, West Germany
Horst Mittelstaedt, Max Planck Institut f. Verhaltensphysiologie, Seewiesen, West Germany
Barbara G. Pickard, Center for the Biology of Natural Systems, Washington University, St. Louis, Missouri 63130 U.S.A.
Ernest C. Pollard, Department of Biophysics, The Pennsylvania State University, University Park 16802 U.S.A.
Orr E. Reynolds, Office of Space Science and Applications, National Aeronautics and Space Administration, Washington, D.C. 20546 U.S.A.
Henry Rufelt, Institute of Physiological Botany, University of Uppsala, Sweden
Joseph F. Saunders, Office of Space Science and Applications, National Aeronautics and Space Administration, Washington, D.C. 20546 U.S.A.
Hermann Schöne, Max Planck Institut f. Verhaltensphysiologie, Seewiesen, West Germany
A. R. Schrank, Department of Zoology, The University of Texas at Austin 78712 U.S.A.
J. Shen-Miller, Division of Biological and Medical Research, Argonne National Laboratory, Argonne, Illinois 60439 U.S.A.
Hiroshi Shimazu, School of Medicine, University of Tokyo, 7-3-1 Hongo, Japan
Andreas Sievers, Abt. f. Cytologie Botanisches Institut der Universität, D-53 Bonn, West Germany
Arthur H. Smith, Department of Animal Physiology, University of California, Davis 95616 U.S.A.
Arnold H. Sparrow, Biology Department, Brookhaven National Laboratory, Upton, New York 11973 U.S.A.
Torkel Weis-Fogh, Department of Zoology, University of Cambridge, U.K.
Gernot Wendler, Max Planck Institut f. Verhaltensphysiologie, 8131 Seewiesen, West Germany
Arthur H. Westing, Department of Biology, Windham College, Putney, Vermont 05346 U.S.A.
Malcolm B. Wilkins, Department of Botany, The University, Glasgow, U.K.
*Donald M. Wilson, Department of Biological Sciences, Stanford University, California 94305 U.S.A.
Charles C. Wunder, Department of Physiology and Biophysics, The University of Iowa, Iowa City 52240 U.S.A.
Richard S. Young, Office of Space Science and Applications, National Aeronautics and Space Administration, Washington, D.C. 20546 U.S.A.

---

*Deceased.